Essential Biochemistry

Essential Biochemistry

Victor Perry

Larsen & Keller
www.larsen-keller.com

Essential Biochemistry
Victor Perry
ISBN: 978-1-64172-610-8 (Hardback)

Larsen & Keller
Published by Larsen and Keller Education,
5 Penn Plaza,
19th Floor,
New York, NY 10001, USA

Cataloging-in-Publication Data

Essential biochemistry / Victor Perry.
p. cm.
Includes bibliographical references and index.
ISBN 978-1-64172-610-8
1. Biochemistry. 2. Biology. 3. Chemistry. I. Perry, Victor.
QH345 .E87 2022
572--dc23

For more information regarding Larsen and Keller Education and its products, please visit the publisher's website www.larsen-keller.com

Table of Contents

Preface

It is with great pleasure that I present this book. It has been carefully written after numerous discussions with my peers and other practitioners of the field. I would like to take this opportunity to thank my family and friends who have been extremely supporting at every step in my life.

The scientific study of chemical processes in living organisms such as plants, animals and microorganisms is known as biochemistry. It is divided into physical biochemistry, molecular genetics, metabolism, clinical chemistry and protein science. This field focuses on the mechanism of biological molecules and how it gives rise to various processes which occurs within living cells. This extensive field studies tissues, organism structure and function, and organs. It also focuses on proteins, nucleic acids, lipids and carbohydrates as they provide cell structure. The concepts and methods of this field are applied in medicine, agriculture, nutrition, and biotechnology. The topics covered in this extensive book deal with the core aspects of biochemistry. While understanding the long-term perspectives of the topics, this book makes an effort in highlighting their impact as a modern tool for the growth of the discipline. Coherent flow of topics, student-friendly language and extensive use of examples make this book an invaluable source of knowledge.

The chapters below are organized to facilitate a comprehensive understanding of the subject:

Chapter - What is Biochemistry?

The study of chemical processes related to living organisms, is known as biochemistry. It deals with the functions, structures and interactions of proteins, carbohydrates, lipids and nucleic acids. This is an introductory chapter which will briefly introduce all these fundamental concepts of biochemistry.

Chapter - Biomolecules: The Molecules of Life

Biomolecules are the large macromolecules and ions present in organisms that are necessary for biological processes like morphogenesis and cell division. Four major classes of biomolecules are carbohydrates, proteins, nucleic acids and lipids. This chapter has been carefully written to provide an easy understanding of these classes of biomolecules.

Chapter - Enzymes and Enzyme Activity

The biological catalysts that accelerate a chemical reaction by lowering the activation energy are known as enzymes. The activities of enzymes can be understood through protein dynamics, enzyme kinetics, enzyme inhibition and enzyme catalysis. The following chapter elucidates these varied activities and mechanisms associated with enzymes.

Chapter - Metabolism and Metabolic Pathway

The set of chemical reactions in organisms that are essential for life and provide energy for vital processes is referred to as metabolism. It includes the conversion of food to energy, proteins, lipids, carbohydrates and nucleic acids. All these different processes of metabolism and metabolic pathway have been carefully analyzed in this chapter.

Chapter - Biochemical Techniques

Various biochemical techniques of immunoprecipitation, western blot, electrophoretic mobility shift assay, northern blot and southern blot are used for the analysis of biomolecules. The topics elaborated in this chapter will help in gaining a better perspective about these biochemical techniques.

Victor Perry

1

What is Biochemistry?

The study of chemical processes related to living organisms, is known as biochemistry. It deals with the functions, structures and interactions of proteins, carbohydrates, lipids and nucleic acids. This is an introductory chapter which will briefly introduce all these fundamental concepts of biochemistry.

Biochemistry is the study of the chemical substances and processes that occur in plants, animals, and microorganisms and of the changes they undergo during development and life. It deals with the chemistry of life, and as such it draws on the techniques of analytical, organic, and physical chemistry, as well as those of physiologists concerned with the molecular basis of vital processes. All chemical changes within the organism—either the degradation of substances, generally to gain necessary energy, or the buildup of complex molecules necessary for life processes—are collectively termed metabolism. These chemical changes depend on the action of organic catalysts known as enzymes, and enzymes, in turn, depend for their existence on the genetic apparatus of the cell. It is not surprising, therefore, that biochemistry enters into the investigation of chemical changes in disease, drug action, and other aspects of medicine, as well as in nutrition, genetics, and agriculture.

The term biochemistry is synonymous with two somewhat older terms: physiological chemistry and biological chemistry. Those aspects of biochemistry that deal with the chemistry and function of very large molecules (e.g., proteins and nucleic acids) are often grouped under the term molecular biology. Biochemistry is a young science, having been known under that term only since about 1900. Its origins, however, can be traced much further back; its early history is part of the early history of both physiology and chemistry.

Areas of Study

A description of life at the molecular level includes a description of all the complexly interrelated chemical changes that occur within the cell—i.e., the processes known as intermediary metabolism. The processes of growth, reproduction, and heredity, also subjects of the biochemist's curiosity, are intimately related to intermediary metabolism and cannot be understood independently of it. The properties and capacities exhibited by a complex multicellular organism can be reduced to the properties of the individual cells of that organism, and the behaviour of each individual cell can be understood in terms of its chemical structure and the chemical changes occurring within that cell.

Chemical Composition of Living Matter

Every living cell contains, in addition to water and salts or minerals, a large number of organic compounds, substances composed of carbon combined with varying amounts of hydrogen and usually also of oxygen. Nitrogen, phosphorus, and sulfur are likewise common constituents. In general, the bulk of the organic matter of a cell may be classified as (1) protein, (2) carbohydrate, and (3) fat, or lipid. Nucleic acids and various other organic derivatives are also important constituents. Each class contains a great diversity of individual compounds. Many substances that cannot be classified in any of the above categories also occur, though usually not in large amounts.

Proteins are fundamental to life, not only as structural elements (e.g., collagen) and to provide defense (as antibodies) against invading destructive forces but also because the essential biocatalysts are proteins. The chemistry of proteins is based on the researches of the German chemist Emil Fischer, whose work from 1882 demonstrated that proteins are very large molecules, or polymers, built up of about 24 amino acids. Proteins may vary in size from small—insulin with a molecular weight of 5,700 (based on the weight of a hydrogen atom as 1)—to very large—molecules with molecular weights of more than 1,000,000. The first complete amino acid sequence was determined for the insulin molecule in the 1950s. By 1963 the chain of amino acids in the protein enzyme ribonuclease (molecular weight 12,700) had also been determined, aided by the powerful physical techniques of X-ray-diffraction analysis. In the 1960s, Nobel Prize winners J.C. Kendrew and M.F. Perutz, utilizing X-ray studies, constructed detailed atomic models of the proteins hemoglobin and myoglobin (the respiratory pigment in muscle), which were later confirmed by sophisticated chemical studies. The abiding interest of biochemists in the structure of proteins rests on the fact that the arrangement of chemical groups in space yields important clues regarding the biological activity of molecules.

Carbohydrates include such substances as sugars, starch, and cellulose. The second quarter of the 20th century witnessed a striking advance in the knowledge of how living cells handle small molecules, including carbohydrates. The metabolism of carbohydrates became clarified during this period, and elaborate pathways of carbohydrate breakdown and subsequent storage and utilization were gradually outlined in terms of cycles (e.g., the Embden–Meyerhof glycolytic cycle and the Krebs cycle). The involvement of carbohydrates in respiration and muscle contraction was well worked out by the 1950s.

Fats, or lipids, constitute a heterogeneous group of organic chemicals that can be extracted from biological material by nonpolar solvents such as ethanol, ether, and benzene. The classic work concerning the formation of body fat from carbohydrates was accomplished during the early 1850s. Those studies, and later confirmatory evidence, have shown that the conversion of carbohydrate to fat occurs continuously in the body. The liver is the main site of fat metabolism. Fat absorption in the intestine, studied as early as the 1930s, still is under investigation by biochemists. The control of fat absorption is known to depend upon a combination action of secretions of the pancreas and bile salts. Abnormalities of fat metabolism, which result in disorders such as obesity and rare clinical conditions, are the subject of much biochemical research. Equally interesting to biochemists is the association between high levels of fat in the blood and the occurrence of arteriosclerosis ("hardening" of the arteries).

Nucleic acids are large, complex compounds of very high molecular weight present in the cells of all organisms and in viruses. They are of great importance in the synthesis of proteins and in the transmission of hereditary information from one generation to the next. Originally discovered as constituents of cell nuclei (hence their name), it was assumed for many years after their isolation in 1869 that they were found nowhere else. This assumption was not challenged seriously until the 1940s, when it was determined that two kinds of nucleic acid exist: deoxyribonucleic acid (DNA), in the nuclei of all cells and in some viruses; and ribonucleic acid (RNA), in the cytoplasm of all cells and in most viruses.

The profound biological significance of nucleic acids came gradually to light during the 1940s and 1950s. Attention turned to the mechanism by which protein synthesis and genetic transmission was controlled by nucleic acids. During the 1960s, experiments were aimed at refinements of the genetic code. Promising attempts were made during the late 1960s and early 1970s to accomplish duplication of the molecules of nucleic acids outside the cell—i.e., in the laboratory. By the mid-1980s genetic engineering techniques had accomplished, among other things, in vitro fertilization and the recombination of DNA (so-called gene splicing).

Nutrition

Biochemists have long been interested in the chemical composition of the food of animals. All animals require organic material in their diet, in addition to water and minerals. This organic matter must be sufficient in quantity to satisfy the caloric, or energy, requirements of the animals. Within certain limits, carbohydrate, fat, and protein may be used interchangeably for this purpose. In addition, however, animals have nutritional requirements for specific organic compounds. Certain essential fatty acids, about ten different amino acids (the so-called essential amino acids), and vitamins are required by many higher animals. The nutritional requirements of various species are similar but not necessarily identical; thus man and the guinea pig require vitamin C, or ascorbic acid, whereas the rat does not.

That plants differ from animals in requiring no preformed organic material was appreciated soon after the plant studies of the late 1700s. The ability of green plants to make all their cellular material from simple substances—carbon dioxide, water, salts, and a source of nitrogen such as ammonia or nitrate—was termed photosynthesis. As the name implies, light is required as an energy source, and it is generally furnished by sunlight. The process itself is primarily concerned with the manufacture of carbohydrate, from which fat can be made by animals that eat plant carbohydrates. Protein can also be formed from carbohydrate, provided ammonia is furnished.

In spite of the large apparent differences in nutritional requirements of plants and animals, the patterns of chemical change within the cell are the same. The plant manufactures all the materials it needs, but these materials are essentially similar to those that the animal cell uses and are often handled in the same way once they are formed. Plants could not furnish animals with their nutritional requirements if the cellular constituents in the two forms were not basically similar.

Digestion

The organic food of animals, including man, consists in part of large molecules. In the digestive tracts of higher animals, these molecules are hydrolyzed, or broken down, to their component building blocks. Proteins are converted to mixtures of amino acids, and polysaccharides are converted to monosaccharides. In general, all living forms use the same small molecules, but many of the large complex molecules are different in each species. An animal, therefore, cannot use the protein of a plant or of another animal directly but must first break it down to amino acids and then recombine the amino acids into its own characteristic proteins. The hydrolysis of food material is necessary also to convert solid material into soluble substances suitable for absorption.

The liquefaction of stomach contents aroused the early interest of observers, long before the birth of modern chemistry, and the hydrolytic enzymes secreted into the digestive tract were among the first enzymes to be studied in detail. Pepsin and trypsin, the proteolytic enzymes of gastric and pancreatic juice, respectively, continue to be intensively investigated.

The products of enzymatic action on the food of an animal are absorbed through the walls of the intestines and distributed to the body by blood and lymph. In organisms without digestive tracts, substances must also be absorbed in some way from the environment. In some instances simple diffusion appears to be sufficient to explain the transfer of a substance across a cell membrane. In other cases, however (e.g., in the case of the transfer of glucose from the lumen of the intestine to the blood), transfer occurs against a concentration gradient. That is, the glucose may move from a place of lower concentration to a place of higher concentration.

In the case of the secretion of hydrochloric acid into gastric juice, it has been shown that active secretion is dependent on an adequate oxygen supply (i.e., on the respiratory metabolism of the tissue), and the same holds for absorption of salts by plant roots. The energy released during the tissue oxidation must be harnessed in some way to provide the energy necessary for the absorption or secretion. This harnessing is achieved by a special chemical coupling system. The elucidation of the nature of such coupling systems has been an objective of the biochemist.

Blood

One of the animal tissues that has always excited special curiosity is blood. Blood has been investigated intensively from the early days of biochemistry, and its chemical composition is known with greater accuracy and in more detail than that of any other tissue in the body. The physician takes blood samples to determine such things as the sugar content, the urea content, or the inorganic-ion composition of the blood, since these show characteristic changes in disease.

The blood pigment hemoglobin has been intensively studied. Hemoglobin is confined within the blood corpuscles and carries oxygen from the lungs to the tissues. It combines with oxygen in the lungs, where the oxygen concentration is high, and releases the oxygen in the tissues, where the oxygen concentration is low. The hemoglobins of higher animals are related but not identical. In invertebrates, other pigments may take the place and function of hemoglobin. The comparative study of these compounds constitutes a fascinating chapter in biochemical investigation.

The proteins of blood plasma also have been extensively investigated. The gamma-globulin fraction of the plasma proteins contains the antibodies of the blood and is of practical value as an immunizing agent. An animal develops resistance to disease largely by antibody production. Antibodies are proteins with the ability to combine with an antigen (i.e., an agent that induces their formation). When this agent is a component of a disease-causing bacterium, the antibody can protect an organism from infection by that bacterium. The chemical study of antigens and antibodies and their interrelationship is known as immunochemistry.

Metabolism and Hormones

The cell is the site of a constant, complex, and orderly set of chemical changes collectively called metabolism. Metabolism is associated with a release of heat. The heat released is the same as that obtained if the same chemical change is brought about outside the living organism. This confirms the fact that the laws of thermodynamics apply to living systems just as they apply to the inanimate world. The pattern of chemical change in a living cell, however, is distinctive and different from anything encountered in nonliving systems. This difference does not mean that any chemical laws are invalidated. It instead reflects the extraordinary complexity of the interrelations of cellular reactions.

Hormones, which may be regarded as regulators of metabolism, are investigated at three levels, to determine (1) their physiological effects, (2) their chemical structure, and (3) the chemical mechanisms whereby they operate. The study of the physiological effects of hormones is properly regarded as the province of the physiologist. Such investigations obviously had to precede the more analytical chemical studies. The chemical structures of thyroxine and adrenaline are known. The chemistry of the sex and adrenal hormones, which are steroids, has also been thoroughly investigated. The hormones of the pancreas—insulin and glucagon—and the hormones of the hypophysis (pituitary gland) are peptides (i.e., compounds composed of chains of amino acids). The structures of most of these hormones has been determined. The chemical structures of the plant hormones, auxin and gibberellic acid, which act as growth-controlling agents in plants, are also known.

The first and second phases of the hormone problem thus have been well, though not completely, explored, but the third phase is still in its infancy. It seems likely that different hormones exert their effects in different ways. Some may act by affecting the permeability of membranes; others appear to control the synthesis of certain enzymes. Evidently some hormones also control the activity of certain genes.

Genes

Genetic studies have shown that the hereditary characteristics of a species are maintained and transmitted by the self-duplicating units known as genes, which are composed of nucleic acids and located in the chromosomes of the nucleus. One of the most fascinating chapters in the history of the biological sciences contains the story of the elucidation, in the mid-20th century, of the chemical structure of the genes, their mode of self-duplication, and the manner in which the deoxyribonucleic acid (DNA) of the nucleus causes the synthesis of ribonucleic acid (RNA), which, among its other activites, causes the synthesis of protein. Thus, the capacity of a protein to behave as an enzyme is determined by the chemical constitution of the gene (DNA) that directs the synthesis of the protein. The relationship of genes to enzymes has been demonstrated in several ways. The first successful experiments, devised by the Nobel Prize winners George W. Beadle and Edward L. Tatum, involved the bread mold Neurospora crassa; the two men were able to collect a variety of strains that differed from the parent strain in nutritional requirements. Such strains had undergone a mutation (change) in the genetic makeup of the parent strain. The mutant strains required a particular amino acid not required for growth by the parent strain. It was then shown that such a mutant had lost an enzyme essential for the synthesis of the amino acid in question. The subsequent development of techniques for the isolation of mutants with specific nutritional requirements led to a special procedure for studying intermediary metabolism.

The exploration of space beginning in the mid-20th century intensified speculation about the possibility of life on other planets. At the same time, man was beginning to understand some of the intimate chemical mechanisms used for the transmission of hereditary characteristics. It was possible, by studying protein structure in different species, to see how the amino acid sequences of functional proteins (e.g., hemoglobin and cytochrome) have been altered during phylogeny (the development of species). It was natural, therefore, that biochemists should look upon the problem of the origin of life as a practical one. The synthesis of a living cell from inanimate material was not regarded as an impossible task for the future.

Applied Biochemistry

An early objective in biochemistry was to provide analytical methods for the determination of various blood constituents because it was felt that abnormal levels might indicate the presence of metabolic diseases. The clinical chemistry laboratory now has become a major investigative arm of the physician in the diagnosis and treatment of disease and is an indispensable unit of every hospital. Some of the older analytical methods directed toward diagnosis of common diseases are still the most commonly used—for example, tests for determining the levels of blood glucose, in diabetes; urea, in kidney disease; uric acid, in gout; and bilirubin, in liver and gallbladder disease. With development of the knowledge of enzymes, determination of certain enzymes in blood plasma has assumed diagnostic value, such as alkaline phosphatase, in bone and liver disease; acid phosphatase, in prostatic cancer; amylase, in pancreatitis; and lactate dehydrogenase and transaminase, in cardiac infarct. Electrophoresis of plasma proteins is commonly employed to aid in the diagnosis of various liver diseases and forms of cancer. Both electrophoresis and ultracentrifugation of serum constituents (lipoproteins) are used increasingly in the diagnosis and examination of therapy of atherosclerosis and heart disease. Many specialized and sophisticated methods have been introduced, and machines have been developed for the simultaneous automated analysis of many different blood constituents in order to cope with increasing medical needs.

Analytical biochemical methods have also been applied in the food industry to develop crops superior in nutritive value and capable of retaining nutrients during the processing and preservation of food. Research in this area is directed particularly to preserving vitamins as well as colour and taste, all of which may suffer loss if oxidative enzymes remain in the preserved food. Tests for enzymes are used for monitoring various stages in food processing.

Biochemical techniques have been fundamental in the development of new drugs. The testing of potentially useful drugs includes studies on experimental animals and man to observe the desired effects and also to detect possible toxic manifestations; such studies depend heavily on many of the clinical biochemistry techniques already described. Although many of the commonly used drugs have been developed on a rather empirical (trial-and-error) basis, an increasing number of therapeutic agents have been designed specifically as enzyme inhibitors to interfere with the metabolism of a host or invasive agent. Biochemical advances in the knowledge of the action of natural hormones and antibiotics promise to aid further in the development of specific pharmaceuticals.

Methods in Biochemistry

Like other sciences, biochemistry aims at quantifying, or measuring, results, sometimes with sophisticated instrumentation. The earliest approach to a study of the events in a living organism was an analysis of the materials entering an organism (foods, oxygen) and those leaving (excretion products, carbon dioxide). This is still the basis of so-called balance experiments conducted on animals, in which, for example, both foods and excreta are thoroughly analyzed. For this purpose many chemical methods involving specific colour reactions have been developed, requiring spectrum-analyzing instruments (spectrophotometers) for quantitative measurement. Gasometric techniques are those commonly used for measurements of oxygen and carbon dioxide, yielding respiratory quotients (the ratio of carbon dioxide to oxygen). Somewhat more detail has been gained by determining the quantities of substances entering and leaving a given organ and also by incubating slices of a tissue in a physiological medium outside the body and analyzing the changes that occur in the medium. Because these techniques yield an overall picture of metabolic capacities, it became necessary to disrupt cellular structure (homogenization) and to isolate the individual parts of the cell—nuclei, mitochondria, lysosomes, ribosomes, membranes—and finally the various enzymes and discrete chemical substances of the cell in an attempt to understand the chemistry of life more fully.

Centrifugation and Electrophoresis

An important tool in biochemical research is the centrifuge, which through rapid spinning imposes high centrifugal forces on suspended particles, or even molecules in solution, and causes separations of such matter on the basis of differences in weight. Thus, red cells may be separated from plasma of blood, nuclei from mitochondria in cell homogenates, and one protein from another in complex mixtures. Proteins are separated by ultracentrifugation—very high speed spinning; with appropriate photography of the protein layers as they form in the centrifugal field, it is possible to determine the molecular weights of proteins.

Another property of biological molecules that has been exploited for separation and analysis is their electrical charge. Amino acids and proteins possess net positive or negative charges according to the acidity of the solution in which they are dissolved. In an electric field, such molecules adopt different rates of migration toward positively (anode) or negatively (cathode) charged poles and permit separation. Such separations can be effected in solutions or when the proteins saturate a stationary medium such as cellulose (filter paper), starch, or acrylamide gels. By appropriate colour reactions of the proteins and scanning of colour intensities, a number of proteins in a mixture may be measured. Separate proteins may be isolated and identified by electrophoresis, and the purity of a given protein may be determined. (Electrophoresis of human hemoglobin revealed the abnormal hemoglobin in sickle-cell anemia, the first definitive example of a "molecular disease.")

Chromatography and Isotopes

The different solubilities of substances in aqueous and organic solvents provide another basis for analysis. In its earlier form, a separation was conducted in complex apparatus by partition of substances in various solvents. A simplified form of the same principle evolved as "paper chromatography," in which small amounts of substances could be separated on filter paper and identified by appropriate colour reactions. In contrast to electrophoresis, this method has been applied to a wide variety of biological compounds and has contributed enormously to research in biochemistry.

The general principle has been extended from filter paper strips to columns of other relatively inert media, permitting larger scale separation and identification of closely related biological substances. Particularly noteworthy has been the separation of amino acids by chromatography in columns of ion-exchange resins, permitting the determination of exact amino acid composition of proteins. Following such determination, other techniques of organic chemistry have been used to elucidate the actual sequence of amino acids in complex proteins. Another technique of column chromatography is based on the relative rates of penetration of molecules into beads of a complex carbohydrate according to size of the molecules. Larger molecules are excluded relative to smaller molecules and emerge first from a column of such beads. This technique not only permits separation of biological substances but also provides estimates of molecular weights.

Perhaps the single most important technique in unravelling the complexities of metabolism has been the use of isotopes (heavy or radioactive elements) in labelling biological compounds and "tracing" their fate in metabolism. Measurement of the isotope-labelled compounds has required considerable technology in mass spectroscopy and radioactive detection devices.

A variety of other physical techniques, such as nuclear magnetic resonance, electron spin spectroscopy, circular dichroism, and X-ray crystallography, have become prominent tools in revealing the relation of chemical structure to biological function.

2

Biomolecules: The Molecules of Life

Biomolecules are the large macromolecules and ions present in organisms that are necessary for biological processes like morphogenesis and cell division. Four major classes of biomolecules are carbohydrates, proteins, nucleic acids and lipids. This chapter has been carefully written to provide an easy understanding of these classes of biomolecules.

Biomolecules are defined as any organic molecule present in a living cell which includes carbohydrates, proteins, fats etc. Each biomolecule is essential for body functions and manufactured within the body. They can vary in nature, type, and structure where some may be straight chains, some may be cyclic rings or both. Also, they can vary in physical properties such as water solubility, melting points.

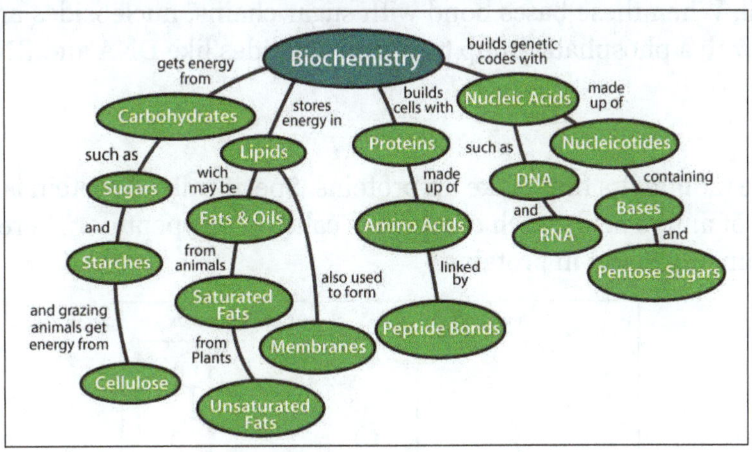

Types of Biomolecules

Biomolecules are primarily classified into 5 types, namely:

Carbohydrates

Polysaccharides, commonly known as carbohydrates are macromolecules. They are made up of monosaccharides (sugar molecules). Majority of living cells are rich in carbohydrates and they are the final products of many metabolisms. For example, Glucose is the final product of photosynthesis. Saccharides can be monosaccharide, disaccharide, polysaccharide etc. based on the number of sugar molecules they are made up of.

Proteins

Proteins are dietary compounds made of monomers called amino acids. Protein is a long chain of amino acid bonded by polypeptide bonds. Hence proteins are also called polypeptides. Amino acids are carbon-containing compounds where a carboxylic acid group and the amino group are present at the two ends. Each amino acid consists of one central carbon surrounded by four substituents. These four substituents include an amino group, carboxylic acid group, hydrogen and a variable group represented by R. The variable group, R decides the nature and type of amino acid.

Lipids

Lipids are a group of water-insoluble compounds which include fats, glycerol, phospholipids, steroids, oils etc. Types of lipids vary according to their constituents. Fatty acids are simple lipids made up of a carboxyl group and a variable group R. They may be saturated or unsaturated fatty acids. Glycerol is trihydroxy propane which combines with fatty acids to give triglycerides. Some lipids consist of a phosphorus group along with the organic chain. Such lipids are called phospholipids, the constituent of the plasma membrane.

Nucleic Acids

Nucleic acids are the genetic materials present in an organism which includes DNA and RNA. Nucleic acids are the combination materials of nitrogenous bases, sugar molecules and phosphate group linked by different bonds in a series of steps. Our body consists of heterocyclic compounds like pyrimidines and purines. These are nitrogenous compounds like adenine, guanine, cytosine, thymine, and uracil. When these bases bond with sugar chains, nucleosides are formed. Nucleosides in turn bond with a phosphate group to give nucleotides like DNA and RNA.

Amino Acids

Amino acids are the monomers that make up proteins. Specifically, a protein is made up of one or more linear chains of amino acids, each of which is called a polypeptide. There are 202020 types of amino acids commonly found in proteins.

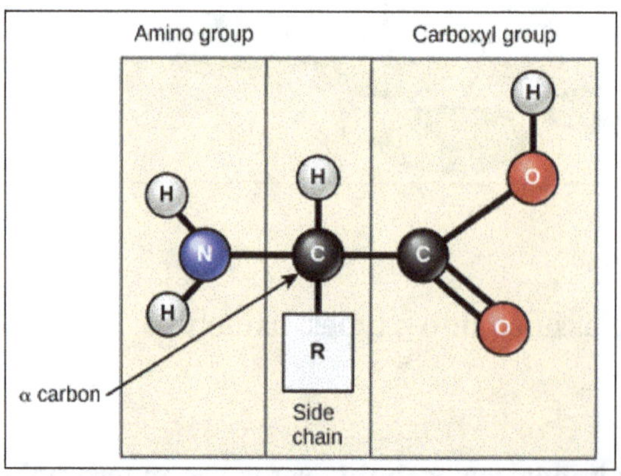

Amino acids share a basic structure, which consists of a central carbon atom, also known as the alpha (α) carbon, bonded to an amino group (NH_2), a carboxyl group (COOH) and a hydrogen atom.

Carbohydrate

Carbohydrate refers to the class of naturally occurring compounds and derivatives formed from them. In the early part of the 19th century, substances such as wood, starch, and linen were found to be composed mainly of molecules containing atoms of carbon (C), hydrogen (H), and oxygen (O) and to have the general formula $C_6H_{12}O_6$; other organic molecules with similar formulas were found to have a similar ratio of hydrogen to oxygen. The general formula $C_x(H_2O)_y$ is commonly used to represent many carbohydrates, which means "watered carbon."

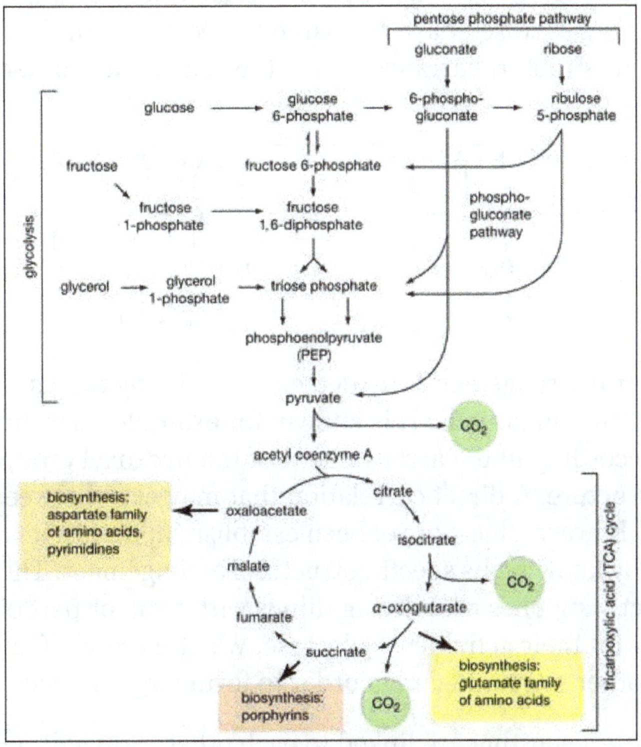

Pathways for the utilization of carbohydrates.

Carbohydrates are probably the most abundant and widespread organic substances in nature, and they are essential constituents of all living things. Carbohydrates are formed by green plants from carbon dioxide and water during the process of photosynthesis. Carbohydrates serve as energy sources and as essential structural components in organisms; in addition, part of the structure of nucleic acids, which contain genetic information, consists of carbohydrate.

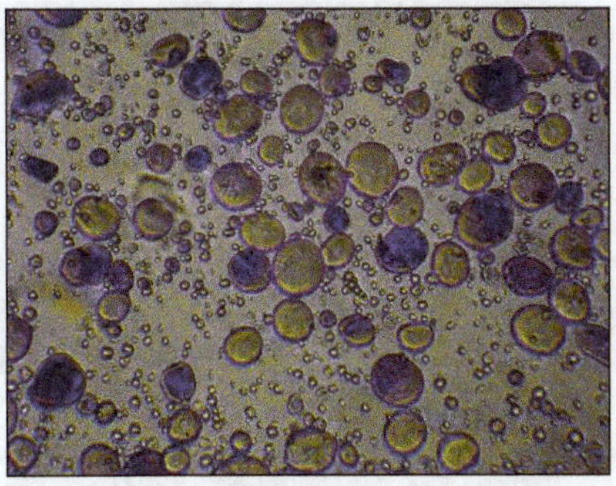

Wheat starch granules stained with iodine.

General Features

Classification and Nomenclature

Although a number of classification schemes have been devised for carbohydrates, the division into four major groups—monosaccharides, disaccharides, oligosaccharides, and polysaccharides—used here is among the most common. Most monosaccharides, or simple sugars, are found in grapes, other fruits, and honey. Although they can contain from three to nine carbon atoms, the most common representatives consist of five or six joined together to form a chainlike molecule. Three of the most important simple sugars—glucose (also known as dextrose, grape sugar, and corn sugar), fructose (fruit sugar), and galactose—have the same molecular formula, ($C_6H_{12}O6$), but, because their atoms have different structural arrangements, the sugars have different characteristics; i.e., they are isomers.

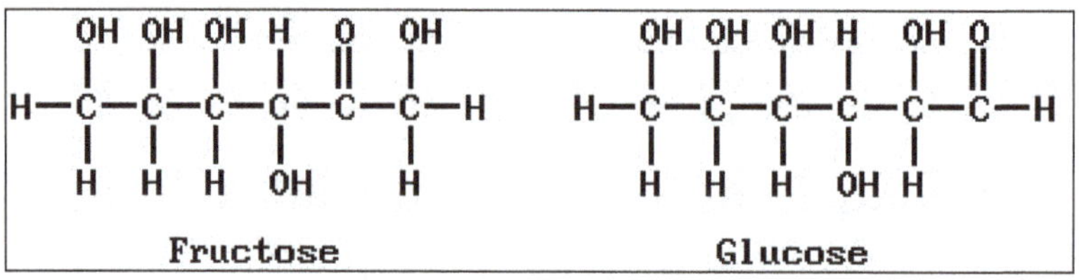

Slight changes in structural arrangements are detectable by living things and influence the biological significance of isomeric compounds. It is known, for example, that the degree of sweetness of various sugars differs according to the arrangement of the hydroxyl groups (−OH) that compose part of the molecular structure. A direct correlation that may exist between taste and any specific structural arrangement, however, has not yet been established; that is, it is not yet possible to predict the taste of a sugar by knowing its specific structural arrangement. The energy in the chemical bonds of glucose indirectly supplies most living things with a major part of the energy that is necessary for them to carry on their activities. Galactose, which is rarely found as a simple sugar, is usually combined with other simple sugars in order to form larger molecules.

Two molecules of a simple sugar that are linked to each other form a disaccharide, or double sugar. The disaccharide sucrose, or table sugar, consists of one molecule of glucose and one molecule of fructose; the most familiar sources of sucrose are sugar beets and cane sugar. Milk sugar, or lactose, and maltose are also disaccharides. Before the energy in disaccharides can be utilized by living things, the molecules must be broken down into their respective monosaccharides. Oligosaccharides, which consist of three to six monosaccharide units, are rather infrequently found in natural sources, although a few plant derivatives have been identified.

Lactose Crystal

Lactose crystals are shown suspended in oil. Their distinct shape allows them to be identified in foods examined for research.

Polysaccharides (the term means many sugars) represent most of the structural and energy-reserve carbohydrates found in nature. Large molecules that may consist of as many as 10,000 monosaccharide units linked together, polysaccharides vary considerably in size, in structural complexity, and in sugar content; several hundred distinct types have thus far been identified. Cellulose, the principal structural component of plants, is a complex polysaccharide comprising many glucose units linked together; it is the most common polysaccharide. The starch found in plants and the glycogen found in animals also are complex glucose polysaccharides. Starch (from the Old English word stercan, meaning "to stiffen") is found mostly in seeds, roots, and stems, where it is stored as an available energy source for plants. Plant starch may be processed into foods such as bread, or it may be consumed directly—as in potatoes, for instance. Glycogen, which consists of branching chains of glucose molecules, is formed in the liver and muscles of higher animals and is stored as an energy source.

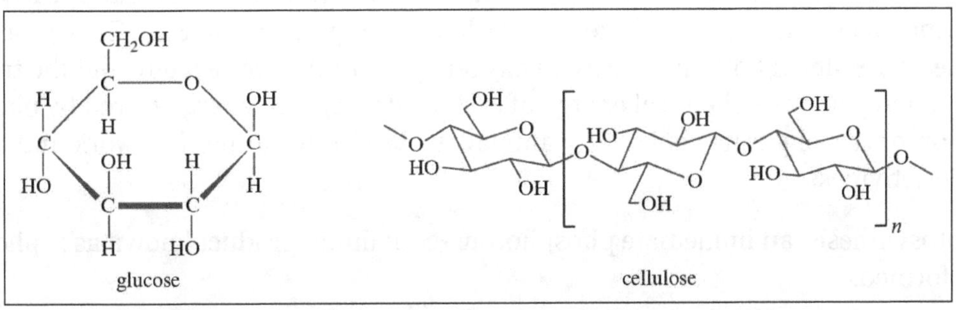

Composition of cellulose and glucose are examples of carbohydrates.

The generic nomenclature ending for the monosaccharides is -ose; thus, the term pentose (pent = five) is used for monosaccharides containing five carbon atoms, and hexose (hex = six) is used for those containing six. In addition, because the monosaccharides contain a chemically reactive group that is either an aldehyde group or a keto group, they are frequently referred to as aldopentoses or ketopentoses or aldohexoses or ketohexoses. The aldehyde group can occur at position 1 of an aldopentose, and the keto group can occur at a further position (e.g., 2) within a ketohexose. Glucose is an aldohexose—i.e., it contains six carbon atoms, and the chemically reactive group is an aldehyde group.

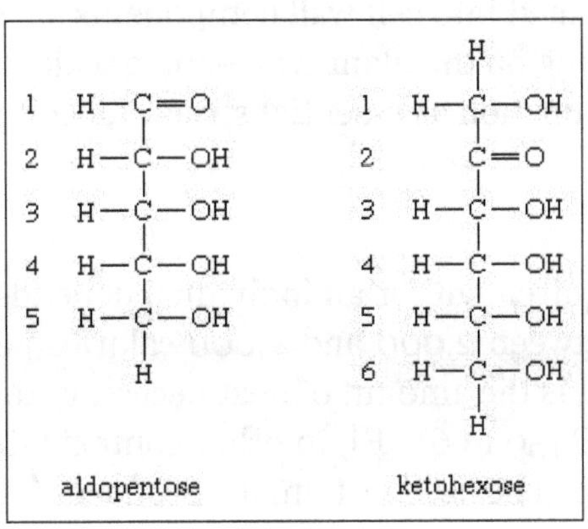

Biological Significance

The importance of carbohydrates to living things can hardly be overemphasized. The energy stores of most animals and plants are both carbohydrate and lipid in nature; carbohydrates are generally available as an immediate energy source, whereas lipids act as a long-term energy resource and tend to be utilized at a slower rate. Glucose, the prevalent uncombined, or free, sugar circulating in the blood of higher animals, is essential to cell function. The proper regulation of glucose metabolism is of paramount importance to survival.

The ability of ruminants, such as cattle, sheep, and goats, to convert the polysaccharides present in grass and similar feeds into protein provides a major source of protein for humans. A number of medically important antibiotics, such as streptomycin, are carbohydrate derivatives. The cellulose in plants is used to manufacture paper, wood for construction, and fabrics.

Role in the Biosphere

The essential process in the biosphere, the portion of Earth in which life can occur, that has permitted the evolution of life as it now exists is the conversion by green plants of carbon dioxide from the atmosphere into carbohydrates, using light energy from the Sun. This process, called photosynthesis, results in both the release of oxygen gas into the atmosphere and the transformation of light energy into the chemical energy of carbohydrates. The energy stored by plants during the formation of carbohydrates is used by animals to carry out mechanical work and to perform biosynthetic activities.

During photosynthesis, an immediate phosphorous-containing product known as 3-phosphoglyceric acid is formed.

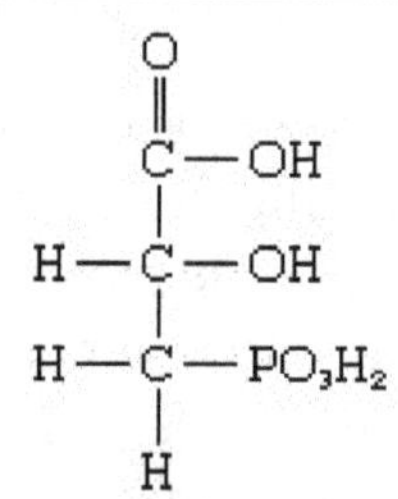

3-phosphoglyceric acid

This compound then is transformed into cell wall components such as cellulose, varying amounts of sucrose, and starch—depending on the plant type—and a wide variety of polysaccharides, other than cellulose and starch, that function as essential structural components.

Role in Human Nutrition

The total caloric, or energy, requirement for an individual depends on age, occupation, and other factors but generally ranges between 2,000 and 4,000 calories per 24-hour period (one calorie, as this term is used in nutrition, is the amount of heat necessary to raise the temperature of 1,000 grams of water from 15 to 16 °C [59 to 61 °F]; in other contexts this amount of heat is called the kilocalorie). Carbohydrate that can be used by humans produces four calories per gram as opposed to nine calories per gram of fat and four per gram of protein. In areas of the world where nutrition is marginal, a high proportion (approximately one to two pounds) of an individual's daily energy requirement may be supplied by carbohydrate, with most of the remainder coming from a variety of fat sources.

Although carbohydrates may compose as much as 80 percent of the total caloric intake in the human diet, for a given diet, the proportion of starch to total carbohydrate is quite variable, depending upon the prevailing customs. In East Asia and in areas of Africa, for example, where rice or tubers such as manioc provide a major food source, starch may account for as much as 80 percent of the total carbohydrate intake. In a typical Western diet, 33 to 50 percent of the caloric intake is in the form of carbohydrate. Approximately half (i.e., 17 to 25 percent) is represented by starch; another third by table sugar (sucrose) and milk sugar (lactose); and smaller percentages by monosaccharides such as glucose and fructose, which are common in fruits, honey, syrups, and certain vegetables such as artichokes, onions, and sugar beets. The small remainder consists of bulk, or indigestible carbohydrate, which comprises primarily the cellulosic outer covering of seeds and the stalks and leaves of vegetables.

Role in Energy Storage

Starches, the major plant-energy-reserve polysaccharides used by humans, are stored in plants in the form of nearly spherical granules that vary in diameter from about three to 100 micrometres (about 0.0001 to 0.004 inch). Most plant starches consist of a mixture of two components: amylose and amylopectin. The glucose molecules composing amylose have a straight-chain, or linear, structure. Amylopectin has a branched-chain structure and is a somewhat more compact molecule. Several thousand glucose units may be present in a single starch molecule. In the diagram, each small circle represents one glucose molecule.

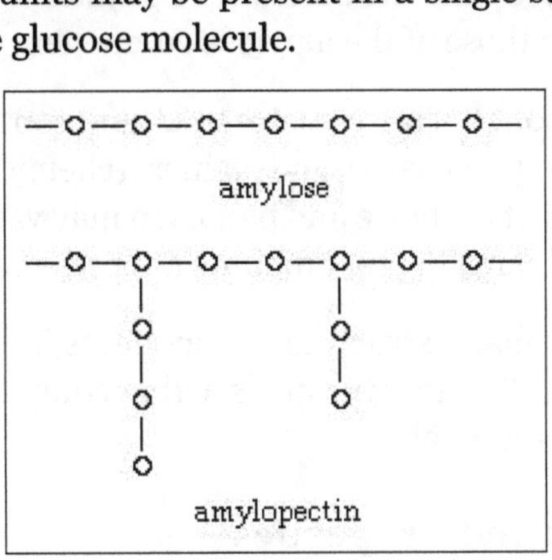

In addition to granules, many plants have large numbers of specialized cells, called parenchymatous cells, the principal function of which is the storage of starch; examples of plants with these cells include root vegetables and tubers. The starch content of plants varies considerably; the highest concentrations are found in seeds and in cereal grains, which contain up to 80 percent of their total carbohydrate as starch. The amylose and amylopectin components of starch occur in variable proportions; most plant species store approximately 25 percent of their starch as amylose and 75 percent as amylopectin. This proportion can be altered, however, by selective-breeding techniques, and some varieties of corn have been developed that produce up to 70 percent of their starch as amylose, which is more easily digested by humans than is amylopectin.

In addition to the starches, some plants (e.g., the Jerusalem artichoke and the leaves of certain grasses, particularly rye grass) form storage polysaccharides composed of fructose units rather than glucose. Although the fructose polysaccharides can be broken down and used to prepare syrups, they cannot be digested by higher animals.

Starches are not formed by animals; instead, they form a closely related polysaccharide, glycogen. Virtually all vertebrate and invertebrate animal cells, as well as those of numerous fungi and protozoans, contain some glycogen; particularly high concentrations of this substance are found in the liver and muscle cells of higher animals. The overall structure of glycogen, which is a highly branched molecule consisting of glucose units, has a superficial resemblance to that of the amylopectin component of starch, although the structural details of glycogen are significantly different. Under conditions of stress or muscular activity in animals, glycogen is rapidly broken down to glucose, which is subsequently used as an energy source. In this manner, glycogen acts as an immediate carbohydrate reserve. Furthermore, the amount of glycogen present at any given time, especially in the liver, directly reflects an animal's nutritional state. When adequate food supplies are available, both glycogen and fat reserves of the body increase, but when food supplies decrease or when the food intake falls below the minimum energy requirements, the glycogen reserves are depleted quite rapidly, while those of fat are used at a slower rate.

Role in Plant and Animal Structure

Whereas starches and glycogen represent the major reserve polysaccharides of living things, most of the carbohydrate found in nature occurs as structural components in the cell walls of plants. Carbohydrates in plant cell walls generally consist of several distinct layers, one of which contains a higher concentration of cellulose than the others. The physical and chemical properties of cellulose are strikingly different from those of the amylose component of starch.

In most plants, the cell wall is about 0.5 micrometre thick and contains a mixture of cellulose, pentose-containing polysaccharides (pentosans), and an inert (chemically unreactive) plastic-like material called lignin. The amounts of cellulose and pentosan may vary; most plants contain between 40 and 60 percent cellulose, although higher amounts are present in the cotton fibre.

Polysaccharides also function as major structural components in animals. Chitin, which is similar to cellulose, is found in insects and other arthropods. Other complex polysaccharides predominate in the structural tissues of higher animals.

Structural Arrangements and Properties

Stereoisomerism

Studies by German chemist Emil Fischer in the late 19th century showed that carbohydrates, such as fructose and glucose, with the same molecular formulas but with different structural arrangements and properties (i.e., isomers) can be formed by relatively simple variations of their spatial, or geometric, arrangements. This type of isomerism, which is called stereoisomerism, exists in all biological systems. Among carbohydrates, the simplest example is provided by the three-carbon aldose sugar glyceraldehyde. There is no way by which the structures of the two isomers of glyceraldehyde, which can be distinguished by the so-called Fischer projection formulas, can be made identical, excluding breaking and reforming the linkages, or bonds, of the hydrogen (–H) and hydroxyl (–OH) groups attached to the carbon at position 2. The isomers are, in fact, mirror images akin to right and left hands; the term enantiomorphism is frequently employed for such isomerism. The chemical and physical properties of enantiomers are identical except for the property of optical rotation.

Optical rotation is the rotation of the plane of polarized light. Polarized light is light that has been separated into two beams that vibrate at right angles to each other; solutions of substances that rotate the plane of polarization are said to be optically active, and the degree of rotation is called the optical rotation of the solution. In the case of the isomers of glyceraldehyde, the magnitudes of the optical rotation are the same, but the direction in which the light is rotated—generally designated as plus, or d for dextrorotatory (to the right), or as minus, or l for levorotatory (to the left)—is opposite; i.e., a solution of D-(d)-glyceraldehyde causes the plane of polarized light to rotate to the right, and a solution of L-(l)-glyceraldehyde rotates the plane of polarized light to the left. Fischer projection formulas for the two isomers of glyceraldehyde are given below.

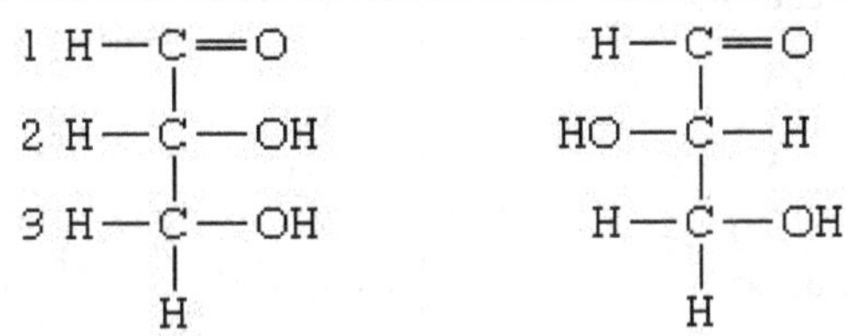

D-(*d*)-glyceraldehyde L-(*l*)-glyceraldehyde

Configuration

Molecules, such as the isomers of glyceraldehyde—the atoms of which can have different structural arrangements—are known as asymmetrical molecules. The number of possible structural arrangements for an asymmetrical molecule depends on the number of centres of asymmetry; i.e., for n (any given number of) centres of asymmetry, 2n different isomers of a molecule are possible. An asymmetrical centre in the case of carbon is defined as a carbon atom to which four different groups are attached. In the three-carbon aldose sugar, glyceraldehyde, the asymmetrical centre is located at the central carbon atom.

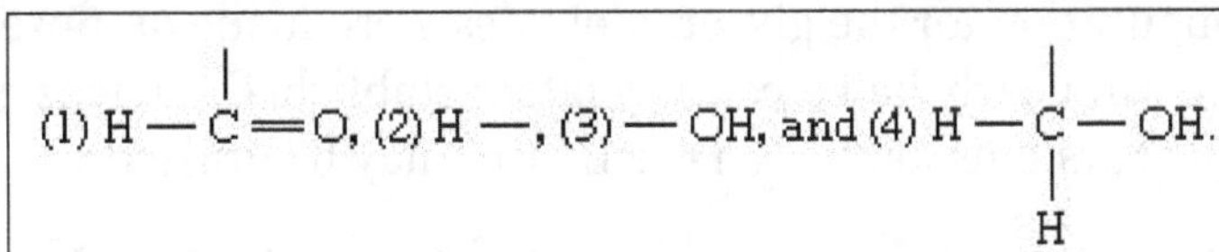

The position of the hydroxyl group (–OH) attached to the central carbon atom–i.e., whether –OH projects from the left or the right—determines whether the molecule rotates the plane of polarized light to the left or to the right. Since glyceraldehyde has one asymmetrical centre, n is one in the relationship 2^n, and there thus are two possible glyceraldehyde isomers. Sugars containing four carbon atoms have two asymmetrical centres; hence, there are four possible isomers (2^2). Similarly, sugars with five carbon atoms have three asymmetrical centres and thus have eight possible isomers (2^3). Keto sugars have one less asymmetrical centre for a given number of carbon atoms than do aldehyde sugars.

A convention of nomenclature, devised in 1906, states that the form of glyceraldehyde whose asymmetrical carbon atom has a hydroxyl group projecting to the right is designated as of the D-configuration; that form, whose asymmetrical carbon atom has a hydroxyl group projecting to the left, is designated as L. All sugars that can be derived from D-glyceraldehyde—i.e., hydroxyl group attached to the asymmetrical carbon atom most remote from the aldehyde or keto end of the molecule projects to the right—are said to be of the D-configuration; those sugars derived from L-glyceraldehyde are said to be of the L configuration.

Representative Disaccharides and Oligosaccharides

aldehydo group (H—C=O)—of the D-configuration.

Common name	Component sugars	Linkages	Sources
Cellobiose	Glucose, glucose	$\beta 1 \rightarrow 4$*	Hydrolysis of cellulose
Gentiobiose	Glucose, glucose	$\beta 1 \rightarrow 6$	Plant glycosides, amygdalin
Isomaltose	Glucose, glucose	$\alpha 1 \rightarrow 6$	Hydrolysis of glycogen, amylopectin
Raffinose**	Galactose, glucose, fructose	$\alpha 1 \rightarrow 6$, $\alpha 1 \rightarrow 2$	Sugarcane, beets, seeds
Stachyose**	Galactose, galactose, glucose, fructose	$\alpha 1 \rightarrow 6$, $\alpha 1 \rightarrow 6$, $\alpha 1 \rightarrow 2$	Soybeans, jasmine, twigs, lentils

*The linkage joins carbon atom 1 (in the β configuration) of one glucose molecule and carbon atom 4 of the second glucose molecule; the linkage may also be abbreviated β-1, 4.

The configurational notation D or L is independent of the sign of the optical rotation of a sugar in solution. It is common, therefore, to designate both, as, for example, D-(l)-fructose or D-(d)-glucose; i.e., both have a D-configuration at the centre of asymmetry most remote from the aldehyde end (in glucose) or keto end (in fructose) of the molecule, but fructose is levorotatory and glucose is dextrorotatory—hence the latter has been given the alternative name dextrose. Although the initial assignments of configuration for the glyceraldehydes were made on purely arbitrary grounds, studies that were carried out nearly half a century later established them as correct in an absolute spatial sense. In biological systems, only the D or L form may be utilized.

When more than one asymmetrical centre is present in a molecule, as is the case with sugars having four or more carbon atoms, a series of DL pairs exists, and they are functionally, physically, and chemically distinct. Thus, although D-xylose and D-lyxose both have five carbon atoms and are of the D-configuration, the spatial arrangement of the asymmetrical centres (at carbon atoms 2, 3, and 4) is such that they are not mirror images.

Hemiacetal and Hemiketal Forms

Although optical rotation has been one of the most frequently determined characteristics of carbohydrates since its recognition in the late 19th century, the rotational Behavior of freshly prepared solutions of many sugars differs from that of solutions that have been allowed to stand. This phenomenon, known as mutarotation, is demonstrable even with apparently identical sugars and is caused by a type of stereoisomerism involving formation of an asymmetrical centre at the first carbon atom (aldehyde carbon) in aldoses and the second one (keto carbon) in ketoses.

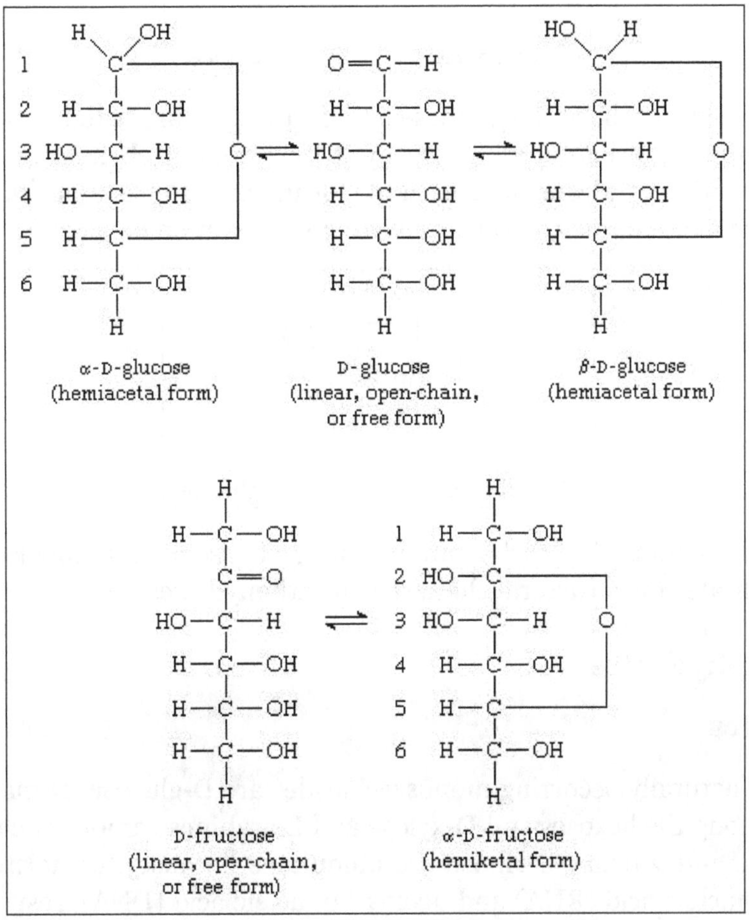

α-D-glucose (hemiacetal form) — D-glucose (linear, open-chain, or free form) — β-D-glucose (hemiacetal form)

D-fructose (linear, open-chain, or free form) — α-D-fructose (hemiketal form)

Most pentose and hexose sugars, therefore, do not exist as linear, or open-chain, structures in solution but form cyclic, or ring, structures in hemiacetal or hemiketal forms, respectively. As illustrated for glucose and fructose, the cyclic structures are formed by the addition of the hydroxyl group (–OH) from either the fourth, fifth, or sixth carbon atom to the carbonyl group ($>C=O$) at position 1 in glucose or 2 in fructose. In the case of five-membered cyclic ketohexose or six-membered cyclic aldohexose, the cyclic forms are in equilibrium with (i.e., the rate of conversion from one form to another is stable) the open-chain structure—a free aldehyde if the solution contains glucose, a free ketone if it contains fructose; each form has a different optical rotation value. Since the forms are in equilibrium with each other, a constant value of optical rotation is measurable; the two cyclic forms represent more than 99.9 percent of the sugar in the case of a glucose solution.

By definition, the carbon atom containing the aldehyde or keto group is called the anomeric carbon atom; similarly, carbohydrate stereoisomers that differ in configuration only at this carbon atom are called anomers. When a cyclic hemiacetal or hemiketal structure forms, the structure with the new hydroxyl group projecting on the same side as that of the oxygeninvolved in forming the ring is called the alpha anomer; that with the hydroxyl group projecting on the opposite side from that of the oxygen ring is called the beta anomer.

alpha anomer beta anomer

The spatial arrangements of the atoms in these cyclic structures are better shown (glucose is used as an example) in the representation devised by British organic chemist Sir Norman Haworth about 1930; they are still in widespread use. In the formulation the asterisk indicates the position of the anomeric carbon atom; the carbon atoms, except at position 6, usually are not labelled.

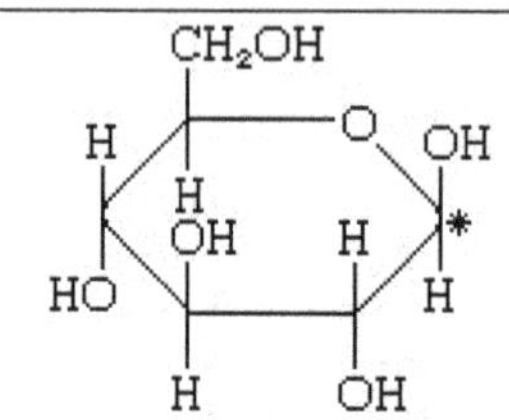

Haworth formulation of β-D-glucose

The large number of asymmetrical carbon atoms and the consequent number of possible isomers considerably complicates the structural chemistry of carbohydrates.

Classes of Carbohydrates

Monosaccharides

The most common naturally occurring monosaccharides are D-glucose, D-mannose, D-fructose, and D-galactose among the hexoses and D-xylose and L-arabinose among the pentoses. In a special sense, D-ribose and 2-deoxy-D-ribose are ubiquitous because they form the carbohydrate component of ribonucleic acid (RNA) and deoxyribonucleic acid (DNA), respectively; these sugars are present in all cells as components of nucleic acids.

Some naturally occurring monosaccharides	
sugar	sources
L-arabinose	mesquite gum, wheat bran
D-ribose	all living cells; as component of ribonucleic acid
D-xylose	corncobs, seed hulls, straw
D-ribulose	as an intermediate in photosynthesis
2-deoxy-D-ribose	as constituent of deoxyribonucleic acid
D-galactose	lactose, agar, gum arabic, brain glycolipids
D-glucose	sucrose, cellulose, starch, glycogen
D-mannose	seeds, ivory nut
D-fructose	sucrose, artichokes, honey
L-fucose	marine algae, seaweed
L-rhamnose	poison-ivy blossom, oak bark
D-mannoheptulose	avocado
D-altroheptulose	numerous plants

D-Xylose, found in most plants in the form of a polysaccharide called xylan, is prepared from corncobs, cottonseed hulls, or straw by chemical breakdown of xylan. D-Galactose, a common constituent of both oligosaccharides and polysaccharides, also occurs in carbohydrate-containing lipids, called glycolipids, which are found in the brain and other nervous tissues of most animals. Galactose is generally prepared by acid hydrolysis (breakdown involving water) of lactose, which is composed of galactose and glucose. Since the biosynthesis of galactose in animals occurs through intermediate compounds derived directly from glucose, animals do not require galactose in the diet. In fact, in most human populations the majority of people do not retain the ability to manufacture the enzyme necessary to metabolize galactose after they reach the age of four, and many individuals possess a hereditary defect known as galactosemiaand never have the ability to metabolize galactose.

D-Glucose, the naturally occurring form, is found in fruits, honey, blood, and, under abnormal conditions, in urine. It is also a constituent of the two most common naturally found disaccharides, sucrose and lactose, as well as the exclusive structural unit of the polysaccharides cellulose, starch, and glycogen. Generally, D-glucose is prepared from either potato starch or cornstarch.

D-Fructose, a ketohexose, is one of the constituents of the disaccharide sucrose and is also found in uncombined form in honey, apples, and tomatoes. Fructose, generally considered the sweetest monosaccharide, is prepared by sucrose hydrolysis and is metabolized by humans.

Chemical Reactions

The reactions of the monosaccharides can be conveniently subdivided into those associated with the aldehyde or keto group and those associated with the hydroxyl groups.

The relative ease with which sugars containing a free or potentially free aldehyde or keto group can be oxidized to form products has been known for a considerable time and once was the basis for the detection of these so-called reducing sugars in a variety of sources. For many years, analyses of blood glucose and urinary glucose were carried out by a procedure involving the use of an alkaline copper compound. Because the reaction has undesirable features—extensive destruction of carbohydrate structure occurs, and the reaction is not very specific (i.e., sugars other than glucose give similar results) and does not result in the formation of readily identifiable products—blood and urinary glucose now are analyzed by using the enzyme glucose oxidase, which catalyzes the oxidation of glucose to products that include hydrogen peroxide. The hydrogen peroxide then is used to oxidize a dye present in the reaction mixture; the intensity of the colour is directly proportional to the amount of glucose initially present. The enzyme, glucose oxidase, is highly specific for β-D-glucose.

In another reaction, the aldehyde group of glucose reacts with alkaline iodine to form a class of compounds called aldonic acids. One important aldonic acid is ascorbic acid (vitamin C), an essential dietary component for humans and guinea pigs. The formation of similar acid derivatives does not occur with the keto sugars.

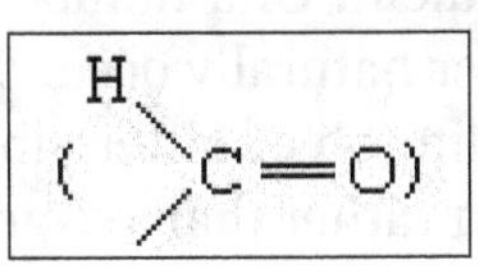

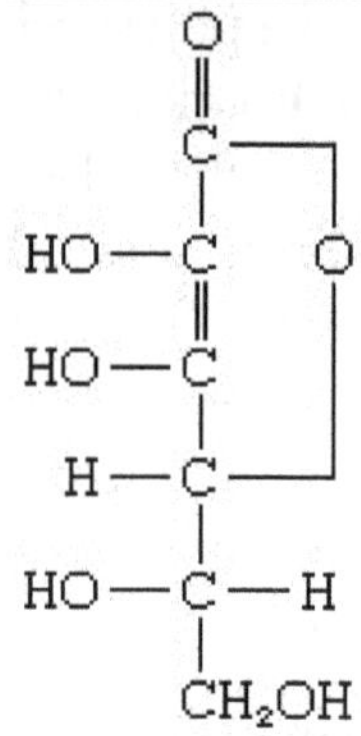

ascorbic acid, vitamin C (L-gulonolactone-2, 3-enediol)

Either the aldehyde or the keto group of a sugar may be reduced (i.e., hydrogen added) to form an alcohol; compounds formed in this way are called alditols, or sugar alcohols. The product formed as a result of the reduction of the aldehyde carbon of D-glucose is called sorbitol (D-glucitol). D-Glucitol also is formed when L-sorbose is reduced. The reduction of mannose results in mannitol, that of galactose in dulcitol.

Sugar alcohols that are of commercial importance include sorbitol (D-glucitol), which is commonly used as a sweetening agent, and D-mannitol, which is also used as a sweetener, particularly in chewing gums, because it has a limited water solubility and remains powdery and granular on long storage.

Formation of Glycosides

The hydroxyl group that is attached to the anomeric carbon atom (i.e., the carbon containing the aldehyde or keto group) of carbohydrates in solution has unusual reactivity, and derivatives, called glycosides, can be formed; glycosides formed from glucose are called glucosides. It is not possible for equilibration between the α- and β-anomers of a glycoside in solution (i.e., mutarotation) to occur. The reaction by which a glycoside is formed involves the hydroxyl group (−OH) of the anomeric carbon atom (numbered 1) of both α and β forms of D-glucose—α and β forms of D-glucose are shown in equilibrium in the reaction sequence—and the hydroxyl group of an alcohol (methyl alcohol in the reaction sequence); methyl α-D-glucosides and β-D-glucosides are formed as products, as is water.

Among the wide variety of naturally occurring glycosides are a number of plant pigments, particularly those red, violet, and blue in colour; these pigments are found in flowers and consist of a pigment molecule attached to a sugar molecule, frequently glucose. Plant indican (from Indigofera species), composed of glucose and the pigment indoxyl, was important in the preparation of indigo dye before synthetic dyes became prevalent. Of a number of heart muscle stimulants that occur as glycosides, digitalis is still used. Other naturally occurring glycosides include vanillin, which is found in the vanilla bean, and amygdalin (oil of bitter almonds); a variety of glycosides found in mustard have a sulfur atom at position 1 rather than oxygen.

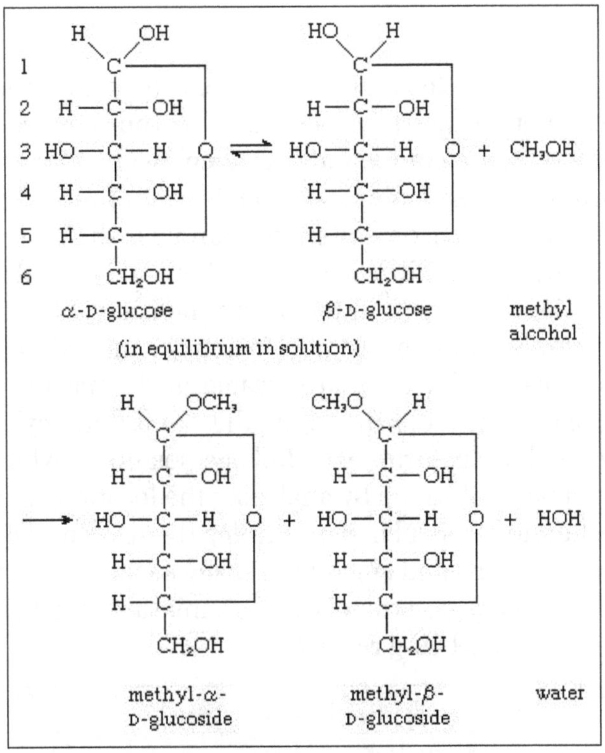

A number of important antibiotics are glycosides; among the best known are streptomycin and erythromycin. Glucosides—i.e., glycosides formed from glucose—in which the anomeric carbon atom (at position 1) has phosphoric acid linked to it, are extremely important biological compounds. For example, α-D-glucose-1-phosphate is an intermediate product in the biosynthesis of cellulose, starch, and glycogen; similar glycosidic phosphate derivatives of other monosaccharides participate in the formation of naturally occurring glycosides and polysaccharides.

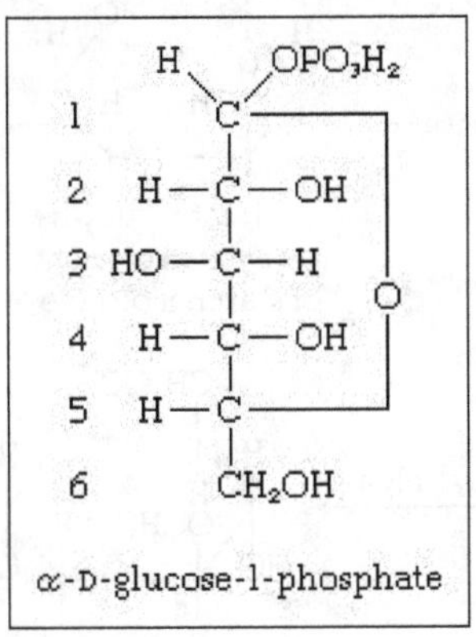

The hydroxyl groups other than the one at the anomeric carbon atom can undergo a variety of reactions. Esterification, which consists of reacting the hydroxyl groups with an appropriate acidic compound, results in the formation of a class of compounds called sugar esters. Among the common ones are the sugar acetates, in which the acid is acetic acid. Esters of phosphoric acid and sulfuric acid are important biological compounds; glucose-6-phosphate, for example, plays a central role in the energy metabolism of most living cells, and D-ribulose 1,5-diphosphate is important in photosynthesis.

Formation of Methyl Ethers

Treatment of a carbohydrate with methyl iodide or similar agents under appropriate conditions results in the formation of compounds in which the hydroxyl groups are converted to methyl groups ($-CH_3$). Called methyl ethers, these compounds are employed in structural studies of oligosaccharides and polysaccharides because their formation does not break the bonds, called glycosidic bonds, that link adjacent monosaccharide units. An example is the etherification of a starch molecule carried out using methyl iodide, in which methyl groups become attached to the glucose molecules, forming a methylated segment in the starch molecule; note that the glycosidic bonds are not broken by the reaction with methyl iodide. When the methylated starch molecule then is broken down (hydrolyzed), hydroxyl groups are located at the positions in the molecule previously involved in linking one sugar molecule to another, and a methylated glucose, in this case named 2,3,6 tri-O-methyl-D-glucose, forms. The linkage positions (which are not methylated) in a complex carbohydrate can be established by analyzing the locations of the methyl groups in the monosaccharides. This technique is useful in determining the structural details of polysaccharides, particularly since the various methylated sugars are easily separated by techniques involving gas chromatography, in which a moving gas stream carries a mixture through a column of a stationary liquid or solid, the components thus being resolved.

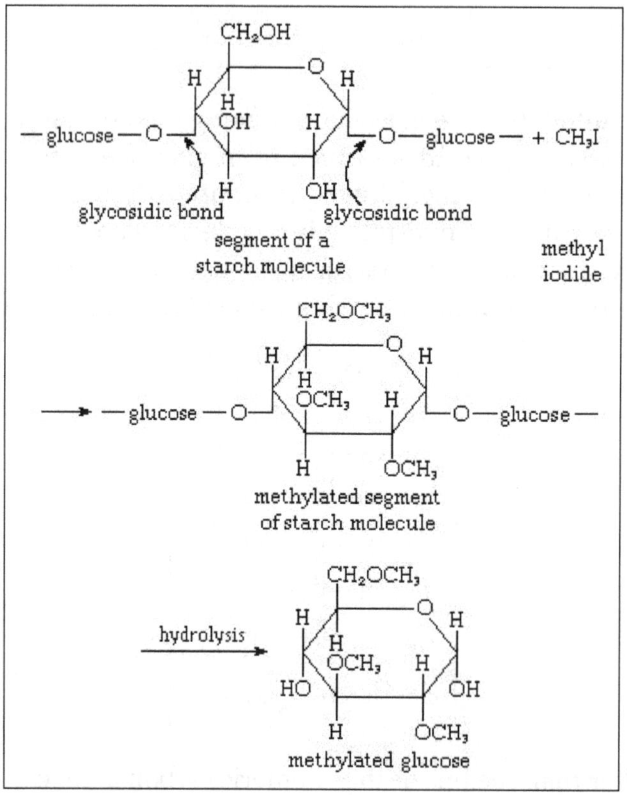

When the terminal group (CH_2OH) of a monosaccharide is oxidized chemically or biologically, a product called a uronic acid is formed. Glycosides that are derived from D-glucuronic acid (the uronic acid formed from D-glucose) and fatty substances called steroids appear in the urine of animals as normal metabolic products; in addition, foreign toxic substances are frequently converted in the liver to glucuronides before excretion in the urine. D-Glucuronic acid also is a major component of connective tissue polysaccharides, and D-galacturonic acid and D-mannuronic acid, formed from D-galactose and D-mannose, respectively, are found in several plant sources.

Other compounds formed from monosaccharides include those in which one hydroxyl group, usually at the carbon at position 2, is replaced by an amino group ($-NH_2$); these compounds, called amino sugars, are widely distributed in nature. The two most important ones are glucosamine (2-amino-2-deoxy-D-glucose) and galactosamine (2-amino-2-deoxy-D-galactose).

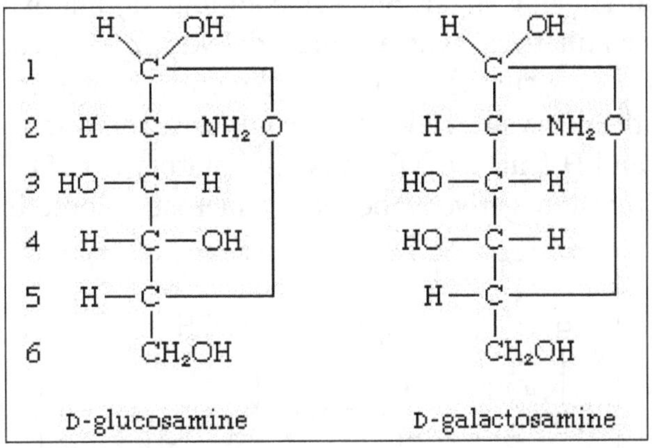

Neither amino sugar is found in the uncombined form. Both occur in animals as components of glycolipids or polysaccharides; e.g., the primary structural polysaccharide (chitin) of insect outer skeletons and various blood group substances.

In a number of naturally occurring sugars, known as deoxy sugars, the hydroxyl group at a particular position is replaced by a hydrogen atom. By far the most important representative is 2-deoxy-D-ribose, the pentose sugar found in deoxyribonucleic acid (DNA); the hydroxyl group at the carbon atom at position 2 has been replaced by a hydrogen atom.

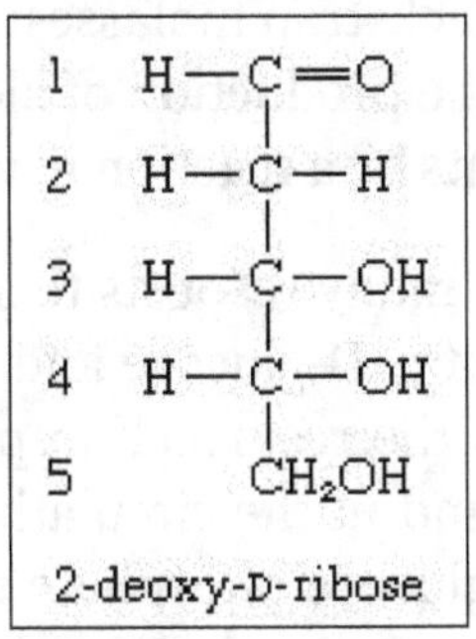

Other naturally occurring deoxy sugars are hexoses, of which L-rhamnose (6-deoxy-L-mannose) and L-fucose (6-deoxy-L-galactose) are the most common; the latter, for example, is present in the carbohydrate portion of blood group substances and on the outer surface of red blood cells.

Disaccharides and Oligosaccharides

Disaccharides are a specialized type of glycoside in which the anomeric hydroxyl group of one sugar has combined with the hydroxyl group of a second sugar with the elimination of the elements of water. Although an enormous number of disaccharide structures are possible, only a limited number are of commercial or biological significance.

Sucrose and Trehalose

Sucrose, or common table sugar, is a major commodity worldwide. By the second decade of the 21st century, its world production had amounted to more than 170 million tons annually. The unusual type of linkage between the two anomeric hydroxyl groups of glucose and fructose means that neither a free aldehyde group (on the glucose moiety) nor a free keto group (on the fructose moiety) is available to react unless the linkage between the monosaccharides is destroyed; for this reason, sucrose is known as a nonreducing sugar. Sucrose solutions do not exhibit mutarotation, which involves formation of an asymmetrical centre at the aldehyde or keto group. If the linkage between the monosaccharides composing sucrose is broken, the optical rotation value of sucrose changes from positive to negative; the new value reflects the composite rotation values for D-glucose, which is dextrorotatory (+52°), and D-fructose, which is levorotatory (−92°). The change in the sign of optical rotation from positive to negative is the reason sucrose is sometimes called invert sugar.

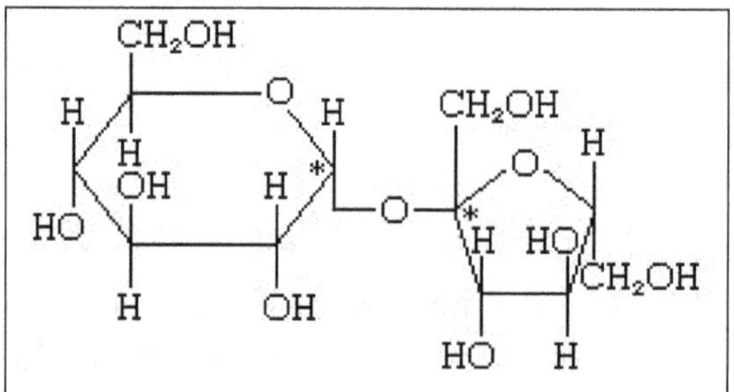

sucrose

The commercial preparation of sucrose takes advantage of the alkaline stability of the sugar, and a variety of impurities are removed from crude sugarcane extracts by treatment with alkali. After this step, syrup preparations are crystallized to form table sugar. Successive "crops" of sucrose crystals are "harvested," and the later ones are known as brown sugar. The residual syrupy material is called either cane final molasses or blackstrap molasses; both are used in the preparation of antibiotics, as sweetening agents, and in the production of alcohol by yeast fermentation. Sucrose is formed following photosynthesis in plants by a reaction in which sucrose phosphate first is formed.

The disaccharide trehalose is similar in many respects to sucrose but is much less widely distributed. It is composed of two molecules of α-D-glucose and is also a nonreducing sugar. Trehalose is present in young mushrooms and in the resurrection plant (Selaginella); it is of considerable biological interest because it is also found in the circulating fluid (hemolymph) of many insects. Since trehalose can be converted to a glucose phosphate compound by an enzyme-catalyzed reaction that does not require energy, its function in hemolymph may be to provide an immediate energy source, a role similar to that of the carbohydrate storage forms (i.e., glycogen) found in higher animals.

Lactose and Maltose

Lactose is one of the sugars (sucrose is another) found most commonly in human diets throughout the world; it constitutes about 7 percent of human milk and about 4–5 percent of the milk of mammals such as cows, goats, and sheep. Lactose consists of two aldohexoses—β-D-galactose and glucose—linked so that the aldehyde group at the anomeric carbon of glucose is free to react; i.e., lactose is a reducing sugar.

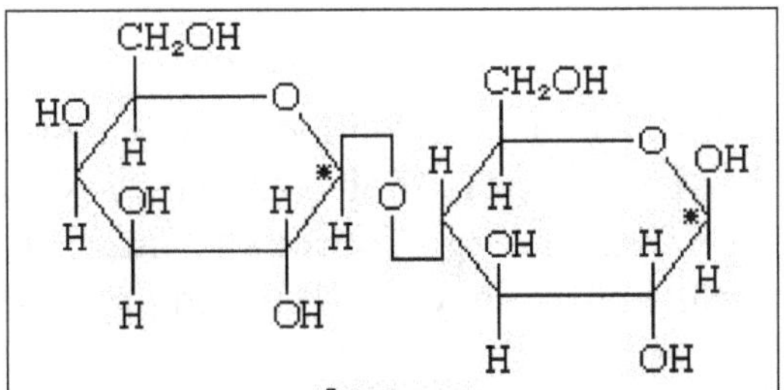

β-lactose

A variety of metabolic disorders related to lactose may occur in infants; in some cases, they are the result of a failure to metabolize properly the galactose portion of the molecule.

Although not found in uncombined form in nature, the disaccharide maltose is biologically important because it is a product of the enzymatic breakdown of starches during digestion. Maltose consists of α-D-glucose linked to a second glucose unit in such a way that maltose is a reducing sugar. Maltose, which is readily hydrolyzed to glucose and can be metabolized by animals, is employed as a sweetening agent and as a food for infants whose tolerance for lactose is limited.

Polysaccharides

Polysaccharides, or glycans, may be classified in a number of ways; the following scheme is frequently used. Homopolysaccharides are defined as polysaccharides formed from only one type of monosaccharide. Homopolysaccharides may be further subdivided into straight-chain and branched-chain representatives, depending upon the arrangement of the monosaccharide units. Heteropolysaccharides are defined as polysaccharides containing two or more different types of monosaccharides; they may also occur in both straight-chain and branched-chain forms. In general, extensive variation of linkage types does not occur within a polysaccharide structure, nor are there many polysaccharides composed of more than three or four different monosaccharides; most contain one or two.

Representative Homopolysaccharides				
Homopolysaccharide	Sugar component	Linkage	Function	Sources
Cellulose	Glucose	$\beta, 1 \rightarrow 4$	Structural	Throughout plant kingdom
Amylose	Glucose	$\alpha, 1 \rightarrow 4$	Food Storage	Starches, especially corn, potatoes, rice
Chitin	N-Acetylglucos-amine	$\beta, 1 \rightarrow 4$	Structural	Insect and crustacean skeleton
Inulin	Fructose	$\beta, 2 \rightarrow 1$	Food Storage	Artichokes, chicory
Xylan	Xylose	$\beta, 1 \rightarrow 4$	Structural	All land plants
Glycogen	Glucose	$\alpha, 1 \rightarrow 4, 6 \rightarrow 1, \alpha$	Food Storage	Liver and muscle cells of all animals
Amylopectin	Glucose	$\alpha, 1 \rightarrow 4, 6 \rightarrow 1, \alpha$	Food Storage	Starches, especially corn, potatoes, rice
Dextran	Glucose	$\alpha, 1 \rightarrow 6, 4 \rightarrow 1, \alpha$	Unknown	Primarily bacterial
Agar	Galactose	$\alpha, 1 \rightarrow 3$	Structural	Seaweeds

May contain sulfate groups.

Homopolysaccharides

In general, homopolysaccharides have a well-defined chemical structure, although the molecular weight of an individual amylose or xylan molecule may vary within a particular range, depending on the source; molecules from a single source also may vary in size, because most polysaccharides are formed biologically by an enzyme-catalyzed process lacking genetic information regarding size.

The basic structural component of most plants, cellulose, is widely distributed in nature. It has been estimated that 50 billion to 100 billion tons of cellulose are synthesized yearly as a result of photosynthesis by higher plants. The proportion of cellulose to total carbohydrate found in plants may vary in various types of woods from 30 to 40 percent, and to more than 98 percent in the seed hair of the cotton plant. Cellulose, a large, linear molecule composed of 3,000 or more β-D-glucose molecules, is insoluble in water.

The chains of glucose units composing cellulose molecules are frequently aligned within the cell-wall structure of a plant to form fibre-like or crystalline arrangements. This alignment permits very tight packing of the chains and promotes their structural stability but also makes structural analysis difficult. The relationships between cellulose and other polysaccharides present in the cell wall are not well established; in addition, the presence of unusual chemical linkages or nonglucose units within the cellulose structure has not yet been established with certainty.

During the preparation of cellulose, raw plant material is treated with hot alkali; this treatment removes most of the lignin, the hemicelluloses, and the mucilaginous components. The cellulose then is processed to produce papers and fibres. The high resistance of cellulose to chemical or enzymatic breakdown is important in the manufacture of paper and cloth. Cellulose also is modified chemically for other purposes; e.g., compounds such as cellulose acetate are used in the plastics industry, in the production of photographic film, and in the rayon-fibre industry. Cellulose nitrate (nitrocellulose) is employed in the lacquer and explosives industries.

The noteworthy biological stability of cellulose is dramatically illustrated by trees, the life-span of which may be several thousand years. Enzymes capable of breaking down cellulose are generally found only among several species of bacteria and molds. The apparent ability of termites to utilize cellulose as an energy source depends on the presence in their intestinal tracts of protozoans that can break it down. Similarly, the single-celled organisms present in the rumina of sheep and cattle are responsible for the ability of these animals to utilize the cellulose present in typical grasses and other feeds.

Xylans are almost as ubiquitous as cellulose in plant cell walls and contain predominantly β-D-xylose units linked as in cellulose. Some xylans contain other sugars, such as L-arabinose, but they form branches and are not part of the main chain. Xylans are of little commercial importance.

The term starch refers to a group of plant reserve polysaccharides consisting almost exclusively of a linear component (amylose) and a branched component (amylopectin). The use of starch as an energy source by humans depends on the ability to convert it completely to individual glucose units; the process is initiated by the action of enzymes called amylases, synthesized by the salivary glands in the mouth, and continues in the intestinal tract. The primary product of amylase action is maltose, which is hydrolyzed to two component glucose units as it is absorbed through the walls of the intestine.

A characteristic reaction of the amylose component of starch is the formation with iodine of a complex compound with a characteristic blue colour. About one iodine molecule is bound for each seven or eight glucose units, and at least five times that many glucose units are needed in an amylose chain to permit the effective development of the colour.

The amylopectin component of starch is structurally similar to glycogen in that both are composed of glucose units linked together in the same way, but the distance between branch points is greater in amylopectin than in glycogen, and the former may be thought of as occupying more space per unit weight.

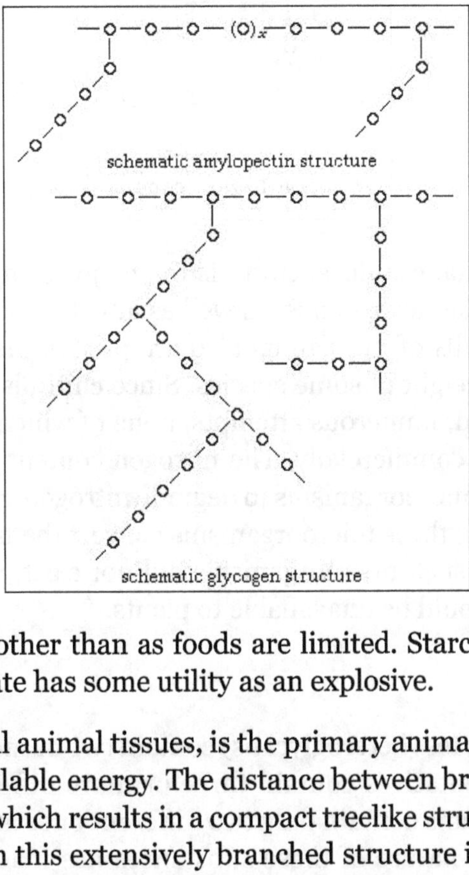

The applications of starches other than as foods are limited. Starches are employed in adhesive manufacture, and starch nitrate has some utility as an explosive.

Glycogen, which is found in all animal tissues, is the primary animal storage form of carbohydrate and, indirectly, of rapidly available energy. The distance between branch points in a glycogen molecule is only five or six units, which results in a compact treelike structure. The ability of higher animals to form and break down this extensively branched structure is essential to their well-being; in conditions known as glycogen storage diseases, these activities are abnormal, and the asymmetrical glycogen molecules that are formed have severe, often fatal, consequences. Glycogen synthesis and breakdown are controlled by substances called hormones.

Large molecules—e.g., pectins and agars—composed of galactose or its uronic-acid derivative (galacturonic acid) are important because they can form gels. Pectins, which are predominantly galacturonans, are produced from citrus fruit rinds; they are used commercially in the preparation of jellies and jams. Agar is widely employed in biological laboratories as a solidifying agent for growth media for microorganisms and in the bakery industry as a gelling agent; it forms a part of the diet of people in several areas of East Asia.

Dextrans, a group of polysaccharides composed of glucose, are secreted by certain strains of bacteria as slimes. The structure of an individual dextran varies with the strain of microorganism. Dextrans can be used as plasma expanders (substitutes for whole blood) in cases of severe shock. In addition, a dextran derivative compound is employed medically as an anticoagulant for blood.

Chitin is structurally similar to cellulose, but the repeating sugar is 2-deoxy-2-acetamido-D-glucose (N-acetyl-D-glucosamine) rather than glucose.

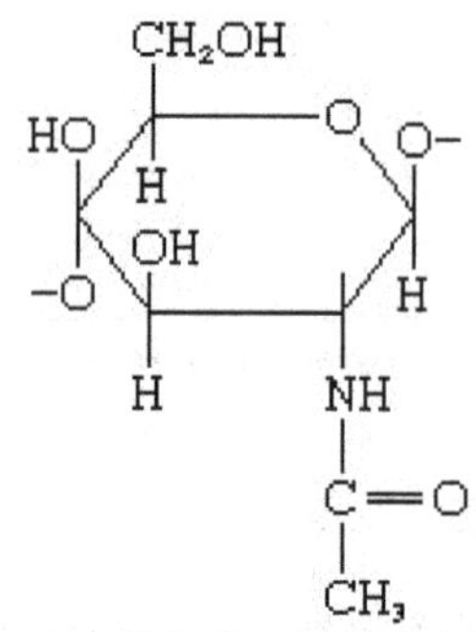

N-acetyl-D-glucosamine

Sometimes referred to as animal cellulose, chitin is the major component of the outer skeletons of insects, crustaceans, and other arthropods, as well as annelid and nematode worms, mollusks, and coelenterates. The cell walls of most fungi also are predominantly chitin, which comprises nearly 50 percent of the dry weight of some species. Since chitin is nearly as chemically inactive as cellulose and easily obtained, numerous attempts, none of which has thus far been successful, have been made to develop it commercially. The nitrogen content of the biosphere, however, is stabilized by the ability of soil microorganisms to degrade nitrogen-containing compounds such as those found in insect skeletons; these microorganisms convert the nitrogen in complex molecules to a form usable by plants. If such microorganisms did not exist, much of the organic nitrogen present in natural materials would be unavailable to plants.

Heteropolysaccharides

In general, heteropolysaccharides (heteroglycans) contain two or more different monosaccharide units. Although a few representatives contain three or more different monosaccharides, most naturally occurring heteroglycans contain only two different ones and are closely associated with lipid or protein. The complex nature of these substances has made detailed structural studies extremely difficult. The major heteropolysaccharides include the connective-tissue polysaccharides, the blood group substances, glycoproteins (combinations of carbohydrates and proteins) such as gamma globulin, and glycolipids (combinations of carbohydrates and lipids), particularly those found in the central nervous system of animals and in a wide variety of plant gums.

Representative Heteropolysaccharides			
Heteropolysaccharide	Component Sugars	Functions	Distribution
Hyaluronic acid	D-glucuronic acid and N-acetyl-D-glucosamine	Lubricant, shock absorber, water binding	Connective tissue, skin

Chondroitin-4-sulfate*	D-glucuronic acid and N-acetyl-D-galactosamine-4-O-sulfate	Calcium accumulation, cartilage and bone formation	Cartilage
Heparin*	D-glucuronic acid, L-iduronic acid, N-sulfo-D-glucosamine	Anticoagulant	Mast cells, blood
Gamma globulin*	N-acetyl-hexosamine, D-mannose, D-galactose	Antibody	Blood
Blood group substance*	D-glucosamine, D-galactosamine, L-fucose, D-galactose	Blood group specificity	Cell surfaces, especially red blood cells

*Covalently linked to protein; the proportion of protein to carbohydrate in such complex molecules varies from about 10% protein in the case of chondroitin-4-sulfate to better than 95% for gamma globulin.

The most important heteropolysaccharides are found in the connective tissues of all animals and include a group of large molecules that vary in size, shape, and interaction with other body substances. They have a structural role, and the structures of individual connective-tissue polysaccharides are related to specific animal functions; hyaluronic acid, for example, the major component of joint fluid in animals, functions as a lubricating agent and shock absorber.

The connective-tissue heteropolysaccharides contain acidic groups (uronic acids or sulfate groups) and can bind both water and inorganic metal ions. They can also play a role in other physiological functions; e.g., in the accumulation of calcium before bone formation. Ion-binding ability also appears to be related to the anticoagulant activity of the heteropolysaccharide heparin.

The size of the carbohydrate portion of glycoproteins such as gamma globulin or hen-egg albumin is usually between five and 10 monosaccharide units; several such units occur in some glycoprotein molecules. The function of the carbohydrate component has not yet been established except for glycoproteins associated with cell surfaces; in this case, they appear to act as antigenic determinants—i.e., they are capable of inducing the formation of specific antibodies.

Lipids

Lipids are any of a diverse group of organic compounds including fats, oils, hormones, and certain components of membranes that are grouped together because they do not interact appreciably with water. One type of lipid, the triglycerides, is sequestered as fat in adipose cells, which serve as the energy-storage depot for organisms and also provide thermal insulation. Some lipids such as steroid hormones serve as chemical messengers between cells, tissues, and organs, and others communicate signals between biochemical systems within a single cell. The membranes of cells and organelles (structures within cells) are microscopically thin structures formed from two layers of phospholipid molecules. Membranes function to separate individual cells from their environments and to compartmentalize the cell interior into structures that carry out special functions. So important is this compartmentalizing function that membranes, and the lipids that form them, must have been essential to the origin of life itself.

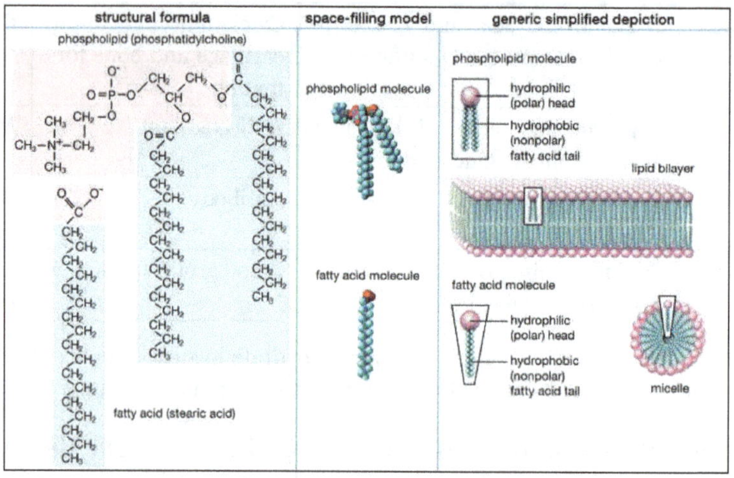

Structure and properties of two representative lipids.

Both stearic acid (a fatty acid) and phosphatidylcholine (a phospholipid) are composed of chemical groups that form polar "heads" and nonpolar "tails." The polar heads are hydrophilic, or soluble in water, whereas the nonpolar tails are hydrophobic, or insoluble in water. Lipid molecules of this composition spontaneously form aggregate structures such as micelles and lipid bilayers, with their hydrophilic ends oriented toward the watery medium and their hydrophobic ends shielded from the water.

Water is the biological milieu—the substance that makes life possible—and almost all the molecular components of living cells, whether they be found in animals, plants, or microorganisms, are soluble in water. Molecules such as proteins, nucleic acids, and carbohydrates have an affinity for water and are called hydrophilic ("water-loving"). Lipids, however, are hydrophobic ("water-fearing"). Some lipids are amphipathic—part of their structure is hydrophilic and another part, usually a larger section, is hydrophobic. Amphipathic lipids exhibit a unique Behavior in water: they spontaneously form ordered molecular aggregates, with their hydrophilic ends on the outside, in contact with the water, and their hydrophobic parts on the inside, shielded from the water. This property is key to their role as the fundamental components of cellular and organelle membranes.

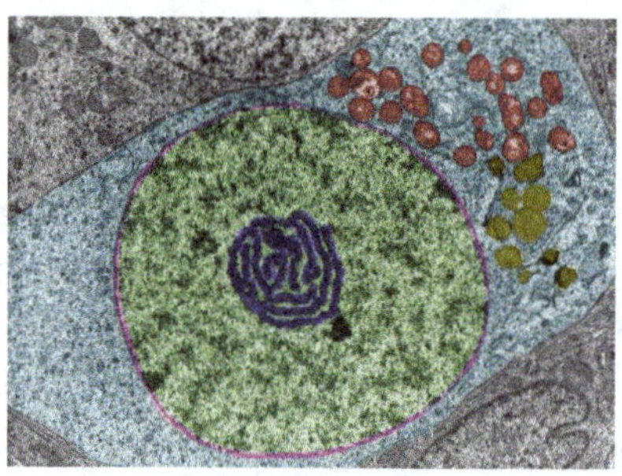

Lipid; oogonium.

A false colour transmission electron microscope micrograph of an oogonium (an egg cell of certain algae and fungi), showing an abundance of lipid droplets (yellow), a nucleus (green), an atypical nucleolus (dark blue), and mitochondria (red).

Although biological lipids are not large macromolecular polymers (e.g., proteins, nucleic acids, and polysaccharides), many are formed by the chemical linking of several small constituent molecules.

Many of these molecular building blocks are similar, or homologous, in structure. The homologies allow lipids to be classified into a few major groups: fatty acids, fatty acid derivatives, cholesterol and its derivatives, and lipoproteins.

Fatty Acids

Fatty acids rarely occur as free molecules in nature but are usually found as components of many complex lipid molecules such as fats (energy-storage compounds) and phospholipids (the primary lipid components of cellular membranes).

Structure

Biological fatty acids, members of the class of compounds known as carboxylic acids, are composed of a hydrocarbon chain with one terminal carboxyl group (COOH). The fragment of a carboxylic acid not including the hydroxyl (OH) group is called an acyl group. Under physiological conditions in water, this acidic group usually has lost a hydrogen ion (H^+) to form a negatively charged carboxylate group (COO^-). Most biological fatty acids contain an even number of carbon atoms because the biosynthetic pathway common to all organisms involves chemically linking two-carbon units together (although relatively small amounts of odd-number fatty acids do occur in some organisms). Although the molecule as a whole is water-insoluble by virtue of its hydrophobic hydrocarbon chain, the negatively charged carboxylate is hydrophilic. This common form for biological lipids—one that contains well-separated hydrophobic and hydrophilic parts—is called amphipathic.

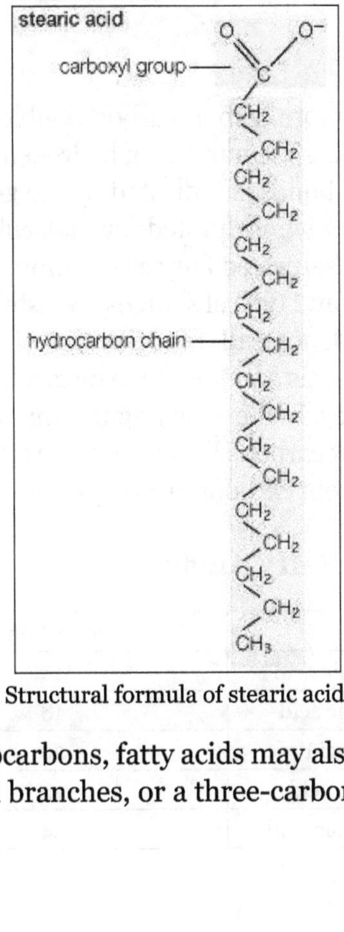

Structural formula of stearic acid.

In addition to straight-chain hydrocarbons, fatty acids may also contain pairs of carbons linked by one or more double bonds, methyl branches, or a three-carbon cyclopropane ring near the centre of the carbon chain.

Saturated Fatty Acids

The simplest fatty acids are unbranched, linear chains of CH_2 groups linked by carbon-carbon single bonds with one terminal carboxylic acid group. The term *saturated* indicates that the maximum possible number of hydrogen atoms are bonded to each carbon in the molecule. Many saturated fatty acids have a trivial or common name as well as a chemically descriptive systematic name. The systematic names are based on numbering the carbon atoms, beginning with the acidic carbon. The table gives the names and typical biological sources of the most common saturated fatty acids. Although the chains are usually between 12 and 24 carbons long, several shorter-chain fatty acids are biochemically important. For instance, butyric acid (C_4) and caproic acid (C_6) are lipids found in milk. Palm kernel oil, an important dietary source of fat in certain areas of the world, is rich in fatty acids that contain 8 and 10 carbons (C_8 and C_{10}).

Common Saturated Fatty Acids

Trivial name	Systematic name	Number of carbons in chain	Typical sources
Lauric acid	n-dodecanoic acid	12	Palm kernel oil, nutmeg
Myristic acid	n-tetradecanoic acid	14	Palm kernel oil, nutmeg
Palmitic acid	n-hexadecanoic acid	16	Olive oil, animal lipids
Stearic acid	n-octadecanoic acid	18	Cocoa butter, animal lipids
Behenic acid	n-docosanoic acid	22	Brain tissue, radish oil
Lignoceric acid	n-tetracosanoic acid	24	Brain tissue, carnauba wax

Unsaturated Fatty Acids

Unsaturated fatty acids have one or more carbon-carbon double bonds. The term unsaturated indicates that fewer than the maximum possible number of hydrogen atoms are bonded to each carbon in the molecule. The number of double bonds is indicated by the generic name—monounsaturated for molecules with one double bond or polyunsaturated for molecules with two or more double bonds. Oleic acid is an example of a monounsaturated fatty acid. Common representative monounsaturated fatty acids together with their names and typical sources are listed in the table. The prefix cis-9 in the systematic name of palmitoleic acid denotes that the position of the double bond is between carbons 9 and 10. Two possible conformations, cis and trans, can be taken by the two CH_2 groups immediately adjacent to the double-bonded carbons. In the cis configuration, the one occurring in all biological unsaturated fatty acids, the two adjacent carbons lie on the same side of the double-bonded carbons. In the trans configuration, the two adjacent carbons lie on opposite sides of the double-bonded carbons.

Common Monounsaturated Fatty Acids

Trivial name	Systematic name	Number of carbons in chain	Typical sources
Palmitoleic acid	cis-9-hexadecenoic acid	16	Marine algae, pine oil
Oleic acid	cis-9-octadecenoic acid	18	Animal tissues, olive oil
Gadoleic acid	cis-9-eicosenoic acid	20	Fish oils (cod, sardine)
Erucic acid	cis-13-docosenoic acid	22	Rapeseed oil
Nervonic acid	cis-15-tetracosenoic acid	24	Sharks, brain tissue

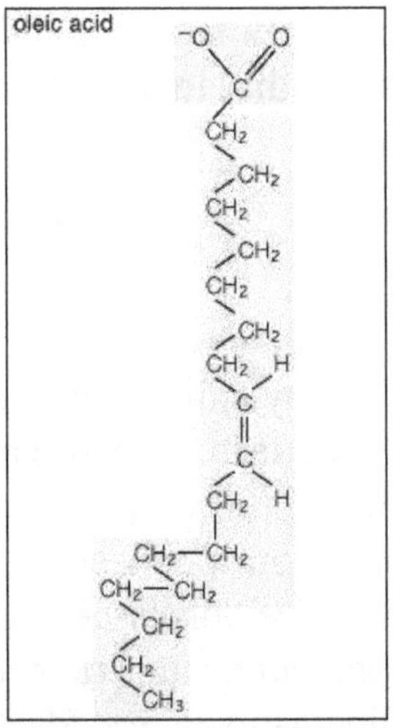

Structural formula of oleic acid.

Fatty acids containing more than one carbon-carbon double bond (polyunsaturated fatty acids) are found in relatively minor amounts. The multiple double bonds are almost always separated by a CH_2 group ($-CH_2-CH=CH-CH_2-CH=CH-CH_2-$), a regular spacing motif that is the result of the biosynthetic mechanism by which the double bonds are introduced into the hydrocarbon chain. The table lists the most common polyunsaturated fatty acids, linoleic and arachidonic, together with several that are less common. Arachidonic acid (C_{20}) is of particular interest as the precursor of a family of molecules, known as eicosanoids, that includes prostaglandins, thromboxanes, and leukotrienes. These compounds, produced by cells under certain conditions, have potent physiological properties. Animals cannot synthesize two important fatty acids, linoleic acid (an omega-6 fatty acid) and alpha-linolenic acid (an omega-3 fatty acid), that are the precursors of the eicosanoids and so must obtain them in the diet from plant sources. For this reason, these precursors are called essential fatty acids.

Common Polyunsaturated Fatty Acids

Trivial name	Systematic name	Number of carbons in chain	Typical sources
Linoleic acid	cis-9-, cis-12-octadecadienoic acid	18	Corn oil, animal tissues, bacteria
Linolenic acid	cis-9-, cis-12-, cis-15-octadecatrienoic acid	18	Animal tissues
	5,8,11-eicosatrienoic acid	20	
	8,11,14-eicosatrienoic acid	20	Brain tissue
	7,10,13-docosatrienoic acid	22	Phospholipids
	8,11,14-docosatrienoic acid	22	
Arachidonic acid	5,8,11,14-eicosatetraenoic acid	20	Liver, brain tissue
	4,7,10,13-docosatetraenoic acid	22	Brain tissue
	4,7,10,13,16,19-docosahexaenoic acid	22	Brain tissue

Trans polyunsaturated fatty acids, although not produced biosynthetically by mammals, are produced by microorganisms in the gut of ruminant animals such as cows and goats, and they are also produced synthetically by partial hydrogenation of fats and oils in the manufacture of margarine (the so-called trans fats). There is evidence that ingestion of trans fats can have deleterious metabolic effects.

Substituent Groups

In addition to the very common fatty acids with straight saturated or unsaturated acyl chains, many fatty acids are chemically modified by substituents on the hydrocarbon chain. For example, the preening gland of ducks secretes a fatty acid 10 carbons long with methyl (CH_3) groups substituted for one of the hydrogens on carbons 2, 4, 6, and 8. Some bacteria produce fatty acids that have a methyl group on the carbon atom farthest from the acidic group or on the penultimate carbon. Other bacteria incorporate a cyclopropane ring near the centre of the acyl chain. The bacterium that causes tuberculosis (*Mycobacterium tuberculosis*) synthesizes a whole family of cyclopropane-containing fatty acids called α-mycolic acids. Similar fatty acids are found in related bacteria. A third common constituent is a hydroxyl group (OH). Monohydroxyl acids are found in both plants and animals in relatively small amounts, but they are more prevalent in bacteria.

Physical Properties

Pure fatty acids form crystals that consist of stacked layers of molecules, with each layer the thickness of two extended molecules. The molecules in a layer are arranged so that the hydrophobic (water-fearing) hydrocarbon chains form the interior of the layer and the hydrophilic (water-loving) carboxylic acid groups form the two faces. For a specific fatty acid the details of the molecular packing may vary, giving rise to different crystal forms known as polymorphs.

The melting temperatures of saturated fatty acids of biological interest are above 27 °C (81 °F) and rise with increasing length of the hydrocarbon chain. Monounsaturated and polyunsaturated molecules melt at substantially lower temperatures than do their saturated analogs, with the lowest melting temperatures occurring when the carbon-carbon double bonds are located near the centre of the hydrocarbon chain, as they are in most biological molecules. As a result, these molecules form viscous liquids at room temperature.

The hydrophobic character of the hydrocarbon chain of most biological fatty acids exceeds the hydrophilic nature of the carboxylic acid group, making the water solubility of these molecules very low. For example, at 25 °C (77 °F) the solubility in grams of fatty acid per gram of solution is 3×10^{-6}. Water solubility decreases exponentially with the addition of each carbon atom to the hydrocarbon chain. This relationship reflects the energy required to transfer the molecule from a pure hydrocarbon solvent to water. With each CH_2 group, for instance, more energy is required to order water molecules around the hydrocarbon chain of the fatty acid, which results in the hydrophobic effect.

In pure water the carboxylate group can dissociate a positively charged hydrogen ion to only a very small degree thus:

$$R{-}COOH \rightarrow RCOO^- + H^+.$$

Here R represents the hydrocarbon chain. The carboxylate ion, bearing a negative charge, is more polar than the undissociated acid. RCOOH can be converted completely to the ion $RCOO^-$ by adding an equal number of molecules of a base such as sodium hydroxide (NaOH). This effectively replaces the H^+ with Na^+ to give the salt of the fatty acid, which is a soap. The very useful detergent property of soaps stems from the fact that the $RCOO^-$ anions in water spontaneously form stable, spherical aggregates called micelles. The interior of these structures, formed by the hydrocarbon chains, is an excellent solvent in which grease and hydrophobic dirt of all sorts can be sequestered. The diameter of each micelle is roughly twice the length of the extended fatty acid. Dispersions of micelles in water can be made quite concentrated and exhibit great cleansing power. These dispersions are stable and generally look very much like pure water. Bubbles and foams on the surface of soap dispersions are the result of the spontaneous adsorption of $RCOO^-$ ions at the interface between the aqueous dispersion and air, with the result that the air-water interfaces are energetically stabilized and can therefore be mechanically expanded.

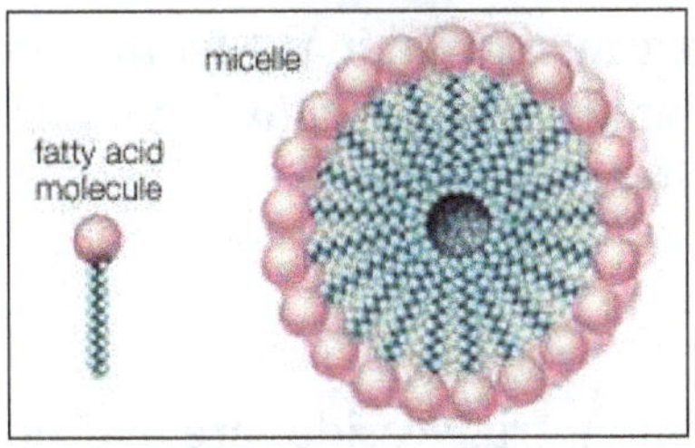

When a soap is dissolved in water, fatty acids in the soap form spherical structures called micelles, in which the hydrophilic "heads" of the fatty acid molecules are turned toward the water and the hydrophobic "tails" are sheltered in the interior.

Chemical Properties

The most chemically reactive portion of fatty acids is the acidic carboxyl group (COOH). It reacts with alcohols (R′OH) to form products known as esters (RCOOR′) and releases water in the process. This ester bond is the principal covalent bond linking fatty acid moieties to other groups in the more-complex lipids discussed in other sections of this article. A second chemical bond, occurring much less frequently in biological lipids involving fatty acids, is the ether bond (R′−O−R). Ether bonds are chemically more stable than ester bonds.

The hydrocarbon part of a fatty acid molecule is quite resistant to chemical attack unless carbon-carbon double bonds are present. A number of different kinds of molecules react with such a double bond. For example, when a catalyst such as platinum is present, hydrogen gas adds to the double bond to give a saturated fatty acid. Halogens (chlorine, bromine, and iodine) and their derivatives such as hydroiodic acid (HI) also react with the double bond to form saturated fatty acids, but in these cases one or two atoms of the halogen replace one or two of the hydrogens normally found in the saturated acyl chain. Carbon-carbon double bonds can also react with oxygen in either nonenzymatic processes or enzymatically catalyzed oxidation reactions. This process generates a variety of products, some of which contribute to the rancid smell in spoiled meat and vegetable products. In general, the more highly unsaturated the fatty acid, the more easily it is oxidized.

Biological Sources

Fatty acids are found in biological systems either as free molecules or as components of more-complex lipids. They are derived from dietary sources or produced by metabolism, as describe below.

Digestion of Dietary Fatty Acids

The main source of fatty acids in the diet is triglycerides, generically called fats. In humans, fat-constitutes an important part of the diet, and in some countries it can contribute as much as 45 percent of energy intake. Triglycerides consist of three fatty acid molecules, each linked by an ester bond to one of the three OH groups of a glycerol molecule. After ingested triglycerides pass through the stomach and into the small intestine, detergents called bile salts are secreted by the liver via the gall bladder and disperse the fat as micelles. Pancreatic enzymes called lipases then hydrolyze the dispersed fats to give monoglycerides and free fatty acids. These products are absorbed into the cells lining the small intestine, where they are resynthesized into triglycerides. The triglycerides, together with other types of lipids, are then secreted by these cells in lipoproteins, large molecular complexes that are transported in the lymph and blood to recipient organs. In detail, the process of triglyceride or fat absorption from dietary sources is quite complex and differs somewhat depending upon the animal species.

Storage

After transport through the circulation, triglycerides are hydrolyzed yet again to fatty acids in the adipose tissue. There they are transported into adipose cells, where once again they are resynthesized into triglycerides and stored as droplets. Fat or adipose tissue essentially consists of cells, whereby the interior of each cell is largely occupied by a fat droplet. The triglyceride in these droplets is available to the body on demand as communicated to adipose tissue by hormone messengers.

Various animals store triglycerides in different ways. In sharks, for example, fat is stored in the liver, but in bony fish it is deposited in and around muscle fibres. Insects store fat in a special organ called the fat body. The development of true adipose tissue is found only in higher animals.

Biosynthesis

In mammals, fatty acids are synthesized in adipose and liver cells from glucose via a fairly complex pathway. In essence, the six carbons of a glucose molecule are oxidized to a pair of two-carbon carboxylic acid fragments called acetate. The starting point for biosynthesis is an acetate group chemically linked to a molecule of CoA (coenzyme A). The process of building up the acyl chain of a fatty acid then begins, basically through the sequential chemical addition of two-carbon fragments from CoA-acetate to generate, for example, the 16-carbon saturated fatty acid palmitate. This process is catalyzed by a complex enzyme known as fatty acid synthase. Elongation of the palmitate carbon chain and the introduction of carbon-carbon double bonds are carried out subsequently by other enzyme systems. The overall process is basically the same in organisms ranging from bacteria to humans.

Fatty Acid Derivatives

Structure of Triglycerides

Triglycerides (chemical name triacylglycerol), the principal means of storing fatty acids in biological systems, are a class of compounds that consist of glycerol (a three-carbon trihydroxy alcohol) with a fatty acid linked to each of the three OH groups by an ester bond. An example of a typical triglyceride is tristearin. Because this molecule contains only one type of fatty acid, it is referred to as a simple triglyceride. Almost all naturally occurring triglyceride molecules, however, contain more than one type of fatty acid; when two or more in a given molecule are different, it is called a mixed triglyceride. For any specific combination of three fatty acids, three different molecules are possible, depending upon which of the three fatty acids is bonded to the central carbon of glycerol. Considering the numbers of common saturated, monounsaturated, and polyunsaturated fatty acids, it is evident that there are a great many different triglycerides.

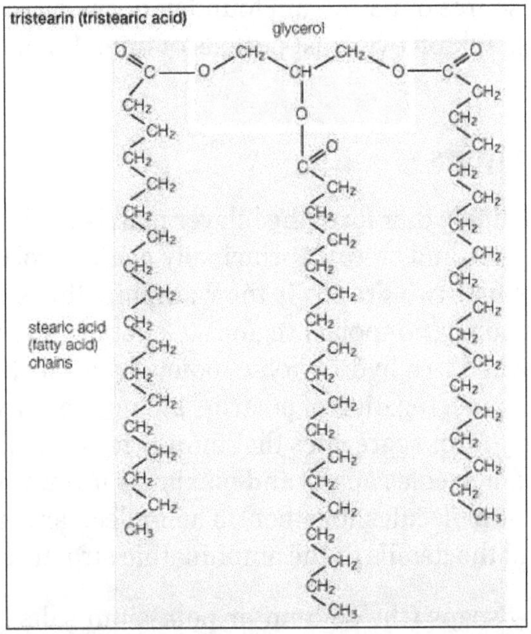

Structural formula of tristearin (tristearic acid).

Physical Properties

Triglycerides are hydrophobic substances that are soluble only in some organic solvents. Unlike many other types of complex lipids, they possess no electric charges and are therefore referred to as neutral lipids. The molecular structure of a few triglycerides that have been studied in crystals indicates that the acyl chains on the 1 and 2 carbons of glycerol, together with the 1 and 2 carbons of glycerol itself, form a straight line. Carbon 3 projects at right angles to this line, but the acyl chain on its glycerol folds over at the carboxyl carbon to lie alongside the acyl chain on carbon 1. Triglyceride molecules look much like a tuning fork and, when packed together, produce layered crystals.

The melting temperatures of mixed triglycerides are roughly an average of the melting temperatures of their constituent fatty acids. In simple triglycerides, melting temperatures rise with increasing acyl chain length but drop with increasing number of double bonds. Melted triglycerides are generally quite viscous oils. From the physiological standpoint, it is important that most stored triglycerides be fluid at body temperature in order to permit their rapid mobilization as an energy source. Liquidity is also important since subcutaneous stored fats perform an insulating function that must not interfere with the mobility of the organism and its parts.

Waxes

A second group of neutral lipids that are of physiological importance, though they are a minor component of biological systems, are waxes. Essentially, waxes consist of a long-chain fatty acid linked through an ester oxygen to a long-chain alcohol. These molecules are completely water-insoluble and generally solid at biological temperatures. Their strongly hydrophobic nature allows them to function as water repellents on the leaves of some plants, on feathers, and on the cuticles of certain insects. Waxes also serve as energy-storage substances in plankton (microscopic aquatic plants and animals) and in higher members of the aquatic food chain. Plankton apparently use the biosynthesis of waxes to adjust their buoyant density and thus their depth in the ocean. It has been suggested that a major source of petroleum found in deep-sea sediments originates from the deposition of wax-rich dead plankton over vast periods of time. Whales and many fishes also store large quantities of waxes.

Biological Membrane Lipids

The three principal classes of lipids that form the bilayer matrix of biological membranes are glycerophospholipids, sphingolipids, and sterols (principally cholesterol). The most important characteristic of molecules in the first two groups is their amphipathic structure—well separated hydrophilic (polar) and hydrophobic (nonpolar) regions. Generally, their shape is elongated, with a hydrophilic end or head attached to a hydrophobic moiety by a short intervening region of intermediate polarity. Because of the segregation of polarity and nonpolarity, amphipathic molecules in any solvent will spontaneously form aggregates that minimize energetically unfavourable contacts (by keeping unlike regions of molecules apart) and maximize favourable contacts with the solvent (by keeping similar regions of molecules together). The molecular arrangement of the aggregate depends upon the solvent and the details of the amphipathic structure of the lipid.

In water, micelles formed by soaps (the sodium or potassium salts of fatty acids) are one such aggregate. The polar or hydrophilic portion of the soap molecules forms the surface of the micelle, while the hydrocarbon chains form its interior and are thus completely protected from the energetically unfavourable contact with water. Biological membrane lipids, however, do not form spherical micelles in water but instead form topologically closed lamellar (layered) structures. The polar heads of the component molecules form the two faces of the lamella, while the hydrophobic moieties form its interior. Each lamella is thus two molecules in thickness, with the long axis of the component molecules perpendicular to the plane of the bilayer.

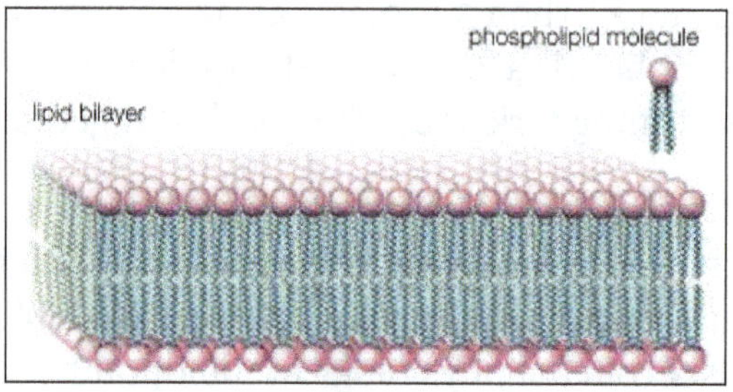

Phospholipid molecules, like molecules of many lipids, are composed of a hydrophilic "head" and one or more hydrophobic "tails." In a water medium, the molecules form a lipid bilayer, or two-layered sheet, in which the heads are turned toward the watery medium and the tails are sheltered inside, away from the water. This bilayer is the basis of the membranes of living cells.

Other types of aggregates are also formed in water by certain amphipathic lipids. For example, liposomes are artificial collections of lipids arranged in a bilayer, having an inside and an outside surface. The lipid bilayers form a sphere that can trap a molecule inside. The liposome structure can be useful for protecting sensitive molecules that are to be delivered orally.

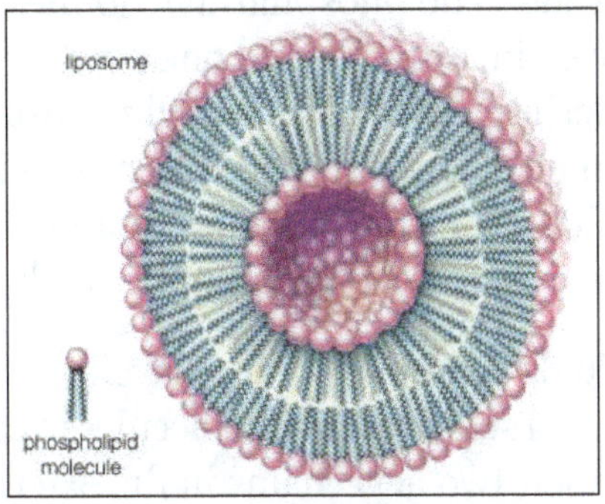

Phospholipids can be used to form artificial structures called liposomes, which are double-walled, hollow spheres useful for encapsulating other molecules such as pharmaceutical drugs.

Glycerophospholipids

Lipids of this class are the most abundant in biological membranes. In glycerophospholipids, fatty acids are linked through an ester oxygen to carbons 1 and 2 of glycerol, the backbone of the molecule. Phosphate is ester-linked to carbon 3, while any one of several possible substituents is also linked to the phosphate moiety. Glycerophospholipids are amphipathic—glycerol and phosphate form the polar end of the molecule, while hydrocarbon chains form the nonpolar end. Although the fatty acids can be any of those common in biological systems, usually those attached to carbon 1 are saturated and those attached to carbon 2 are unsaturated. The various combinations of two fatty acids give rise to many different molecules bearing the same substituent group. Since this is true for each head group, there are altogether about a thousand possible types of glycerophospholipids. The great majority are found in biological membranes.

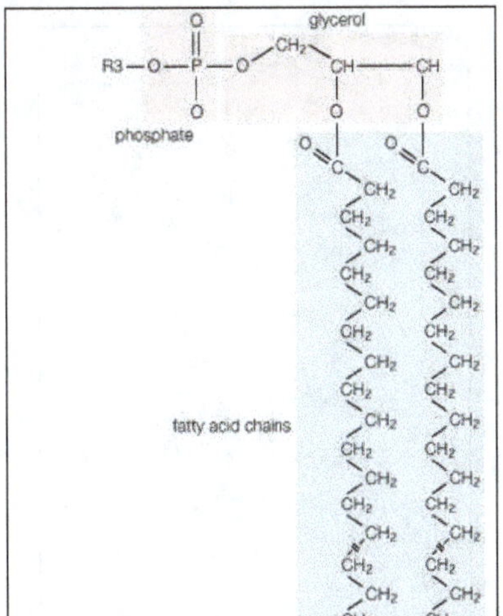

General structural formula of a glycerophospholipid. The composition of the specific molecule depends on the chemical group (designated R3 in the diagram) linked to the phosphate and glycerol "head" and also on the lengths of the fatty acid "tails" (R1 and R2).

From the standpoint of physical properties, the greatest difference among various molecules lies in the particular substituent. This is due in part to the different sizes of the various types and in part to differences in their electric charges. The phosphatidylcholines and phosphatidylethanolamines are zwitterionic, meaning they have one negative and one positive charge on the substituent group. Phosphatidic acid, phosphatidylserine, and phosphatidylinositol have a net negative charge. Differences in fatty acid composition also contribute to differences in physical properties of a series of molecules with the same substituent. In the presence of an excess of water, the molecules form aggregates with a variety of geometries, the most common of which is the bilayer.

In bilayers many glycerophospholipids as well as sphingomyelin can be in either one of two states, gel or liquid-crystalline. In the solidlike gel state, the lipid molecules in each half of the bilayer are arranged in a two-dimensional lattice, with their two acyl chains in the extended form. With the application of heat, the gel converts into the liquid-crystalline state at some temperature characteristic of the lipid mixture. In this state the molecules in each half of the bilayer remain in a fairly regular two-dimensional lattice but are free to rotate about their long axes and slide laterally through the layer. Their acyl chains now undergo considerable motion, leading to transiently kinked conformations. These motions give the bilayer a quasi-liquid Behavior that is characteristic of the bilayers in all biological membranes.

Sphingolipids

A second major class of lipids usually associated with the membranes surrounding cells is sphingolipids. Sphingolipids are based on an 18-carbon amine alcohol, sphingosine, and to a much lesser extent on a 20-carbon analog, phytosphingosine. All but one generic member of this class have a simple or complex sugar linked to the alcohol on carbon 1. The single deviant member is sphingomyelin, a molecule with a phosphorylcholine group (the same polar head group as in phosphatidylcholine) instead of the sugar moiety, making it an analog of phosphatidylcholine. All sphingolipids have, in addition to the sugar, a fatty acid attached to the amino group of sphingosine. Among the sphingolipids, only sphingomyelin, a phospholipid, is a major component of biological membranes.

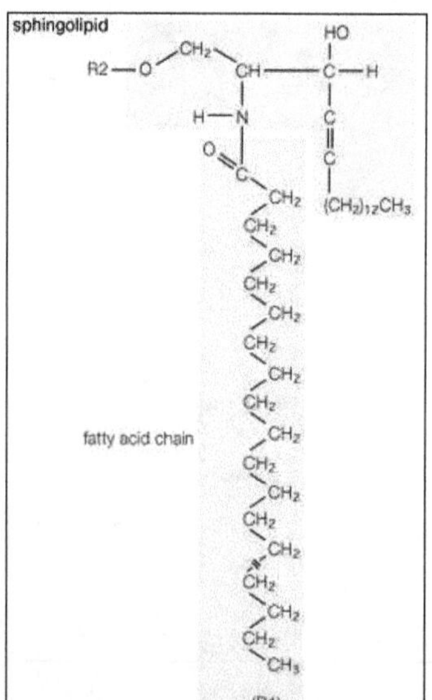

General structural formula of a sphingolipid. The composition of the specific molecule depends on the chemical group (designated R2 in the diagram) linked to the alcohol "head" and also on the length of the fatty acid "tail" (R1).

The principal factor determining the physical properties of sphingolipids is the substituent group attached to carbon 1 of sphingosine. Minor variations in properties depend upon the particular fatty acid component. The glycosphingolipids, all containing a sugar attached to carbon 1 of sphingosine, have physical properties that depend primarily on the complexity and composition of this substituent. Two generic types of glycosphingolipids are recognized: neutral glycosphingolipids, which contain only neutral sugars, and gangliosides, which contain one or more sialic acid residues linked to the sugar. Many hundreds of different glycosphingolipids have been isolated, and many more unidentified types probably exist. Glycosphingolipids are found exclusively on the external surface of the cell membrane, where their sugar moieties often act as antigens and as receptors for hormones and other signaling molecules.

Lipoproteins

Lipoproteins are lipid-protein complexes that allow all lipids derived from food or synthesized in specific organs to be transported throughout the body by the circulatory system. The basic structure of these aggregates is that of an oil droplet made up of triglycerides and cholesteryl esters surrounded by a layer of proteins and amphipathic lipids—very similar to that of a micelle, a spherical structure described in the section Fatty acids. If the concentration of one or another lipoprotein becomes too high, then a fraction of the complex becomes insoluble and is deposited on the walls of arteries and capillaries. This buildup of deposits is called atherosclerosis and ultimately results in blockage of critical arteries to cause a heart attack or stroke. Because of the gravity of this condition, much research is focused on lipoproteins and their functions. The emphasis in the following discussion is therefore placed on human lipoproteins.

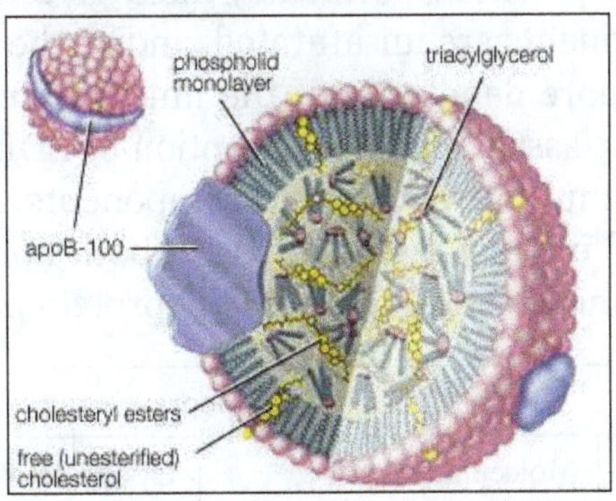

Cutaway view of a low-density lipoprotein (LDL) complex.

The LDL complex is essentially a droplet of triacylglycerols and cholesteryl esters encased in a sphere made up of phospholipid, free cholesterol, and protein molecules known as apoprotein B-100 (ApoB-100). The LDL complex is the principal vehicle for delivering cholesterol to body tissues through the blood.

Classification and Formation

There are four major classes of circulating lipoproteins, each with its own characteristic protein and lipid composition. They are chylomicrons, very low-density lipoproteins (VLDL), low-density lipoproteins (LDL), and high-density lipoproteins (HDL). Within all these classes of complexes, the various molecular components are not chemically linked to each other but are simply associated in such a way as to minimize hydrophobic contacts with water. The most distinguishing feature of each class is the relative amounts of lipid and protein. Because the lipid and protein composition is reflected in the density of each lipoprotein (lipid molecules being less dense than proteins), density, an easily measured attribute, forms the operational basis of defining the lipoprotein classes. Measuring density also provides the basis of separating and purifying lipoproteins from plasma for study and diagnosis. The table gives a summary of the characteristics of the lipoprotein classes and shows the correlation between composition and density.

Human Plasma Lipoproteins					
	Chylomicron	VLDL	IDL	LDL	HDL
Density (g/ml)	<0.95	0.950–1.006	1.006–1.019	1.019–1.063	1.063–1.210
Components (% dry weight)					
Protein	2	7	15	20	40–55
Triglycerides	83	50	31	10	8
Free cholesterol	2	7	7	8	4
Cholesteryl esters	3	12	23	42	12–20
Phospholipids	7	20	22	22	22
Apoprotein composition	A-I, A-II, B-48, C-I, C-II, C-III	B-100, C-I, C-II, C-III, E	B-100, C-I, C-II, C-III, E	B-100	A-I, A-II, C-I, C-II, C-III, D, E

The principal lipid components are triglycerides, cholesterol, cholesteryl esters, and phospholipids. The hydrophobic core of the particle is formed by the triglycerides and cholesteryl esters. The fatty acyl chains of these components are unsaturated, and so the core structure is liquid at body temperature. The table gives more details about the nine different protein components, called apoproteins, of the lipoprotein classes. With the exception of LDL, which contains only one type of apoprotein, all classes have multiple apoprotein components. All the apoproteins, like phospholipids, are amphipathic and interact favourably with both lipids and water. A more-detailed consideration of the character and functions of these lipoprotein particles follows.

Human Plasma Apolipoproteins		
Apolipoprotein	Molecular weight	Lipoprotein distribution
apoA-I	28,331	HDL
apoA-II	17,380	HDL
apoB-48	241,000	chylomicrons
apoB-100	500,000	VLDL, LDL
apoC-I	7,000	HDL, VLDL
apoC-II	8,837	chylomicrons, VLDL, HDL
apoC-III	8,750	chylomicrons, VLDL, HDL
apoD	33,000	HDL
apoE	34,145	chylomicrons, VLDL, HDL

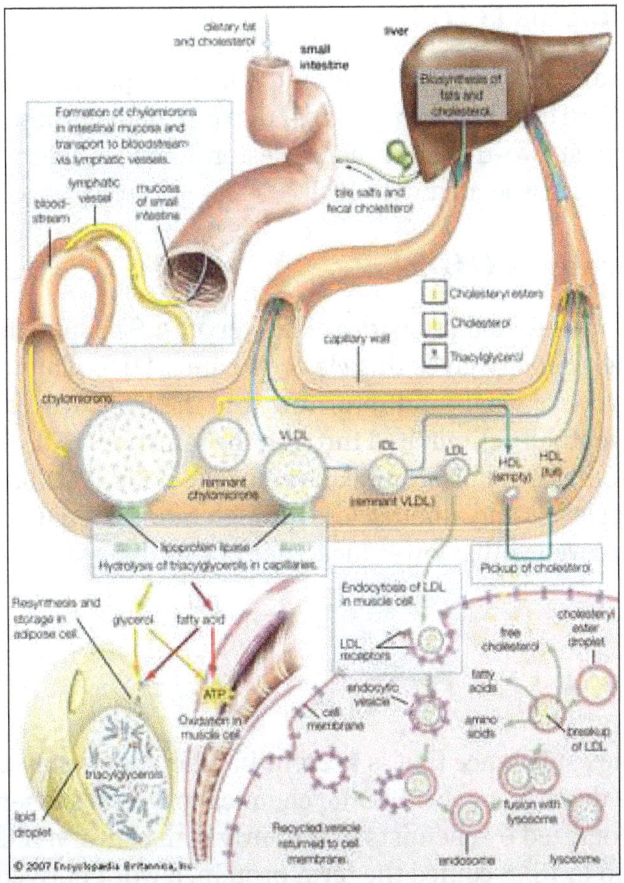

Synthesis of lipoprotein complexes in the small intestine, liver, and blood plasma and their delivery to peripheral tissues of the body.

Chylomicrons

Chylomicrons are the largest lipoproteins, with diameters of 75–600 nanometres (nm; 1 nm = 10^{-9} metre). They have the lowest protein-to-lipid ratio (being about 90 percent lipid) and therefore the lowest density. Chylomicrons are synthesized by the absorptive cells of the intestinal lining and are secreted by these cells into the lymphatic system, which joins the blood circulation at the subclavian vein. The triglyceride, cholesteryl ester, and free cholesterol content of these particles is derived from the digestion of dietary fat. Their primary destinations in peripheral areas are heart muscle, skeletal muscle, adipose tissue, and lactating mammary tissue. The transfer of triglycerides and cholesteryl esters to the tissues depletes the lipid-protein aggregates of these substances and leaves remnant chylomicrons, which are eventually taken up by the liver. The lipid and protein remnants are used to form VLDL and LDL.

Very Low-density Lipoproteins (VLDL)

VLDL is a lipoprotein class synthesized by the liver that is analogous to the chylomicrons secreted by the intestine. Its purpose is also to deliver triglycerides, cholesteryl esters, and cholesterol to peripheral tissues. VLDL is largely depleted of its triglyceride content in these tissues and gives rise to an intermediate-density lipoprotein (IDL) remnant, which is returned to the liver in the bloodstream. As might be expected, the same proteins are present in both VLDL and IDL.

Low-density Lipoproteins (LDL)

Low-density lipoproteins are derived from VLDL and IDL in the plasma and contain a large amount of cholesterol and cholesteryl esters. Their principal role is to deliver these two forms of cholesterol to peripheral tissues. Almost two-thirds of the cholesterol and its esters found in plasma (blood free of red and white cells) is associated with LDL.

High-density Lipoproteins (HDL)

Lipoproteins of this class are the smallest, with a diameter of 10.8 nm and the highest protein-to-lipid ratio. The resulting high density gives this class its name. HDL plays a primary role in the removal of excess cholesterol from cells and returning it to the liver, where it is metabolized to bile acids and salts that are eventually eliminated through the intestine. LDL and HDL together are the major factors in maintaining the cholesterol balance of the body. Because of the high correlation between blood cholesterol levels and atherosclerosis, high ratios of HDL to cholesterol (principally as found in LDL) correlate well with a lower incidence of this disease in humans.

Proteins

Protein is a highly complex substance that is present in all living organisms. Proteins are of great nutritional value and are directly involved in the chemical processes essential for life. The importance of proteins was recognized by chemists in the early 19th century, including Swedish chemist Jöns Jacob Berzelius, who in 1838 coined the term protein, a word derived from the Greek proteios, meaning "holding first place." Proteins are species-specific; that is, the proteins of one species differ from those of another species. They are also organ-specific; for instance, within a single organism, muscle proteins differ from those of the brain and liver.

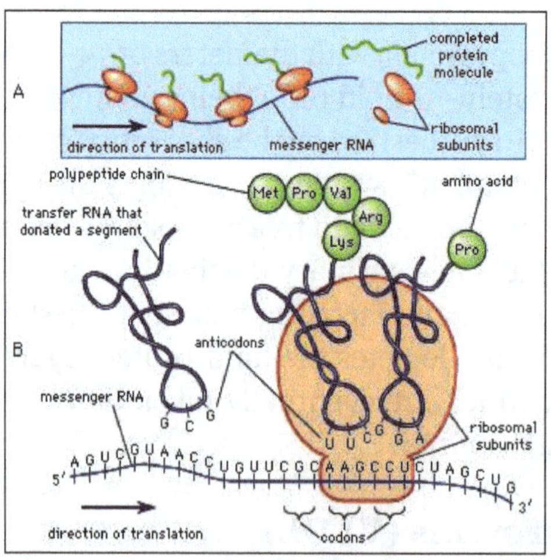

Synthesis of protein.

A protein molecule is very large compared with molecules of sugar or salt and consists of many amino acids joined together to form long chains, much as beads are arranged on a string. There are about 20 different amino acids that occur naturally in proteins. Proteins of similar function have similar amino acid composition and sequence. Although it is not yet possible to explain all of the functions of a protein from its amino acid sequence, established correlations between structure and function can be attributed to the properties of the amino acids that compose proteins.

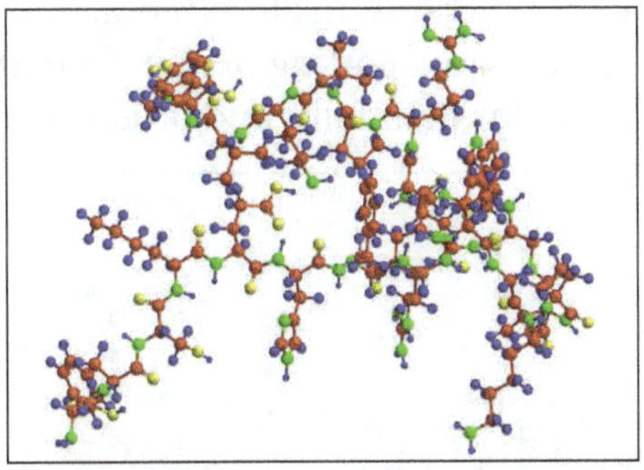

Peptide: The molecular structure of a peptide (a small protein) consists of a sequence of amino acids.

Plants can synthesize all of the amino acids; animals cannot, even though all of them are essential for life. Plants can grow in a medium containing inorganic nutrients that provide nitrogen, potassium, and other substances essential for growth. They utilize the carbon dioxide in the air during the process of photosynthesis to form organic compounds such as carbohydrates. Animals, however, must obtain organic nutrients from outside sources. Because the protein content of most plants is low, very large amounts of plant material are required by animals, such as ruminants (e.g., cows), that eat only plant material to meet their amino acid requirements. Nonruminant animals, including humans, obtain proteins principally from animals and their products—e.g., meat, milk, and eggs. The seeds of legumes are increasingly being used to prepare inexpensive protein-rich food.

Legume; amino acid.

Legumes—such as beans, lentils, and peas—are high in protein and contain many essential amino acids.

The protein content of animal organs is usually much higher than that of the blood plasma. Muscles, for example, contain about 30 percent protein, the liver 20 to 30 percent, and red blood cells 30 percent. Higher percentages of protein are found in hair, bones, and other organs and tissues with a low water content. The quantity of free amino acids and peptides in animals is much smaller than the amount of protein; protein molecules are produced in cells by the stepwise alignment of amino acids and are released into the body fluids only after synthesis is complete.

The high protein content of some organs does not mean that the importance of proteins is related to their amount in an organism or tissue; on the contrary, some of the most important proteins, such as enzymes and hormones, occur in extremely small amounts. The importance of proteins is related principally to their function. All enzymes identified thus far are proteins. Enzymes, which are the catalysts of all metabolic reactions, enable an organism to build up the chemical substances necessary for life—proteins, nucleic acids, carbohydrates, and lipids—to convert them into other substances, and to degrade them. Life without enzymes is not possible. There are several protein hormones with important regulatory functions. In all vertebrates, the respiratory protein hemoglobin acts as oxygen carrier in the blood, transporting oxygen from the lung to body organs and tissues. A large group of structural proteins maintains and protects the structure of the animal body.

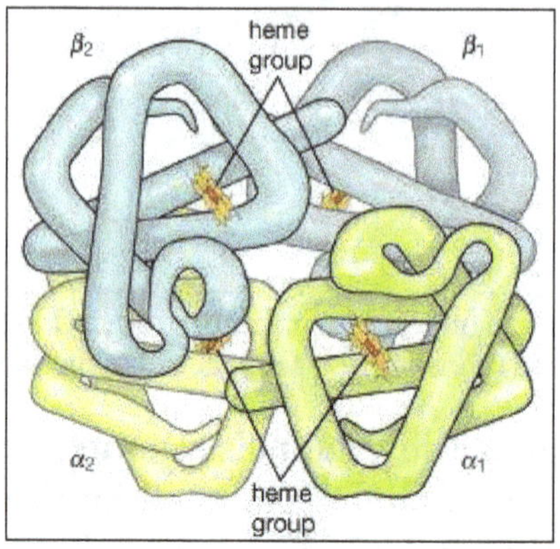

Hemoglobin is a protein made up of four polypeptide chains (α_1, α_2, β_1, and β_2). Each chain is attached to a heme group composed of porphyrin (an organic ringlike compound) attached to an iron atom. These iron-porphyrin complexes coordinate oxygen molecules reversibly, an ability directly related to the role of hemoglobin in oxygen transport in the blood.

General Structure and Properties of Proteins

Amino Acid Composition of Proteins

The common property of all proteins is that they consist of long chains of α-amino (alpha amino) acids. The general structure of α-amino acids is shown in. The α-amino acids are so called because the α-carbon atom in the molecule carries an amino group ($-NH_2$); the α-carbon atom also carries a carboxyl group ($-COOH$).

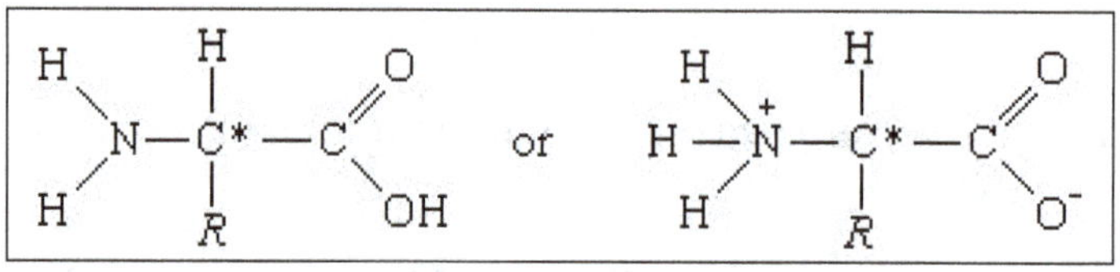

In acidic solutions, when the pH is less than 4, the $-COO$ groups combine with hydrogen ions (H^+) and are thus converted into the uncharged form ($-COOH$). In alkaline solutions, at pH above 9, the ammonium groups ($-NH^+_3$) lose a hydrogen ion and are converted into amino groups ($-NH_2$). In the pH range between 4 and 8, amino acids carry both a positive and a negative charge and therefore do not migrate in an electrical field. Such structures have been designated as dipolar ions, or zwitterions (i.e., hybrid ions).

Although more than 100 amino acids occur in nature, particularly in plants, only 20 types are commonly found in most proteins. In protein molecules the α-amino acids are linked to each other by peptide bonds between the amino group of one amino acid and the carboxyl group of its neighbour.

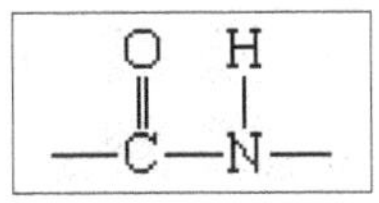

The condensation (joining) of three amino acids yields the tripeptid.

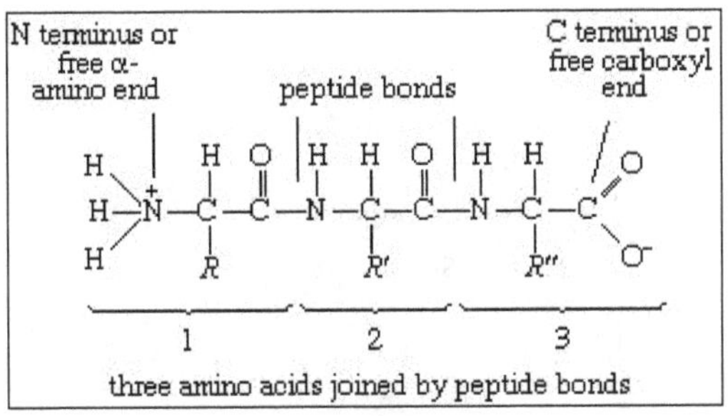

It is customary to write the structure of peptides in such a way that the free α-amino group (also called the N terminus of the peptide) is at the left side and the free carboxyl group (the C terminus) at the right side. Proteins are macromolecular polypeptides—i.e., very large molecules composed of many peptide-bonded amino acids. Most of the common ones contain more than 100 amino acids linked to each other in a long peptide chain. The average molecular weight (based on the weight of a hydrogen atom as 1) of each amino acid is approximately 100 to 125; thus, the molecular weights of proteins are usually in the range of 10,000 to 100,000 daltons (one dalton is the weight of one hydrogen atom). The species-specificity and organ-specificity of proteins result from differences in the number and sequences of amino acids. Twenty different amino acids in a chain 100 amino acids long can be arranged in far more than 10100 ways (10100 is the number one followed by 100 zeroes).

Structures of Common Amino Acids

The amino acids present in proteins differ from each other in the structure of their side (*R*) chains. The simplest amino acid is glycine, in which *R* is a hydrogen atom. In a number of amino acids, *R* represents straight or branched carbon chains. One of these amino acids is alanine, in which *R* is the methyl group ($-CH_3$). Valine, leucine, and isoleucine, with longer *R* groups, complete the alkyl side-chain series. The alkyl side chains (*R* groups) of these amino acids are nonpolar; this means that they have no affinity for water but some affinity for each other. Although plants can form all of the alkyl amino acids, animals can synthesize only alanine and glycine; thus valine, leucine, and isoleucine must be supplied in the diet.

Two amino acids, each containing three carbon atoms, are derived from alanine; they are serine and cysteine. Serine contains an alcohol group ($-CH_2OH$) instead of the methyl group of alanine, and cysteine contains a mercapto group ($-CH_2SH$). Animals can synthesize serine but not cysteine or cystine. Cysteine occurs in proteins predominantly in its oxidized form (oxidation in this sense meaning the removal of hydrogen atoms), called cystine. Cystine consists of two cysteine molecules linked by the disulfide bond ($-S-S-$) that results when a hydrogen atom is removed from the mercapto group of each of the cysteines. Disulfide bonds are important in protein structure because they allow the linkage of two different parts of a protein molecule to—and thus the formation of loops in—the otherwise straight chains. Some proteins contain small amounts of cysteine with free sulfhydryl ($-SH$) groups.

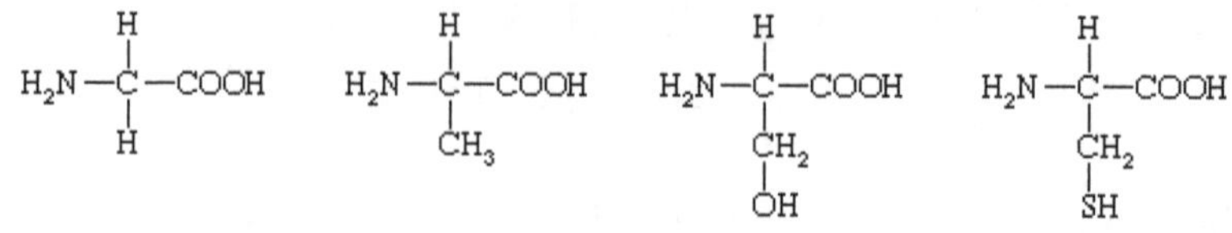

glycine (Gly, G) | alanine (Ala, A) | serine (Ser, S) | cysteine (CysH, C)

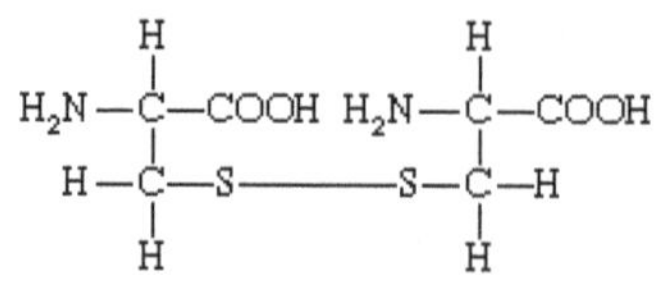

cystine (Cys-S-S-Cys, C-C)

Four amino acids, each consisting of four carbon atoms, occur in proteins; they are aspartic acid, asparagine, threonine, and methionine. Aspartic acid and asparagine, which occur in large amounts, can be synthesized by animals. Threonine and methionine cannot be synthesized and thus are essential amino acids; i.e., they must be supplied in the diet. Most proteins contain only small amounts of methionine.

Proteins also contain an amino acid with five carbon atoms (glutamic acid) and a secondary amine (in proline), which is a structure with the amino group ($-NH_2$) bonded to the alkyl side chain, forming a ring. Glutamic acid and aspartic acid are dicarboxylic acids; that is, they have two carboxyl groups ($-COOH$).

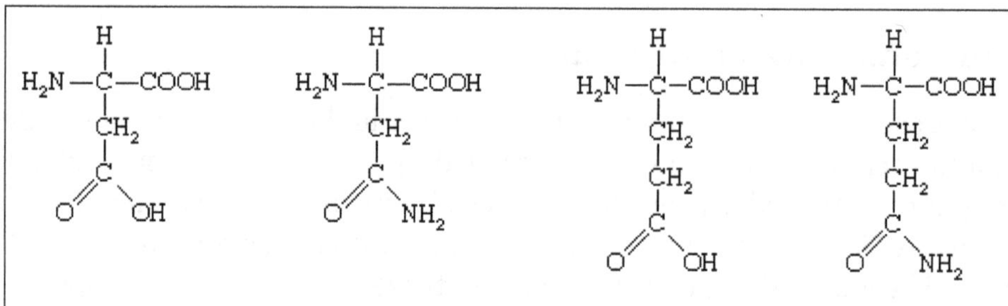

aspartic acid (Asp, D; Asx or B) | asparagine ($AspNH_2$ or Asn, N; Asx or B) | glutamic acid (Glu, E; Glx or Z) | glutamine ($GluNH_2$, GluN, or Gln, Q; Glx or Z)

Glutamine is similar to asparagine in that both are the amides of their corresponding dicarboxylic acid forms; i.e., they have an amide group ($-CONH_2$) in place of the carboxyl ($-COOH$) of the side chain. Glutamic acid and glutamine are abundant in most proteins; e.g., in plant proteins they sometimes comprise more than one-third of the amino acids present. Both glutamic acid and glutamine can be synthesized by animals.

	Protein					
Amino Acid	Alpha-Casein	Gliadin	Edestin	Collagen (Ox Hide)	Keratin (Wool)	Myosin
Lysine	60.9	4.45	19.9	27.4	6.2	85
Histidine	18.7	11.7	18.6	4.5	19.7	15

Arginine	24.7	15.7	99.2	47.1	56.9	41
Aspartic Acid**	63.1	10.1	99.4	51.9	51.5	85
Threonine	41.2	17.6	31.2	19.3	55.9	41
Serine	63.1	46.7	55.7	41.0	79.5	41
Glutamic Acid**	153.1	311.0	144.9	76.2	99.0	155
Proline	71.3	117.8	32.9	125.2	58.3	22
Glycine	37.3	—	68.0	354.6	78.0	39
Alanine	41.5	23.9	57.7	115.7	43.8	78
Half-Cystine	3.6	21.3	10.9	0.0	105.0	86
Valine	53.8	22.7	54.6	21.4	46.6	42
Methionine	16.8	11.3	16.4	6.5	4.0	22
Isoleucine	48.8	90.8***	41.9	14.5	29.0	42
Leucine	60.3		60.0	28.2	59.9	79
Tyrosine	44.7	17.7	26.9	5.5	28.7	18
Phenylalanine	27.9	39.0	38.4	13.9	22.4	27
Tryptophan	7.8	3.2	6.6	0.0	9.6	—
Hydroxyproline	0.0	0.0	0.0	97.5	12.2	—
Hydroxylysine	—	—	—	8.0	1.2	—
Total	839	765	883	1,058	863	832
Average Residual Weight	119	131	113	95	117	120

*Number of gram molecules of amino acid per 100,000 grams of protein.

**The values for aspartic acid and glutamic acid include asparagine and glutamine, respectively.

***Isoleucine Plus Leucine.

The amino acids proline and hydroxyproline occur in large amounts in collagen, the protein of the connective tissue of animals. Proline and hydroxyproline lack free amino ($-NH_2$) groups because the amino group is enclosed in a ring structure with the side chain; they thus cannot exist in a zwitterion form. Although the nitrogen-containing group (>NH) of these amino acids can form a peptide bond with the carboxyl group of another amino acid, the bond so formed gives rise to a kink in the peptide chain; i.e., the ring structure alters the regular bond angle of normal peptide bonds.

Proteins usually are almost neutral molecules; that is, they have neither acidic nor basic properties. This means that the acidic carboxyl ($-COO^-$) groups of aspartic and glutamic acid are about equal in number to the amino acids with basic side chains. Three such basic amino acids, each containing six carbon atoms, occur in proteins. The one with the simplest structure, lysine, is synthesized by plants but not by animals. Even some plants have a low lysine content. Arginine is found in all proteins; it occurs in particularly high amounts in the strongly basic protamines (simple proteins composed of relatively few amino acids) of fish sperm. The third basic amino acid is histidine. Both arginine and histidine can be synthesized by animals. Histidine is a weaker base than either lysine or arginine. The imidazole ring, a five-membered ring structure containing two nitrogen atoms in the side chain of histidine, acts as a buffer (i.e., a stabilizer of hydrogen ion concentration) by binding hydrogen ions (H^+) to the nitrogen atoms of the imidazole ring.

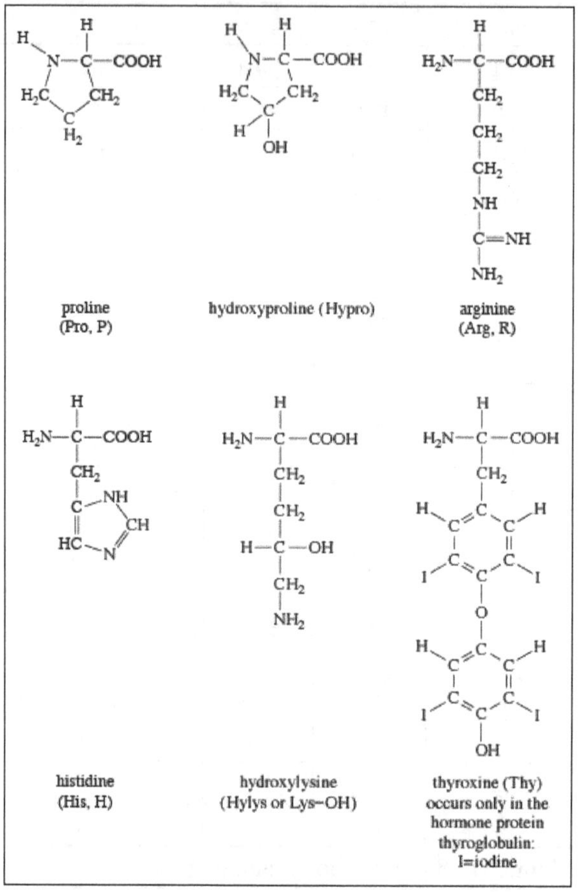

The remaining amino acids–phenylalanine, tyrosine, and tryptophan–have in common an aromatic structure; i.e., a benzene ring is present. These three amino acids are essential, and, while animals cannot synthesize the benzene ring itself, they can convert phenylalanine to tyrosine.

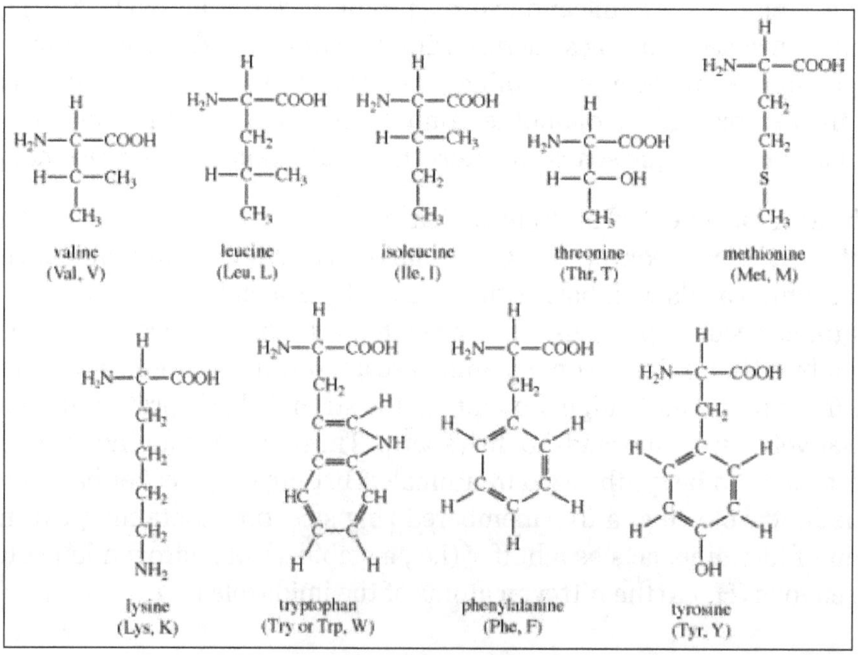

Because these amino acids contain benzene rings, they can absorb ultraviolet light at wavelengths between 270 and 290 nanometres (nm; 1 nanometre = 10^{-9} metre = 10 angstrom units). Phenylalanine absorbs very little ultraviolet light; tyrosine and tryptophan, however, absorb it strongly and are responsible for the absorption band most proteins exhibit at 280–290 nanometres. This absorption is often used to determine the quantity of protein present in protein samples.

Most proteins contain only the amino acids described above; however, other amino acids occur in proteins in small amounts. For example, the collagen found in connective tissue contains, in addition to hydroxyproline, small amounts of hydroxylysine. Other proteins contain some monomethyl-, dimethyl-, or trimethyllysine—i.e., lysine derivatives containing one, two, or three methyl groups ($-CH_3$). The amount of these unusual amino acids in proteins, however, rarely exceeds 1 or 2 percent of the total amino acids.

Physicochemical Properties of the Amino Acids

The physicochemical properties of a protein are determined by the analogous properties of the amino acids in it.

The α-carbon atom of all amino acids, with the exception of glycine, is asymmetric; this means that four different chemical entities (atoms or groups of atoms) are attached to it. As a result, each of the amino acids, except glycine, can exist in two different spatial, or geometric, arrangements (i.e., isomers), which are mirror images akin to right and left hands.

These isomers exhibit the property of optical rotation. Optical rotation is the rotation of the plane of polarized light, which is composed of light waves that vibrate in one plane, or direction, only. Solutions of substances that rotate the plane of polarization are said to be optically active, and the degree of rotation is called the optical rotation of the solution. The direction in which the light is rotated is generally designed as plus, or *d*, for dextrorotatory (to the right), or as minus, or *l*, for levorotatory (to the left). Some amino acids are dextrorotatory, others are levorotatory. With the exception of a few small proteins (peptides) that occur in bacteria, the amino acids that occur in proteins are l-amino acids.

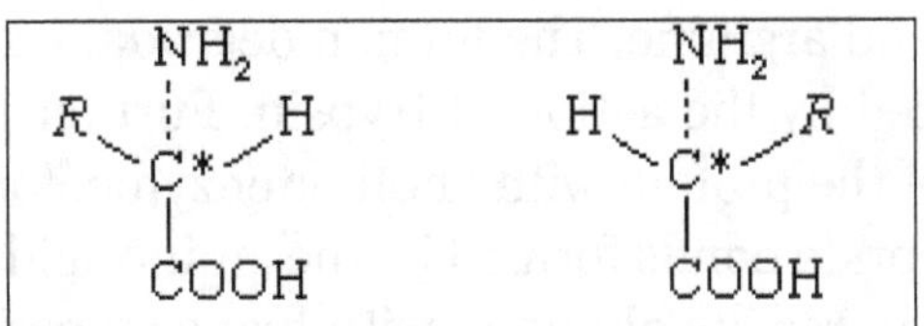

L-amino acid D-amino acid

In bacteria, d-alanine and some other d-amino acids have been found as components of gramicidin and bacitracin. These peptides are toxic to other bacteria and are used in medicine as antibiotics. The d-alanine has also been found in some peptides of bacterial membranes.

In contrast to most organic acids and amines, the amino acids are insoluble in organic solvents. In aqueous solutions they are dipolar ions (zwitterions, or hybrid ions) that react with strong acids or bases in a way that leads to the neutralization of the negatively or positively charged ends, respectively. Because of their reactions with strong acids and strong bases, the amino acids act as buffers—stabilizers of hydrogen ion (H^+) or hydroxide ion (OH^-) concentrations. In fact, glycine is frequently used as a buffer in the pH range from 1 to 3 (acid solutions) and from 9 to 12 (basic solutions).

In acid solutions, glycine has a positive charge and therefore migrates to the cathode (negative electrode of a direct-current electrical circuit with terminals in the solution). Its charge, however, is negative in alkaline solutions, in which it migrates to the anode (positive electrode). At pH 6.1 glycine does not migrate, because each molecule has one positive and one negative charge. The pH at which an amino acid does not migrate in an electrical field is called the isoelectric point. Most of the monoamino acids (i.e., those with only one amino group) have isoelectric points similar to that of glycine. The isoelectric points of aspartic and glutamic acids, however, are close to pH 3, and those of histidine, lysine, and arginine are at pH 7.6, 9.7, and 10.8, respectively.

Amino Acid Sequence in Protein Molecules

Since each protein molecule consists of a long chain of amino acid residues, linked to each other by peptide bonds, the hydrolytic cleavage of all peptide bonds is a prerequisite for the quantitative determination of the amino acid residues. Hydrolysis is most frequently accomplished by boiling the protein with concentrated hydrochloric acid. The quantitative determination of the amino acids is based on the discovery that amino acids can be separated from each other by chromatography on filter paper and made visible by spraying the paper with ninhydrin. The amino acids of the protein hydrolysate are separated from each other by passing the hydrolysate through a column of adsorbents, which adsorb the amino acids with different affinities and, on washing the column with buffer solutions, release them in a definite order. The amount of each of the amino acids can be determined by the intensity of the colour reaction with ninhydrin.

To obtain information about the sequence of the amino acid residues in the protein, the protein is degraded stepwise, one amino acid being split off in each step. This is accomplished by coupling the free α-amino group ($-NH_2$) of the N-terminal amino acid with phenyl isothiocyanate; subsequent mild hydrolysis does not affect the peptide bonds. The procedure, called the Edman degradation, can be applied repeatedly; it thus reveals the sequence of the amino acids in the peptide chain.

Unavoidable small losses that occur during each step make it impossible to determine the sequence of more than about 30 to 50 amino acids by this procedure. For this reason the protein is usually first hydrolyzed by exposure to the enzyme trypsin, which cleaves only peptide bonds formed by the carboxyl groups of lysine and arginine. The Edman degradation is then applied to each of the few resulting peptides produced by the action of trypsin. Further information can be gained by hydrolyzing another portion of the protein with another enzyme, for instance with chymotrypsin, which splits predominantly peptide bonds formed by the amino acids tyrosine, phenylalanine, and tryptophan. The combination of results obtained with two or more different proteolytic (protein degrading) enzymes was first applied by English biochemist Frederick Sanger, and it enabled him to elucidate the amino acid sequence of insulin. The amino acid sequences of many other proteins subsequently were determined in the same manner.

Levels of Structural Organization in Proteins

Primary Structure

Analytical and synthetic procedures reveal only the primary structure of the proteins—that is, the amino acid sequence of the peptide chains. They do not reveal information about the conformation (arrangement in space) of the peptide chain—that is, whether the peptide chain is present as a long straight thread or is irregularly coiled and folded into a globule. The configuration, or conformation, of a protein is determined by mutual attraction or repulsion of polar or nonpolar groups in the side chains (*R* groups) of the amino acids. The former have positive or negative charges in their side chains; the latter repel water but attract each other. Some parts of a peptide chain containing 100 to 200 amino acids may form a loop, or helix; others may be straight or form irregular coils.

The terms *secondary, tertiary*, and *quaternary structure* are frequently applied to the configuration of the peptide chain of a protein. A nomenclature committee of the International Union of Biochemistry (IUB) has defined these terms as follows: The primary structure of a protein is determined by its amino acid sequence without any regard for the arrangement of the peptide chain in space. The secondary structure is determined by the spatial arrangement of the main peptide chain without any regard for the conformation of side chains or other segments of the main chain. The tertiary structure is determined by both the side chains and other adjacent segments of the main chain, without regard for neighbouring peptide chains. Finally, the term *quaternary structure* is used for the arrangement of identical or different subunits of a large protein in which each subunit is a separate peptide chain.

Secondary Structure

The nitrogen and carbon atoms of a peptide chain cannot lie on a straight line, because of the magnitude of the bond angles between adjacent atoms of the chain; the bond angle is about 110°. Each of the nitrogen and carbon atoms can rotate to a certain extent, however, so that the chain has a limited flexibility. Because all of the amino acids, except glycine, are asymmetric L-amino acids, the peptide chain tends to assume an asymmetric helical shape; some of the fibrous proteins consist of elongated helices around a straight screw axis. Such structural features result from properties common to all peptide chains. The product of their effects is the secondary structure of the protein.

Tertiary Structure

The tertiary structure is the product of the interaction between the side chains (*R*) of the amino acids composing the protein. Some of them contain positively or negatively charged groups, others are polar, and still others are nonpolar. The number of carbon atoms in the side chain varies from zero in glycine to nine in tryptophan. Positively and negatively charged side chains have the tendency to attract each other; side chains with identical charges repel each other. The bonds formed by the forces between the negatively charged side chains of aspartic or glutamic acid on the one hand, and the positively charged side chains of lysine or arginine on the other hand, are called salt bridges. Mutual attraction of adjacent peptide chains also results from the formation of numerous hydrogen bonds.

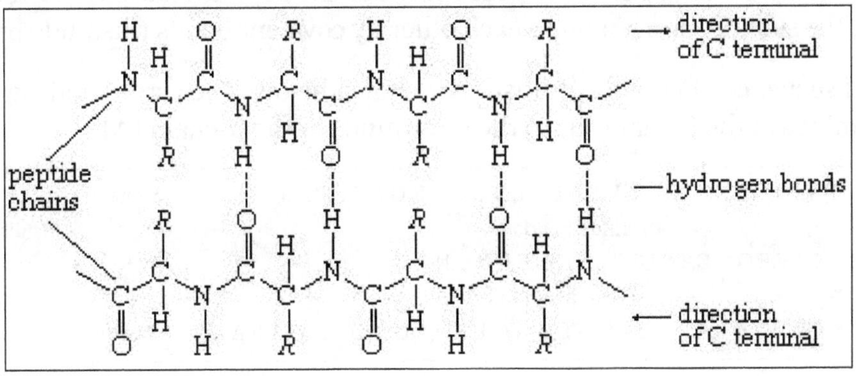

Hydrogen bonds form as a result of the attraction between the nitrogen-bound hydrogen atom (the imide hydrogen) and the unshared pair of electrons of the oxygen atom in the double bonded carbon–oxygen group (the carbonyl group). The result is a slight displacement of the imide hydrogen toward the oxygen atom of the carbonyl group. Although the hydrogen bond is much weaker than a covalent bond (i.e., the type of bond between two carbon atoms, which equally share the pair of bonding electrons between them), the large number of imide and carbonyl groups in peptide chains results in the formation of numerous hydrogen bonds. Another type of attraction is that between nonpolar side chains of valine, leucine, isoleucine, and phenylalanine; the attraction results in the displacement of water molecules and is called hydrophobic interaction.

In proteins rich in cystine, the conformation of the peptide chain is determined to a considerable extent by the disulfide bonds (–S–S–) of cystine. The halves of cystine may be located in different parts of the peptide chain and thus may form a loop closed by the disulfide bond.

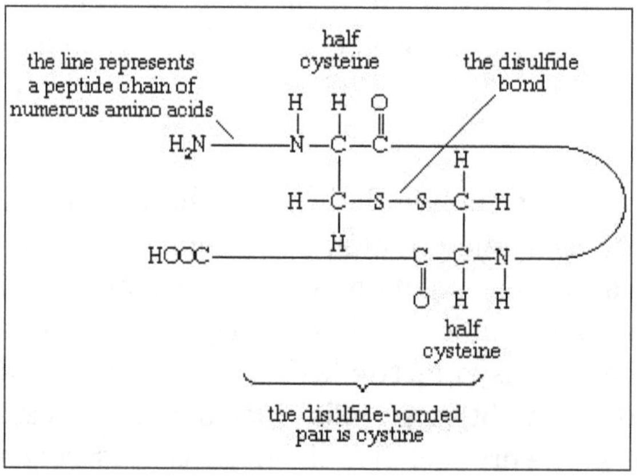

If the disulfide bond is reduced (i.e., hydrogen is added) to two sulfhydryl (–SH) groups, the tertiary structure of the protein undergoes a drastic change—closed loops are broken and adjacent disulfide-bonded peptide chains separate.

Quaternary Structure

The nature of the quaternary structure is demonstrated by the structure of hemoglobin. Each molecule of human hemoglobin consists of four peptide chains, two α-chains and two β-chains; i.e., it is a tetramer. The four subunits are linked to each other by hydrogen bonds and hydrophobic interaction. Because the four subunits are so closely linked, the hemoglobin tetramer is called a molecule, even though no covalent bonds occur between the peptide chains of the four subunits. In other proteins, the subunits are bound to each other by covalent bonds (disulfide bridges).

The amino acid sequence of porcine proinsulin is shown below. The arrows indicate the direction from the N terminus of the β-chain (B) to the C terminus of the α-chain (A).

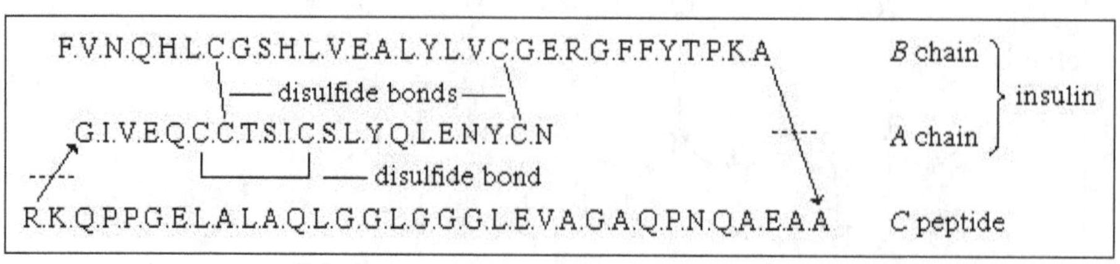

Isolation and Determination of Proteins

Animal material usually contains large amounts of protein and lipids and small amounts of carbohydrate; in plants, the bulk of the dry matter is usually carbohydrate. If it is necessary to determine the amount of protein in a mixture of animal foodstuffs, a sample is converted to ammonium salts by boiling with sulfuric acid and a suitable inorganic catalyst, such as copper sulfate (Kjeldahl method). The method is based on the assumption that proteins contain 16 percent nitrogen, and that nonprotein nitrogen is present in very small amounts. The assumption is justified for most tissues from higher animals but not for insects and crustaceans, in which a considerable portion of the body nitrogen is present in the form of chitin, a carbohydrate. Large amounts of nonprotein nitrogen are also found in the sap of many plants. In such cases, the precise quantitative analyses are made after the proteins have been separated from other biological compounds.

Proteins are sensitive to heat, acids, bases, organic solvents, and radiation exposure; for this reason, the chemical methods employed to purify organic compounds cannot be applied to proteins. Salts and molecules of small size are removed from protein solutions by dialysis—i.e., by placing the solution into a sac of semipermeable material, such as cellulose or acetylcellulose, which will allow small molecules to pass through but not large protein molecules, and immersing the sac in water or a salt solution. Small molecules can also be removed either by passing the protein solution through a column of resin that adsorbs only the protein or by gel filtration. In gel filtration, the large protein molecules pass through the column, and the small molecules are adsorbed to the gel.

Groups of proteins are separated from each other by salting out—i.e., the stepwise addition of sodium sulfate or ammonium sulfate to a protein solution. Some proteins, called globulins, become insoluble and precipitate when the solution is half-saturated with ammonium sulfate or when its sodium sulfate content exceeds about 12 percent. Other proteins, the albumins, can be precipitated from the supernatant solution (i.e., the solution remaining after a precipitation has taken place) by saturation with ammonium sulfate. Water-soluble proteins can be obtained in a dry state by freeze-drying (lyophilization), in which the protein solution is deep-frozen by lowering the temperature below −15 °C (5 °F) and removing the water; the protein is obtained as a dry powder.

Most proteins are insoluble in boiling water and are denatured by it—i.e., irreversibly converted into an insoluble material. Heat denaturation cannot be used with connective tissue because the principal structural protein, collagen, is converted by boiling water into water-soluble gelatin.

Fractionation (separation into components) of a mixture of proteins of different molecular weight can be accomplished by gel filtration. The size of the proteins retained by the gel depends upon the properties of the gel. The proteins retained in the gel are removed from the column by solutions of a suitable concentration of salts and hydrogen ions.

Many proteins were originally obtained in crystalline form, but crystallinity is not proof of purity; many crystalline protein preparations contain other substances. Various tests are used to determine whether a protein preparation contains only one protein. The purity of a protein solution can be determined by such techniques as chromatography and gel filtration. In addition, a solution of pure protein will yield one peak when spun in a centrifuge at very high speeds (ultracentrifugation) and will migrate as a single band in electrophoresis (migration of the protein in an electrical field). After these methods and others (such as amino acid analysis) indicate that the protein solution is pure, it can be considered so. Because chromatography, ultracentrifugation, and electrophoresis cannot be applied to insoluble proteins, little is known about them; they may be mixtures of many similar proteins.

Very small (microheterogeneous) differences in some of the apparently pure proteins are known to occur. They are differences in the amino acid composition of otherwise identical proteins and are transmitted from generation to generation; i.e., they are genetically determined. For example, some humans have two hemoglobins, hemoglobin A and hemoglobin S, which differ in one amino acid at a specific site in the molecule. In hemoglobin A the site is occupied by glutamic acid and in hemoglobin S by valine. Refinement of the techniques of protein analysis has resulted in the discovery of other instances of microheterogeneity.

The quantity of a pure protein can be determined by weighing or by measuring the ultraviolet absorbancy at 280 nanometres. The absorbency at 280 nanometres depends on the content of tyrosine and tryptophan in the protein. Sometimes the slightly less sensitive biuret reaction, a purple colour given by alkaline protein solutions upon the addition of copper sulfate, is used; its intensity depends only on the number of peptide bonds per gram, which is similar in all proteins.

Physicochemical Properties of Proteins

Molecular Weight of Proteins

The molecular weight of proteins cannot be determined by the methods of classical chemistry (e.g., freezing-point depression), because they require solutions of a higher concentration of protein than can be prepared.

If a protein contains only one molecule of one of the amino acids or one atom of iron, copper, or another element, the minimum molecular weight of the protein or a subunit can be calculated; for example, the protein myoglobin contains 0.34 gram of iron in 100 grams of protein. The atomic weight of iron is 56; thus the minimum molecular weight of myoglobin is $(56 \times 100)/0.34$ = about 16,500. Direct measurements of the molecular weight of myoglobin yield the same value. The molecular weight of hemoglobin, however, which also contains 0.34 percent iron, has been found to be 66,000 or $4 \times 16{,}500$; thus hemoglobin contains four atoms of iron.

The method most frequently used to determine the molecular weight of proteins is ultracentrifugation—i.e., spinning in a centrifuge at velocities up to about 60,000 revolutions per minute. Centrifugal forces of more than 200,000 times the gravitational force on the surface of Earth are achieved at such velocities. The first ultracentrifuges, built in 1920, were used to determine the molecular weight of proteins. The molecular weights of a large number of proteins have been determined. Most consist of several subunits, the molecular weight of which is usually less than 100,000 and frequently ranges from 20,000 to 30,000. Proteins of very high molecular weights are found among hemocyanins, the copper-containing respiratory proteins of invertebrates;

some range as high as several million. Although there is no definite lower limit for the molecular weight of proteins, short amino acid sequences are usually called peptides.

The Shape of Protein Molecules

In the technique of X-ray diffraction, the X-rays are allowed to strike a protein crystal. The X-rays, diffracted (bent) by the crystal, impinge on a photographic plate, forming a pattern of spots. This method reveals that peptide chains can assume very complicated, apparently irregular shapes. Two extremes in shape include the closely folded structure of the globular proteins and the elongated, unidimensional structure of the threadlike fibrous proteins; both were recognized many years before the technique of X-ray diffraction was developed. Solutions of fibrous proteins are extremely viscous (i.e., sticky); those of the globular proteins have low viscosity (i.e., they flow easily). A 5 percent solution of a globular protein—ovalbumin, for example—easily flows through a narrow glass tube; a 5 percent solution of gelatin, a fibrous protein, however, does not flow through the tube, because it is liquid only at high temperatures and solidifies at room temperature. Even solutions containing only 1 or 2 percent of gelatin are highly viscous and flow through a narrow tube either very slowly or only under pressure.

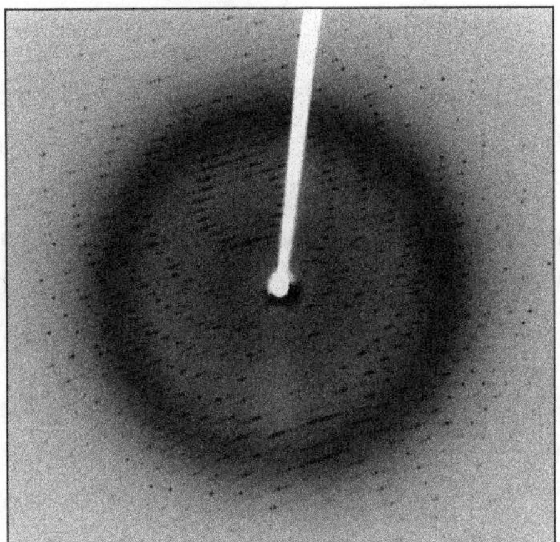

X-ray diffraction: X-ray diffraction pattern of a crystallized enzyme.

The elongated peptide chains of the fibrous proteins can be imagined to become entangled not only mechanically but also by mutual attraction of their side chains, and in this way they incorporate large amounts of water. Most of the hydrophilic (water-attracting) groups of the globular proteins, however, lie on the surface of the molecules, and, as a result, globular proteins incorporate only a few water molecules. If a solution of a fibrous protein flows through a narrow tube, the elongated molecules become oriented parallel to the direction of the flow, and the solution thus becomes birefringent like a crystal; i.e., it splits a light ray into two components that travel at different velocities and are polarized at right angles to each other. Globular proteins do not show this phenomenon, which is called flow birefringence. Solutions of myosin, the contractile protein of muscles, show very high flow birefringence; other proteins with very high flow birefringence include solutions of fibrinogen, the clotting material of blood plasma, and solutions of tobacco mosaic virus. The gamma-globulins of the blood plasma show low flow birefringence, and none can be observed in solutions of serum albumin and ovalbumin.

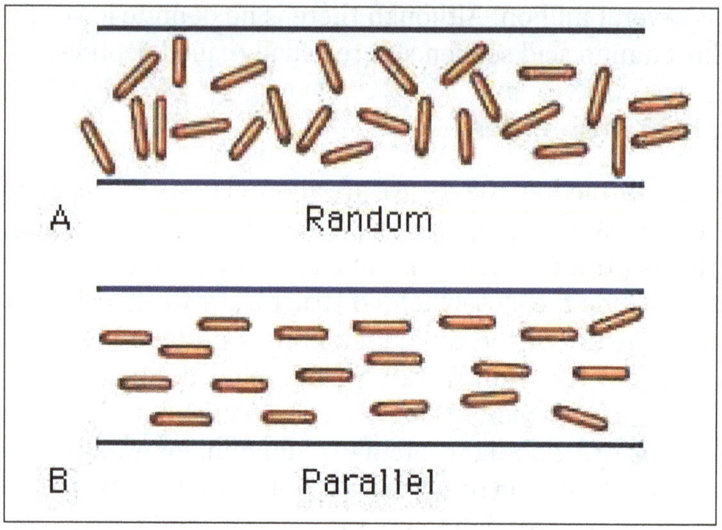

Flow birefringence. Orientation of elongated, rodlike macromolecules (A) in resting solution, or (B) during flow through a horizontal tube.

Hydration of Proteins

When dry proteins are exposed to air of high water content, they rapidly bind water up to a maximum quantity, which differs for different proteins; usually it is 10 to 20 percent of the weight of the protein. The hydrophilic groups of a protein are chiefly the positively charged groups in the side chains of lysine and arginine and the negatively charged groups of aspartic and glutamic acid. Hydration (i.e., the binding of water) may also occur at the hydroxyl (–OH) groups of serine and threonine or at the amide ($-CONH_2$) groups of asparagine and glutamine.

The binding of water molecules to either charged or polar (partly charged) groups is explained by the dipolar structure of the water molecule; that is, the two positively charged hydrogen atoms form an angle of about 105°, with the negatively charged oxygen atom at the apex. The centre of the positive charges is located between the two hydrogen atoms; the centre of the negative charge of the oxygen atom is at the apex of the angle. The negative pole of the dipolar water molecule binds to positively charged groups; the positive pole binds negatively charged ones. The negative pole of the water molecule also binds to the hydroxyl and amino groups of the protein.

The water of hydration is essential to the structure of protein crystals; when they are completely dehydrated, the crystalline structure disintegrates. In some proteins this process is accompanied by denaturation and loss of the biological function.

In aqueous solutions, proteins bind some of the water molecules very firmly; others are either very loosely bound or form islands of water molecules between loops of folded peptide chains. Because the water molecules in such an island are thought to be oriented as in ice, which is crystalline water, the islands of water in proteins are called icebergs. Water molecules may also form bridges between the carbonyl and imino groups of adjacent peptide chains, resulting in structures similar to those of the pleated sheet but with a water molecule in the position of the hydrogen bonds of that configuration. The extent of hydration of protein molecules in aqueous solutions is important, because some of the methods used to determine the molecular weight of proteins yield the molecular weight of the hydrated protein. The amount of water bound to one gram of a globular protein in solution varies from 0.2 to 0.5 gram. Much larger amounts of water are mechanically immobilized between the elongated peptide chains of fibrous proteins; for example, one gram of gelatin can immobilize at room temperature 25 to 30 grams of water.

Hydration of proteins is necessary for their solubility in water. If the water of hydration of a protein dissolved in water is reduced by the addition of a salt such as ammonium sulfate, the protein is no longer soluble and is salted out, or precipitated. The salting-out process is reversible because the protein is not denatured (i.e., irreversibly converted to an insoluble material) by the addition of such salts as sodium chloride, sodium sulfate, or ammonium sulfate. Some globulins, called euglobulins, are insoluble in water in the absence of salts; their insolubility is attributed to the mutual interaction of polar groups on the surface of adjacent molecules, a process that results in the formation of large aggregates of molecules. Addition of small amounts of salt causes the euglobulins to become soluble. This process, called salting in, results from a combination between anions (negatively charged ions) and cations (positively charged ions) of the salt and positively and negatively charged side chains of the euglobulins. The combination prevents the aggregation of euglobulin molecules by preventing the formation of salt bridges between them. The addition of more sodium or ammonium sulfate causes the euglobulins to salt out again and to precipitate.

Electrochemistry of Proteins

Because the α-amino group and α-carboxyl group of amino acids are converted into peptide bonds in the protein molecule, there is only one α-amino group (at the N terminus) and one α-carboxyl group (at the C terminus) in a given protein molecule. The electrochemical character of a protein is affected very little by these two groups. Of importance, however, are the numerous positively charged ammonium groups ($-NH_3^+$) of lysine and arginine and the negatively charged carboxyl groups ($-COO^-$) of aspartic acid and glutamic acid. In most proteins, the number of positively and negatively charged groups varies from 10 to 20 per 100 amino acids.

Electrometric Titration

When measured volumes of hydrochloric acid are added to a solution of protein in salt-free water, the pH decreases in proportion to the amount of hydrogen ions added until it is about 4. Further addition of acid causes much less decrease in pH because the protein acts as a buffer at pH values of 3 to 4. The reaction that takes place in this pH range is the protonation of the carboxyl group—i.e., the conversion of $-COO^-$ into $-COOH$. Electrometric titration of an isoelectric protein with potassium hydroxide causes a very slow increase in pH and a weak buffering action of the protein at pH 7; a very strong buffering action occurs in the pH range from 9 to 10. The buffering action at pH 7, which is caused by loss of protons (positively charged hydrogen) from the imidazolium groups (i.e., the five-member ring structure in the side chain) of histidine, is weak because the histidine content of proteins is usually low. The much stronger buffering action at pH values from 9 to 10 is caused by the loss of protons from the hydroxyl group of tyrosine and from the ammonium groups of lysine. Finally, protons are lost from the guanidinium groups (i.e., the nitrogen-containing terminal portion of the arginine side chains) of arginine at pH 12. Electrometric titrations of proteins yield similar curves. Electrometric titration makes possible the determination of the approximate number of carboxyl groups, ammonium groups, histidines, and tyrosines per molecule of protein.

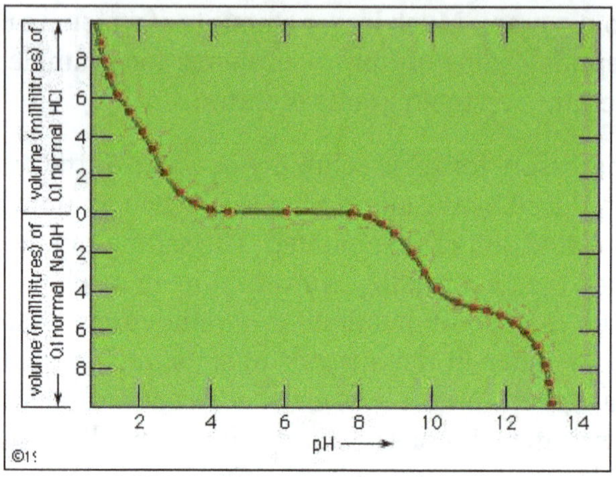

Electrometric titration of glycine.

Electrophoresis

The positively and negatively charged side chains of proteins cause them to behave like amino acids in an electrical field; that is, they migrate during electrophoresis at low pH values to the cathode (negative terminal) and at high pH values to the anode (positive terminal). The isoelectric point, the pH value at which the protein molecule does not migrate, is in the range of pH 5 to 7 for many proteins. Proteins such as lysozyme, cytochrome c, histone, and others rich in lysine and arginine, however, have isoelectric points in the pH range between 8 and 10. The isoelectric point of pepsin, which contains very few basic amino acids, is close to 1.

Number of amino acids per protein molecule:

	Protein*						
Amino acid	Cyto	Hb alpha	Hb beta	RNase	Lys	Chgen	Fdox
Lysine	18	11	11	10	6	14	4
Histidine	3	10	9	4	1	2	1
Arginine	2	3	3	4	11	4	1
Aspartic acid**	8	12	13	15	21	23	13
Threonine	7	9	7	10	7	23	8
Serine	2	11	5	15	10	28	7
Glutamic acid**	10	5	11	12	5	15	13
Proline	4	7	7	4	2	9	4
Glycine	13	7	13	3	12	23	6
Alanine	6	21	15	12	12	22	9
Half-Cystine	2	1	2	8	8	10	5
Valine	3	13	18	9	6	23	7
Methionine	3	2	1	4	2	2	0
Isoleucine	8	0	0	3	6	10	4
Leucine	6	18	18	2	8	19	8
Tyrosine	5	3	3	6	3	4	4
Phenylalanine	3	7	8	3	3	6	2

Tryptophan	1	1	2	0	6	8	1
Total	104	141	146	124	129	245	97
*Cyto = Human Cytochrome C; Hb Alpha = Human Hemoglobin A, Alpha-Chain; Hb Beta = Human Hemoglobin A, Beta-Chain; Rnase = Bovine Ribonuclease; Lys = Chicken Lysozyme; Chgen = Bovine Chymotrypsinogen; Fdox = Spinach Ferredoxin.							
**The values recorded for aspartic acid and glutamic acid include asparagine and glutamine, respectively.							

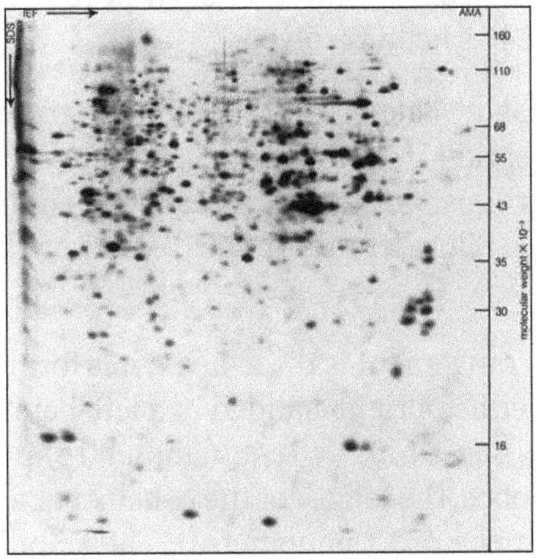

Two-dimensional Gel Electrophoresis

In two-dimensional gel electrophoresis, proteins are separated based on charge and size. Approaches commonly employed include isoelectric focusing (IEF) sodium dodecyl sulfate (SDS) polyacrylamide gel electrophoresis (PAGE) and immobilized pH gradient (IPG-Dalt) SDS-PAGE.

Free-boundary electrophoresis, the original method of determining electrophoretic migration, has been replaced in many instances by zone electrophoresis, in which the protein is placed in either a gel of starch, agar, or polyacrylamide or in a porous medium such as paper or cellulose acetate. The migration of hemoglobin and other coloured proteins can be followed visually. Colourless proteins are made visible after the completion of electrophoresis by staining them with a suitable dye.

Conformation of Globular Proteins

Results of X-ray Diffraction Studies

Most knowledge concerning secondary and tertiary structure of globular proteins has been obtained by the examination of their crystals using X-ray diffraction. In this technique, X-rays are allowed to strike the crystal; the X-rays are diffracted by the crystal and impinge on a photographic plate, forming a pattern of spots. The measured intensity of the diffraction pattern, as recorded on a photographic film, depends particularly on the electron density of the atoms in the protein crystal. This density is lowest in hydrogen atoms, and they do not give a visible diffraction pattern. Although carbon, oxygen, and nitrogen atoms yield visible diffraction patterns, they are present in such great number—about 700 or 800 per 100 amino acids—that the resolution of the structure of a protein containing more than 100 amino acids is almost impossible. Resolution is considerably improved by substituting into the side chains of certain amino acids very heavy atoms, particularly those of heavy metals. Mercury ions, for example, bind to the sulfhydryl (−SH) groups of cysteine. Platinum chloride has been used in other proteins. In the iron-containing proteins, the iron atom already in the molecule is adequate.

Although the X-ray diffraction technique cannot resolve the complete three-dimensional conformation (that is, the secondary and tertiary structure of the peptide chain), complete resolution has been obtained by combination of the results of X-ray diffraction with those of amino acid sequence analysis. In this way the complete conformation of such proteins as myoglobin, chymotrypsinogen, lysozyme, and ribonuclease has been resolved.

The X-ray diffraction method has revealed regular structural arrangements in proteins; one is an extended form of antiparallel peptide chains that are linked to each other by hydrogen bonds between the carbonyl and imino groups. This conformation, called the pleated sheet, or β-structure, is found in some fibrous proteins. Short strands of the β-structure have also been detected in some globular proteins.

A second important structural arrangement is the α-helix; it is formed by a sequence of amino acids wound around a straight axis in either a right-handed or a left-handed spiral. Each turn of the helix corresponds to a distance of 5.4 angstroms (= 0.54 nanometre) in the direction of the screw axis and contains 3.7 amino acids. Hence, the length of the α-helix per amino acid residue is 5.4 divided by 3.7, or 1.5 angstroms (1 angstrom = 0.1 nanometre). The stability of the α-helix is maintained by hydrogen bonds between the carbonyl and imino groups of neighbouring turns of the helix. It was once thought, based on data from analyses of the myoglobin molecule, more than half of which consists of α-helices, that the α-helix is the predominant structural element of the globular proteins; it is now known that myoglobin is exceptional in this respect. The other globular proteins for which the structures have been resolved by X-ray diffraction contain only small regions of α-helix. In most of them the peptide chains are folded in an apparently random fashion, for which the term random coil has been used. The term is misleading, however, because the folding is not random; rather, it is dictated by the primary structure and modified by the secondary and tertiary structures.

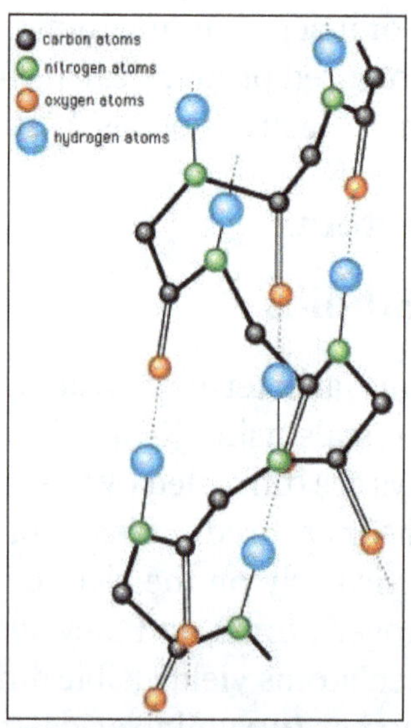

Protein structure; α-helix.

The α-helix in the structural arrangement of a protein.

The first proteins for which the internal structures were completely resolved are the iron-containing proteins myoglobin and hemoglobin. The investigation of the hydrated crystals of these proteins by Austrian-born British biochemist Max Perutz and British biochemist John C. Kendrew, who won the 1962 Nobel Prize for Chemistry for their work, revealed that the folding of the peptide chains is so tight that most of the water is displaced from the centre of the globular molecules. The amino acids that carry the ammonium ($-NH_3^+$) and carboxyl ($-COO^-$) groups were found to be shifted to the surface of the globular molecules, and the nonpolar amino acids were found to be concentrated in the interior.

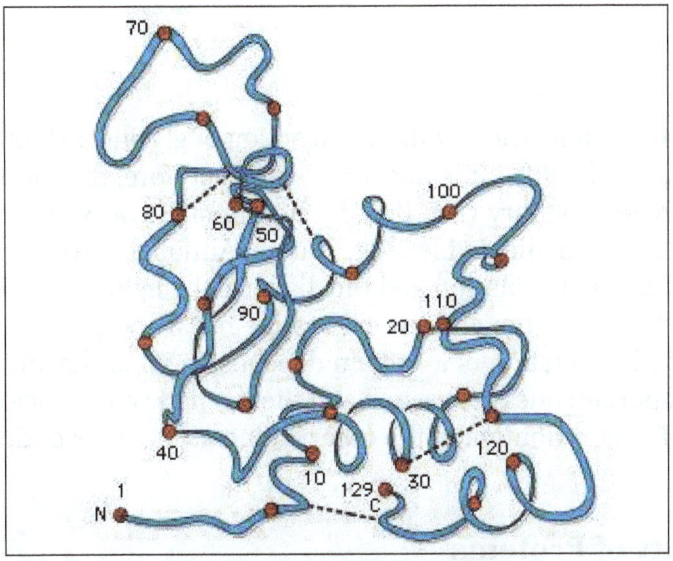

Lysozyme; protein conformation.

The simplified structure of lysozyme from hen's egg white has a single peptide chain of 129 amino acids. The amino acid residues are numbered from the terminal α group (N) to the terminal carboxyl group (C). Circles indicate every fifth residue, and every tenth residue is numbered. Broken lines indicate the four disulfide bridges. Alpha-helices are visible in the ranges 25 to 35, 90 to 100, and 120 to 125.

Other Approaches to the Determination of Protein Structure

None of the several other physical methods that have been used to obtain information on the secondary and tertiary structure of proteins provides as much direct information as the X-ray diffraction technique. Most of the techniques, however, are much simpler than X-ray diffraction, which requires, for the resolution of the structure of one protein, many years of work and equipment such as electronic computers. Some of the simpler techniques are based on the optical properties of proteins—refractivity, absorption of light of different wavelengths, rotation of the plane polarized light at different wavelengths, and luminescence.

Spectrophotometric Behavior

Spectrophotometry of protein solutions (the measurement of the degree of absorbance of light by a protein within a specified wavelength) is useful within the range of visible light only with proteins that contain coloured prosthetic groups (the nonprotein components). Examples of such proteins include the red heme proteins of the blood, the purple pigments of the retina of the eye, green and yellow proteins that contain bile pigments, blue copper-containing proteins, and dark brown proteins called melanins. Peptide bonds, because of their carbonyl groups, absorb light energy at very short wavelengths (185–200 nanometres). The aromatic rings of phenylalanine, tyrosine, and tryptophan, however, absorb ultraviolet light between wavelengths of 280 and 290 nanometres. The absorbance of ultraviolet light by tryptophan is greatest, that of tyrosine is less, and that of phenylalanine is least. If the tyrosine or tryptophan content of the protein is known, therefore, the concentration of the protein solution can be determined by measuring its absorbance between 280 and 290 nanometres.

Optical Activity

It will be recalled that the amino acids, with the exception of glycine, exhibit optical activity (rotation of the plane of polarized light;). It is not surprising, therefore, that proteins also are optically active. They are usually levorotatory (i.e., they rotate the plane of polarization to the left) when polarized light of wavelengths in the visible range is used. Although the specific rotation (a function of the concentration of a protein solution and the distance the light travels in it) of most L-amino acids varies from −30° to +30°, the amino acid cystine has a specific rotation of approximately −300°. Although the optical rotation of a protein depends on all of the amino acids of which it is composed, the most important ones are cystine and the aromatic amino acids phenylalanine, tyrosine, and tryptophan. The contribution of the other amino acids to the optical activity of a protein is negligibly small.

Chemical Reactivity of Proteins

Information on the internal structure of proteins can be obtained with chemical methods that reveal whether certain groups are present on the surface of the protein molecule and thus able to react or whether they are buried inside the closely folded peptide chains and thus are unable to react. The chemical reagents used in such investigations must be mild ones that do not affect the structure of the protein.

The reactivity of tyrosine is of special interest. It has been found, for example, that only three of the six tyrosines found in the naturally occurring enzyme ribonuclease can be iodinated (i.e., reacted to accept an iodine atom). Enzyme-catalyzed breakdown of iodinated ribonuclease is used to identify the peptides in which the iodinated tyrosines are present. The three tyrosines that can be iodinated lie on the surface of ribonuclease; the others, assumed to be inaccessible, are said to be buried in the molecule. Tyrosine can also be identified by using other techniques—e.g., treatment with diazonium compounds or tetranitromethane. Because the compounds formed are coloured, they can easily be detected when the protein is broken down with enzymes.

Cysteine can be detected by coupling with compounds such as iodoacetic acid or iodoacetamide; the reaction results in the formation of carboxymethylcysteine or carbamidomethylcysteine, which can be detected by amino acid determination of the peptides containing them. The imidazole groups of certain histidines can also be located by coupling with the same reagents under different conditions. Unfortunately, few other amino acids can be labelled without changes in the secondary and tertiary structure of the protein.

Association of Protein Subunits

Many proteins with molecular weights of more than 50,000 occur in aqueous solutions as complexes: dimers, tetramers, and higher polymers—i.e., as chains of two, four, or more repeating basic structural units. The subunits, which are called monomers or protomers, usually are present as an even number. Less than 10 percent of the polymers have been found to have an odd number of monomers. The arrangement of the subunits is thought to be regular and may be cyclic, cubic, or tetrahedral. Some of the small proteins also contain subunits. Insulin, for example, with a molecular weight of about 6,000, consists of two peptide chains linked to each other by disulfide bridges (−S−S−). Similar interchain disulfide bonds have been found in the immunoglobulins. In other proteins, hydrogen bonds and hydrophobic bonds (resulting from the interaction between the amino acid side chains of valine, leucine, isoleucine, and phenylalanine) cause the formation of aggregates of the subunits. The subunits of some proteins are identical; those of others differ. Hemoglobin is a tetramer consisting of two α-chains and two β-chains.

Protein Denaturation

When a solution of a protein is boiled, the protein frequently becomes insoluble—i.e., it is denatured—and remains insoluble even when the solution is cooled. The denaturation of the proteins of egg white by heat—as when boiling an egg—is an example of irreversible denaturation. The denatured protein has the same primary structure as the original, or native, protein. The weak forces between charged groups and the weaker forces of mutual attraction of nonpolar groups are disrupted at elevated temperatures, however; as a result, the tertiary structure of the protein is lost. In some instances the original structure of the protein can be regenerated; the process is called renaturation.

Denaturation can be brought about in various ways. Proteins are denatured by treatment with alkaline or acid, oxidizing or reducing agents, and certain organic solvents. Interesting among denaturing agents are those that affect the secondary and tertiary structure without affecting the primary structure. The agents most frequently used for this purpose are urea and guanidinium chloride. These molecules, because of their high affinity for peptide bonds, break the hydrogen bonds and the salt bridges between positive and negative side chains, thereby abolishing the tertiary structure of the peptide chain. When denaturing agents are removed from a protein solution, the native protein re-forms in many cases. Denaturation can also be accomplished by reduction of the disulfide bonds of cystine—i.e., conversion of the disulfide bond (−S−S−) to two sulfhydryl groups (−SH). This, of course, results in the formation of two cysteines. Reoxidation of the cysteines by exposure to air sometimes regenerates the native protein. In other cases, however, the wrong cysteines become bound to each other, resulting in a different protein. Finally, denaturation can also be accomplished by exposing proteins to organic solvents such as ethanol or acetone. It is believed that the organic solvents interfere with the mutual attraction of nonpolar groups.

Some of the smaller proteins, however, are extremely stable, even against heat; for example, solutions of ribonuclease can be exposed for short periods of time to temperatures of 90 °C (194 °F) without undergoing significant denaturation. Denaturation does not involve identical changes in protein molecules. A common property of denatured proteins, however, is the loss of biological activity—e.g., the ability to act as enzymes or hormones.

Although denaturation had long been considered an all-or-none reaction, it is now thought that many intermediary states exist between native and denatured protein. In some instances, however, the breaking of a key bond could be followed by the complete breakdown of the conformation of the native protein.

Although many native proteins are resistant to the action of the enzyme trypsin, which breaks down proteins during digestion, they are hydrolyzed by the same enzyme after denaturation. The peptide bonds that can be split by trypsin are inaccessible in the native proteins but become accessible during denaturation. Similarly, denatured proteins give more intense colour reactions for tyrosine, histidine, and arginine than do the same proteins in the native state. The increased accessibility of reactive groups of denatured proteins is attributed to an unfolding of the peptide chains.

If denaturation can be brought about easily and if renaturation is difficult, how is the native conformation of globular proteins maintained in living organisms, in which they are produced stepwise, by incorporation of one amino acid at a time? Experiments on the biosynthesis of proteins from amino acids containing radioactive carbon or heavy hydrogen reveal that the protein molecule grows stepwise from the N terminus to the C terminus; in each step a single amino acid residue is incorporated. As soon as the growing peptide chain contains six or seven amino acid residues, the side chains interact with each other and thus cause deviations from the straight or β-chain configuration. Depending on the nature of the side chains, this may result in the formation of an α-helix or of loops closed by hydrogen bonds or disulfide bridges. The final conformation is probably frozen when the peptide chain attains a length of 50 or more amino acid residues.

Conformation of Proteins in Interfaces

Like many other substances with both hydrophilic and hydrophobic groups, soluble proteins tend to migrate into the interface between air and water or oil and water; the term oil here means a hydrophobic liquid such as benzene or xylene. Within the interface, proteins spread, forming thin films. Measurements of the surface tension, or interfacial tension, of such films indicate that tension is reduced by the protein film. Proteins, when forming an interfacial film, are present as a monomolecular layer—i.e., a layer one molecule in height. Although it was once thought that globular protein molecules unfold completely in the interface, it has now been established that many proteins can be recovered from films in the native state. The application of lateral pressure on a protein film causes it to increase in thickness and finally to form a layer with a height corresponding to the diameter of the native protein molecule. Protein molecules in an interface, because of Brownian motions (molecular vibrations), occupy much more space than do those in the film after the application of pressure. The Brownian motion of compressed molecules is limited to the two dimensions of the interface, since the protein molecules cannot move upward or downward.

The motion of protein molecules at the air–water interface has been used to determine the molecular weight of proteins. The technique involves measuring the force exerted by the protein layer on a barrier.

When a protein solution is vigorously shaken in air, it forms a foam, because the soluble proteins migrate into the air–water interface and persist there, preventing or slowing the reconversion of the foam into a homogeneous solution. Some of the unstable, easily modified proteins are denatured when spread in the air–water interface. The formation of a permanent foam when egg white is vigorously stirred is an example of irreversible denaturation by spreading in a surface.

Classification of Proteins

Classification by Solubility

After two German chemists, Emil Fischer and Franz Hofmeister, independently stated in 1902 that proteins are essentially polypeptides consisting of many amino acids, an attempt was made to classify proteins according to their chemical and physical properties, because the biological function of proteins had not yet been established. (The protein character of enzymes was not proved until the 1920s.) Proteins were classified primarily according to their solubility in a number of solvents. This classification is no longer satisfactory, however, because proteins of quite different structure and function sometimes have similar solubilities; conversely, proteins of the same function and similar structure sometimes have different solubilities. The terms associated with the old classification, however, are still widely used.

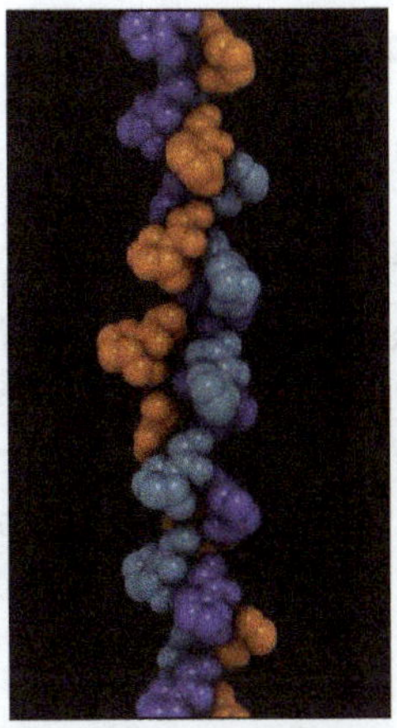

Albumins are proteins that are soluble in water and in water half-saturated with ammonium sulfate. On the other hand, globulins are salted out (i.e., precipitated) by half-saturation with ammonium sulfate. Globulins that are soluble in salt-free water are called pseudoglobulins; those insoluble in salt-free water are euglobulins. Both prolamins and glutelins, which are plant proteins, are insoluble in water; the prolamins dissolve in 50 to 80 percent ethanol, the glutelins in acidified or alkaline solution. The term protamine is used for a number of proteins in fish sperm that consist of approximately 80 percent arginine and therefore are strongly alkaline. Histones, which are less alkaline, apparently occur only in cell nuclei, where they are bound to nucleic acids. The term scleroproteins has been used for the insoluble proteins of animal organs. They include keratin, the insoluble protein of certain epithelial tissues such as the skin or hair, and collagen, the protein of the connective tissue. A large group of proteins has been called conjugated proteins, because they are complex molecules of protein consisting of protein and nonprotein moieties. The nonprotein portion is called the prosthetic group. Conjugated proteins can be subdivided into mucoproteins, which, in addition to protein, contain carbohydrate; lipoproteins, which contain lipids; phosphoproteins, which are rich in phosphate; chromoproteins, which contain pigments such as iron-porphyrins, carotenoids, bile pigments, and melanin; and finally, nucleoproteins, which contain nucleic acid.

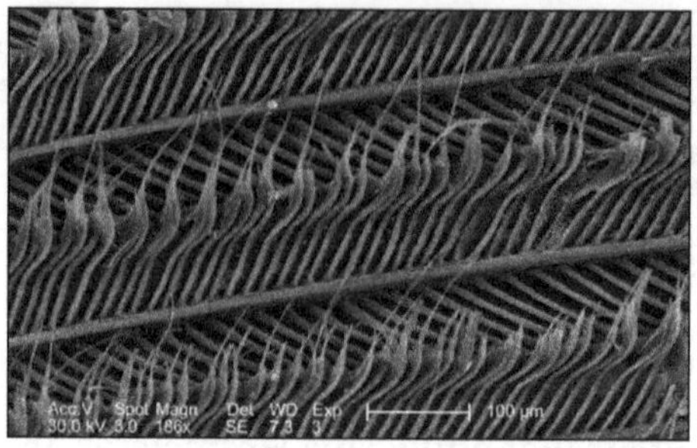

Keratin: Scanning electron micrograph showing strands of keratin in a feather, magnified 186×.

The weakness of the above classification lies in the fact that many, if not all, globulins contain small amounts of carbohydrate; thus there is no sharp borderline between globulins and mucoproteins. Moreover, the phosphoproteins do not have a prosthetic group that can be isolated; they are merely proteins in which some of the hydroxyl groups of serine are phosphorylated (i.e., contain phosphate). Finally, the globulins include proteins with quite different roles—enzymes, antibodies, fibrous proteins, and contractile proteins.

Classification by Biological Functions

In view of the unsatisfactory state of the old classification, it is preferable to classify the proteins according to their biological function. Such a classification is far from ideal, however, because one protein can have more than one function. The contractile protein myosin, for example, also acts as an ATPase (adenosine triphosphatase), an enzyme that hydrolyzes adenosine triphosphate (removes a phosphate group from ATP by introducing a water molecule). Another problem with functional classification is that the definite function of a protein frequently is not known. A protein cannot be called an enzyme as long as its substrate (the specific compound upon which it acts) is not known. It cannot even be tested for its enzymatic action when its substrate is not known.

Protein Hormones

Some hormones that are products of endocrine glands are proteins or peptides, others are steroids. None of the hormones has any enzymatic activity. Each has a target organ in which it elicits some biological action—e.g., secretion of gastric or pancreatic juice, production of milk, production of steroid hormones. The mechanism by which the hormones exert their effects is not fully understood. Cyclic adenosine monophosphate is involved in the transmittance of the hormonal stimulus to the cells whose activity is specifically increased by the hormone.

Hormones of the Thyroid Gland

Thyroglobulin, the active groups of which are two molecules of the iodine-containing compound thyroxine, has a molecular weight of 670,000. Thyroglobulin also contains thyroxine with two and three iodine atoms instead of four and tyrosine with one and two iodine atoms. Injection of the hormone causes an increase in metabolism; lack of it results in a slowdown.

Another hormone, calcitonin, which lowers the calcium level of the blood, occurs in the thyroid gland. The amino acid sequences of calcitonin from pig, beef, and salmon differ from humancalcitonin in some amino acids. All of them, however, have the half-cystines (C) and the prolinamide (P) in the same position.

C.S.N.L.S.T.C.V.L.S.A.Y.W.K.D.L.N.N.Y.H.R.F.S.G.M.G.F.G.P.E.T.P($CONH_2$)

Parathyroid hormone (parathormone), produced in small glands that are embedded in or lie behind the thyroid gland, is essential for maintaining the calcium level of the blood. A decrease in its production results in hypocalcemia (a reduction of calcium levels in the bloodstream below the normal range). Bovine parathormone has a molecular weight of 8,500; it contains no cystine or cysteine and is rich in aspartic acid, glutamic acid, or their amides.

Hormones of the Pancreas

Although the amino acid structure of insulin has been known since 1949, repeated attempts to synthesize it gave very poor yields because of the failure of the two peptide chains to combine forming the correct disulfide bridge. The ease of the biosynthesis of insulin is explained by the discovery in the pancreas of proinsulin, from which insulin is formed. The single peptide chain of proinsulin loses a peptide consisting of 33 amino acids and called the connecting peptide, or C peptide, during its conversion to insulin. The disulfide bridges of proinsulin connect the A and B chains.

F.V.N.Q.H.L.C.G.S.H.L.V.E.A.L.Y.L.V.C.G.E.R.G.F.F.Y.T.P.K.A — *B* chain (insulin)
— disulfide bonds —
G.I.V.E.Q.C.C.T.S.I.C.S.L.Y.Q.L.E.N.Y.C.N — *A* chain (insulin)
— disulfide bond
R.K.Q.P.P.G.E.L.A.L.A.Q.L.G.G.L.G.G.G.L.E.V.A.G.A.Q.P.N.Q.A.E.A.A — *C* peptide

In aqueous solutions, insulin exists predominantly as a complex of six subunits, each of which contains an A and a B chain. The insulins of several species have been isolated and analyzed; their amino acid sequences have been found to differ somewhat, but all apparently contain the same disulfide bridges between the two chains.

Although the injection of insulin lowers the blood sugar, administration of glucagon, another pancreas hormone, raises the blood sugar level. Glucagon consists of a straight peptide chain of 29 amino acids. It has been synthesized; the synthetic product has the full biological activity of natural glucagon. The structure of glucagon is free of cystine and isoleucine.

The pituitary gland has an anterior lobe, a posterior lobe, and an intermediate portion; they differ in cellular structure and in the structure and action of the hormones they form. The posterior lobe produces two similar hormones, oxytocin and vasopressin. The former causes contraction of the pregnant uterus; the latter raises the blood pressure. Both are octapeptides formed by a ring of five amino acids (the two cystine halves count as one amino acid) and a side chain of three amino acids. The two cystine halves are linked to each other by a disulfide bond, and the C terminal amino acid is glycinamide. The structure has been established and confirmed. Human vasopressin differs from oxytocin in that isoleucine is replaced by phenylalanine and leucine by arginine.

A Cys.Tyr.Ile.GluN.Asn.Cys.Pro.Leu.Gly($CONH_2$)

B Cys.Tyr.Phe.GluN.Asn.Cys.Pro.Arg.Gly($CONH_2$)

The intermediate part of the pituitary gland produces the melanocyte-stimulating hormone(MSH), which causes expansion of the pigmented melanophores (cells) in the skin of frogs and other batrachians. Two hormones, called α-MSH and β-MSH, have been prepared from hog pituitary glands. The first, α-MSH, consists of 13 amino acids; its N terminal serine is acetylated (i.e., the acetyl group, CH_3CO, of acetic acid is attached), and its C terminal valine residue is present as valinamide. The second, β-MSH, contains in its 18 amino acids many of those occurring in α-MSH.

(CH_3CO)S.Y.S.M.E.H.F.R.W.G.K.P.V($CONH_2$)	porcine α-MSH, melanocyte-stimulating hormone
D.S.G.P.Y.K.M.E.H.F.R.W.G.S.P.P.K.D	porcine β-MSH
A.E.K.K.D.E.G.P.Y.K.M.E.H.F.R.W.G.S.P.P.K.D	human β-MSH
S.Y.S.M.E.H.F.R.W.G.K.P.V.G.K.K.R.R.P.V.K.V.Y.P.D.G.A.E.D.Q.L.A.E.A.F.P.L.E.F	porcine β-corticotropin

The anterior pituitary lobe produces several protein hormones—a thyroid-stimulating hormone (thyrotropin), molecular weight 28,000; a lactogenic hormone, molecular weight 22,500; a growth hormone, molecular weight 21,500; a luteinizing hormone, molecular weight 30,000; and a follicle-stimulating hormone, molecular weight 29,000. The thyroid-stimulating hormone consists of α and β subunits with a composition similar to the subunits of luteinizing hormone. When separated, neither of the two subunits has hormonal activity; when combined, however, they regain about 50 percent of the original activity. The lactogenic hormone (prolactin) from sheep pituitary glands contains 190 amino acids. Their sequence has been elucidated; a similar peptide chain of 188 amino acids that has been synthesized not only has 10 percent of the biological activity of the natural hormone but also some activity of the growth hormone. The amino acid sequence of the growth hormone (somatotropic hormone) is also known; it seems to stimulate the synthesis of RNA and in this way to accelerate growth. The luteinizing hormone, a mucoprotein containing about 12 percent carbohydrate, consists of two subunits, each with a molecular weight of approximately 15,000; when separated, the subunits recombine spontaneously. The urine of pregnant women contains chorionic gonadotropin, the presence of which makes possible early diagnosis of pregnancy. The amino acid sequence is known. The sequence of 160 of its 190 amino acids is identical with those of the growth hormone; 100 of these also occur in the same sequence as in lactogenic hormone. The different pituitary hormones and the chorionic gonadotropin thus may have been derived from a common substance that, during evolution, underwent differentiation.

Peptides with Hormonelike Activity

Small peptides have been discovered that, like hormones, act on certain target organs. One peptide, angiotensin (angiotonin or hypertensin), is formed in the blood from angiotensinogen by the action of renin, an enzyme of the kidney. It is an octapeptide and increases blood pressure. Similar peptides include bradykinin, which stimulates smooth muscles; gastrin, which stimulates secretion of hydrochloric acid and pepsin in the stomach; secretin, which stimulates the flow of pancreatic juice; and kallikrein, the activity of which is similar to bradykinin.

Immunoglobulins and Antibodies

Antibodies, proteins that combat foreign substances in the body, are associated with the globulin fraction of the immune serum. As stated previously, when the serum globulins are separated into α-, β-, and γ- fractions, antibodies are associated with the γ-globulins. Antibodies can be purified by precipitation with the antigen (i.e., the foreign substance) that caused their formation, followed by separation of the antigen-antibody complex. Antibodies prepared in this way consist of a mixture of many similar antibody molecules, which differ in molecular weight, amino acid composition, and other properties. The same differences are found in the γ-globulins of normal blood serums. The γ-globulin of normal blood serum is thought to consist of a mixture of hundreds of different γ-globulins, each of which occurs in amounts too small for isolation. Because the physical and chemical properties of normal γ-globulins are the same as those of antibodies, the γ-globulins are frequently called immunoglobulins. They may be considered to be antibodies against unknown antigens. If solutions of γ-globulin are resolved by gel filtration through dextran, the first fraction has a molecular weight of 900,000. This fraction is called IgM or γM; Ig is an abbreviation for immunoglobulin and M for macroglobulin. The next two fractions are IgA (γA) and IgG (γG), with molecular weights of about 320,000 and 150,000 respectively. Two other immunoglobulins, known as IgD and IgE, have also been detected in much smaller amounts in some immune sera.

The bulk of the immunoglobulins is found in the IgG fraction, which also contains most of the antibodies. The IgM molecules are apparently pentamers—aggregates of five of the IgG molecules. Electron microscopy shows their five subunits to be linked to each other by disulfide bonds in the form of a pentagon. The IgA molecules are found principally in milk and in secretions of the intestinal mucosa. Some of them contain, in addition to a dimer of IgG, a "secretory piece" that enables the passage of IgA molecules between tissue and fluid; the structure of the secretory piece is not yet known. The IgM and IgA immunoglobulins and antibodies contain 10 to 15 percent carbohydrate; the carbohydrate content of the IgG molecules is 2 to 3 percent.

IgG molecules treated with the enzyme papain split into three fragments of almost identical molecular weight of 50,000. Two of these, called Fab fragments, are identical; the third is abbreviated Fc. Reduction to sulfhydryl groups of some of the disulfide bonds of IgG results in the formation of two heavy, or H, chains (molecular weight 55,000) and two light, or L, chains (molecular weight 22,000). They are linked by disulfide bonds in the order L–H–H–L. Each H chain contains four intrachain disulfide bonds, and each L chain contains two.

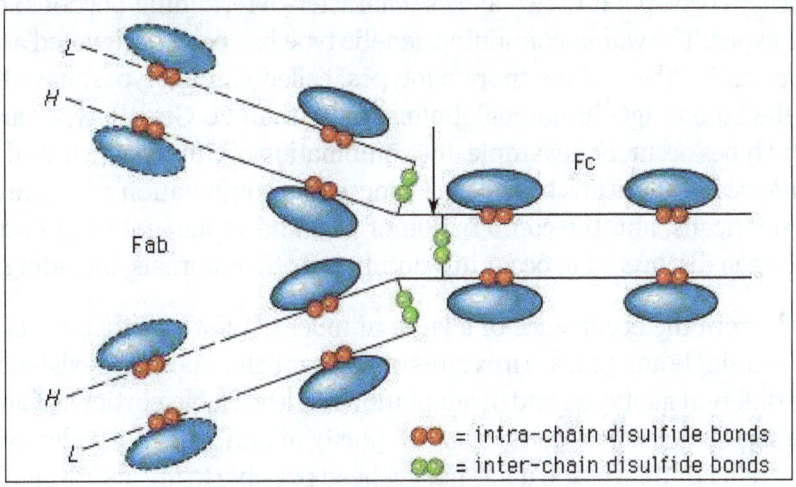

Diagram of an IgG immunoglobulin.

Two heavy chains (H) and two light chains (L) are linked to each other by inter-chain disulfide bonds. Intra-chain disulfide bonds cause loops to from in the 12 peptide portions, each of which contains about 110 amino acid residues. The 12 peptide regions have oystine residues at similar positions and other similarities in their amino acid sequences. The broken lines represent variable portions and the solid lines represent constant portions of the chains. Specific sites that bind antigens are formed by the variable portions. The vertical arrow indicates cleavage of the IgG molecule into two Fab fragments and one Fc fragment by the action of the enzyme papain.

Antibody preparations of the IgG type, even after removal of IgM and IgA antibodies, are heterogeneous. The H and L chains consist of a large number of different L chains and a variety of H chains. Pure IgG, IgM, and IgA immunoglobulins, however, occur in the blood serum of patients suffering from myelomas, which are malignant tumours of the bone marrow. The tumours produce either an IgG, an IgM, or an IgA protein, but rarely more than one class. A protein called the Bence-Jones protein, which is found in the urine of patients suffering from myeloma tumours, is identical with the L chains of the myeloma protein. Each patient has a different Bence-Jones protein; no two of the more than 100 Bence-Jones proteins that have been analyzed thus far are identical. It is thought that one lymphoid cell among hundreds of thousands becomes malignant and multiplies rapidly, forming the mass of a myeloma tumour that produces one γ-globulin.

Analyses of the Bence-Jones proteins have revealed that the L chains of humans and other mammals are of two quite different types, kappa (κ) and lambda (λ). Both consist of approximately 220 amino acids. The N–terminal halves of κ- and λ-chains are variable, differing in each Bence-Jones protein. The C–terminal halves of these same L chains have a constant amino acid sequence of either the κ- or the λ-type. The fact that one half of a peptide chain is variable and the other half invariant is contradictory to the view that the amino acid sequence of each peptide chain is determined by one gene. Evidently, two genes, one of them variable, the other invariant, fuse to form the gene for the single peptide chain of the L chains. Whereas the normal human L chains are always mixtures of the κ- and λ-types, the H chains of IgG, IgM, and IgA are different. They have been designated as gamma (γ), mu (μ), and alpha (α) chains, respectively. The N-terminal quarter of the H chains has a variable amino acid sequence; the C-terminal three-quarters of the H chains have a constant amino acid sequence.

Some of the amino acid sequences in the L and H chains are transmitted from generation to generation. As a result, the constant portion of the human L chains of the κ-type has in position 191 either valine or leucine. They correspond to two alleles (character-determining portions) of a gene; the two types are called allotypes. The valine-containing genetic type has been designated as InV(a+), the leucine-containing type as InV(b+). Many more allotypes, called Gm allotypes, have been found in the gamma chains of the human IgG immunoglobulins; more than 20 Gm allotypes are known. Certain combinations of Gm types occur. For example, the combination of Gm types 5, 6, and 11 has been found in Caucasians and African Americans but not in Chinese; the combination of 1, 2, and 17 has not been found in African Americans; and the combination of 1, 4, and 17 has not been found in Caucasians. Allotypes have also been discovered to occur in a number of other animals, including rabbits and mice.

It is understandable from the occurrence of a large number of allotypes that antibodies, even if produced in response to a single antigen, are mixtures of different allotypes. The existence of several classes of antibodies, of different allotypes, and of adaptation of the variable portions of antibodies to different regions of an antigen molecule results in a multiplicity of antibody molecules even if only a single antigen is administered. For this reason it has not yet been possible to unravel the amino acid sequence in the variable portion of antibody molecules. Much of the amino acid sequence in the constant regions of the L and H chains of humans and rabbit immunoglobulins, however, has been resolved.

Nucleic Acid

Nucleic acids are the most important macromolecules for the continuity of life. They carry the genetic blueprint of a cell and carry instructions for the functioning of the cell.

The two main types of nucleic acids are deoxyribonucleic acid (DNA) and ribonucleic acid (RNA). DNA is the genetic material found in all living organisms, ranging from single-celled bacteria to multicellular mammals. It is found in the nucleus of eukaryotes and in the organelles, chloroplasts, and mitochondria. In prokaryotes, the DNA is not enclosed in a membranous envelope.

The entire genetic content of a cell is known as its genome, and the study of genomes is genomics. In eukaryotic cells but not in prokaryotes, DNA forms a complex with histone proteins to form chromatin, the substance of eukaryotic chromosomes. A chromosome may contain tens of thousands of genes. Many genes contain the information to make protein products; other genes code for RNA products. DNA controls all of the cellular activities by turning the genes "on" or "off."

The other type of nucleic acid, RNA, is mostly involved in protein synthesis. The DNA molecules never leave the nucleus but instead use an intermediary to communicate with the rest of the cell. This intermediary is the messenger RNA (mRNA). Other types of RNA—like rRNA, tRNA, and microRNA—are involved in protein synthesis and its regulation.

DNA and RNA are made up of monomers known as nucleotides. The nucleotides combine with each other to form a polynucleotide, DNA or RNA. Each nucleotide is made up of three components: a nitrogenous base, a pentose (five-carbon) sugar, and a phosphate group . Each nitrogenous base in a nucleotide is attached to a sugar molecule, which is attached to one or more phosphate groups.

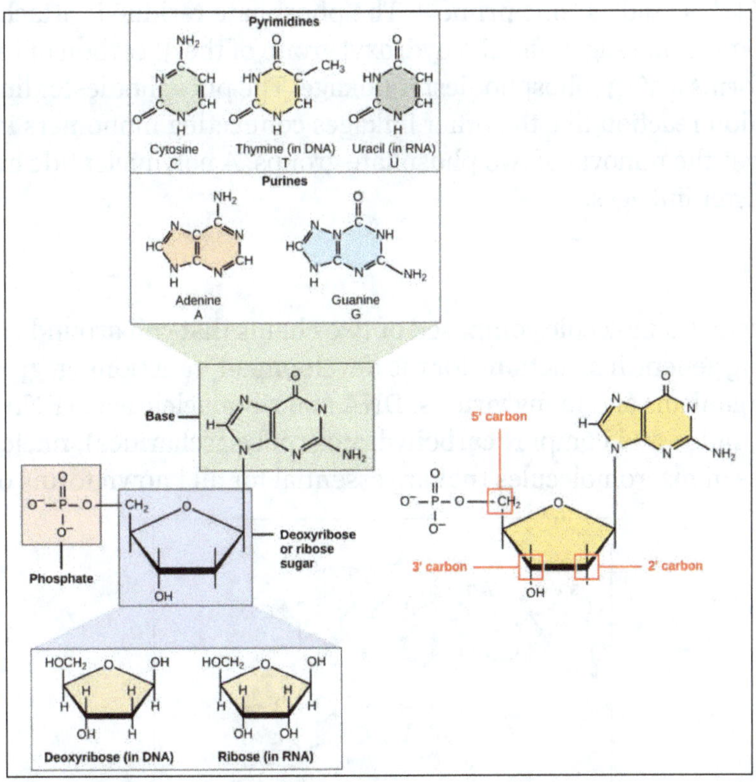

In figure a nucleotide is made up of three components: a nitrogenous base, a pentose sugar, and one or more phosphate groups. Carbon residues in the pentose are numbered 1′ through 5′ (the prime distinguishes these residues from those in the base, which are numbered without using a prime notation). The base is attached to the 1′ position of the ribose, and the phosphate is attached to the 5′ position. When a polynucleotide is formed, the 5′ phosphate of the incoming nucleotide attaches to the 3′ hydroxyl group at the end of the growing chain. Two types of pentose are found in nucleotides, deoxyribose (found in DNA) and ribose (found in RNA). Deoxyribose is similar in structure to ribose, but it has an H instead of an OH at the 2′ position. Bases can be divided into two categories: purines and pyrimidines. Purines have a double ring structure, and pyrimidines have a single ring.

The nitrogenous bases, important components of nucleotides, are organic molecules and are so named because they contain carbon and nitrogen. They are bases because they contain an amino group that has the potential of binding an extra hydrogen, and thus, decreases the hydrogen ion concentration in its environment, making it more basic. Each nucleotide in DNA contains one of four possible nitrogenous bases: adenine (A), guanine (G) cytosine (C), and thymine (T). RNA nucleotides also contain one of four possible bases: adenine, guanine, cytosine, and uracil (U) rather than thymine.

Adenine and guanine are classified as purines. The primary structure of a purine is two carbon-nitrogen rings. Cytosine, thymine, and uracil are classified as pyrimidines which have a single carbon-nitrogen ring as their primary structure. Each of these basic carbon-nitrogen rings has different functional groups attached to it. In molecular biology shorthand, the nitrogenous bases are simply known by their symbols A, T, G, C, and U. DNA contains A, T, G, and C whereas RNA contains A, U, G, and C.

The pentose sugar in DNA is deoxyribose, and in RNA, the sugar is ribose. The difference between the sugars is the presence of the hydroxyl group on the second carbon of the ribose and hydrogen on the second carbon of the deoxyribose. The carbon atoms of the sugar molecule are numbered as 1′, 2′, 3′, 4′, and 5′ (1′ is read as "one prime"). The phosphate residue is attached to the hydroxyl group of the 5′ carbon of one sugar and the hydroxyl group of the 3′ carbon of the sugar of the next nucleotide, which forms a 5′–3′ phosphodiester linkage. The phosphodiester linkage is not formed by simple dehydration reaction like the other linkages connecting monomers in macromolecules: its formation involves the removal of two phosphate groups. A polynucleotide may have thousands of such phosphodiester linkages.

DNA

Deoxyribonucleic acid is a molecule composed of two chains that coil around each other to form a double helix carrying genetic instructions for the development, functioning, growth and reproduction of all known organisms and many viruses. DNA and ribonucleic acid (RNA) are nucleic acids; alongside proteins, lipids and complex carbohydrates (polysaccharides), nucleic acids are one of the four major types of macromolecules that are essential for all known forms of life.

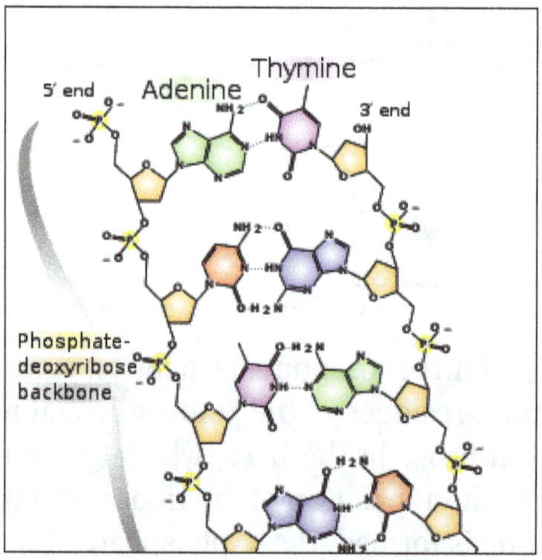

The structure of the DNA double helix. The atoms in the structure are colour-coded by element and the detailed structures of two base pairs are shown in the bottom right.

The two DNA strands are also known as polynucleotides as they are composed of simpler monomeric units called nucleotides. Each nucleotide is composed of one of four nitrogen-containing nucleobases (cytosine [C], guanine [G], adenine [A] or thymine [T]), a sugar called deoxyribose, and a phosphate group. The nucleotides are joined to one another in a chain by covalent bonds between the sugar of one nucleotide and the phosphate of the next, resulting in an alternating sugar-phosphate backbone. The nitrogenous bases of the two separate polynucleotide strands are bound together, according to base pairing rules (A with T and C with G), with hydrogen bonds to make double-stranded DNA. The complementary nitrogenous bases are divided into two groups, pyrimidines and purines. In DNA, the pyrimidines are thymine and cytosine; the purines are adenine and guanine.

Both strands of double-stranded DNA store the same biological information. This information is replicated as and when the two strands separate. A large part of DNA (more than 98% for humans) is non-coding, meaning that these sections do not serve as patterns for protein sequences. The two strands of DNA run in opposite directions to each other and are thus antiparallel. Attached to each sugar is one of four types of nucleobases (informally, *bases*). It is the sequence of these four nucleobases along the backbone that encodes genetic information. RNA strands are created using DNA strands as a template in a process called transcription. Under the genetic code, these RNA strands specify the sequence of amino acids within proteins in a process called translation.

Within eukaryotic cells, DNA is organized into long structures called chromosomes. Before typical cell division, these chromosomes are duplicated in the process of DNA replication, providing a complete set of chromosomes for each daughter cell. Eukaryotic organisms (animals, plants, fungi and protists) store most of their DNA inside the cell nucleus as nuclear DNA, and some in the mitochondria as mitochondrial DNA or in chloroplasts as chloroplast DNA. In contrast, prokaryotes (bacteria and archaea) store their DNA only in the cytoplasm, in circular chromosomes. Within eukaryotic chromosomes, chromatin proteins, such as histones, compact and organize DNA. These compacting structures guide the interactions between DNA and other proteins, helping control which parts of the DNA are transcribed.

DNA was first isolated by Friedrich Miescher in 1869. Its molecular structure was first identified by Francis Crick and James Watson at the Cavendish Laboratory within the University of Cambridge in 1953, whose model-building efforts were guided by X-ray diffraction data acquired by Raymond Gosling, who was a post-graduate student of Rosalind Franklin. DNA is used by researchers as a molecular tool to explore physical laws and theories, such as the ergodic theorem and the theory of elasticity. The unique material properties of DNA have made it an attractive molecule for material scientists and engineers interested in micro- and nano-fabrication. Among notable advances in this field are DNA origami and DNA-based hybrid materials.

Properties

DNA is a long polymer made from repeating units called nucleotides. The structure of DNA is dynamic along its length, being capable of coiling into tight loops and other shapes. In all species it is composed of two helical chains, bound to each other by hydrogen bonds. Both chains are coiled around the same axis, and have the same pitch of 34 angstroms (Å) (3.4 nanometres). The pair of chains has a radius of 10 angstroms (1.0 nanometre). According to another study, when measured in a different solution, the DNA chain measured 22 to 26 angstroms wide (2.2 to 2.6 nanometres), and one nucleotide unit measured 3.3 Å (0.33 nm) long. Although each individual nucleotide is very small, a DNA polymer can be very large and contain hundreds of millions, such as in chromosome 1. Chromosome 1 is the largest human chromosome with approximately 220 million base pairs, and would be 85 mm long if straightened.

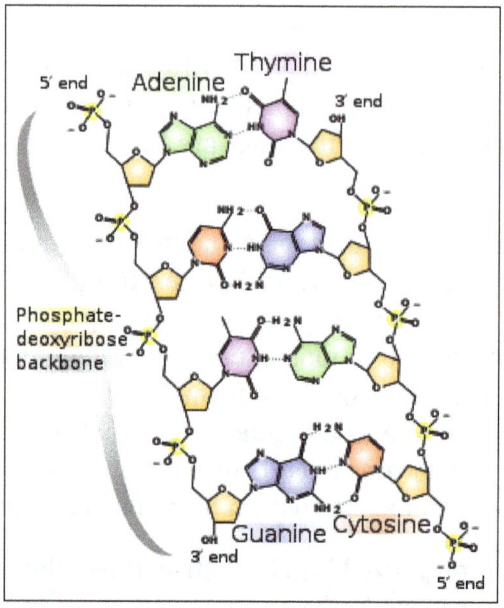

Chemical structure of DNA; hydrogen bonds shown as dotted lines.

DNA does not usually exist as a single strand, but instead as a pair of strands that are held tightly together. These two long strands coil around each other, in the shape of a double helix. The nucleotide contains both a segment of the backbone of the molecule (which holds the chain together) and a nucleobase (which interacts with the other DNA strand in the helix). A nucleobase linked to a sugar is called a nucleoside, and a base linked to a sugar and to one or more phosphate groups is called a nucleotide. A biopolymer comprising multiple linked nucleotides (as in DNA) is called a polynucleotide.

The backbone of the DNA strand is made from alternating phosphate and sugar residues. The sugar in DNA is 2-deoxyribose, which is a pentose (five-carbon) sugar. The sugars are joined together by phosphate groups that form phosphodiester bonds between the third and fifth carbon atoms of adjacent sugar rings. These are known as the 3′-end (three prime end), and 5′-end (five prime end) carbons, the prime symbol being used to distinguish these carbon atoms from those of the base to which the deoxyribose forms a glycosidic bond. When imagining DNA, each phosphoryl is normally considered to "belong" to the nucleotide whose 5′ carbon forms a bond therewith. Any DNA strand therefore normally has one end at which there is a phosphoryl attached to the 5′ carbon of a ribose (the 5′ phosphoryl) and another end at which there is a free hydroxyl attached to the 3′ carbon of a ribose (the 3′ hydroxyl). The orientation of the 3′ and 5′ carbons along the sugar-phosphate backbone confers directionality (sometimes called polarity) to each DNA strand. In a nucleic acid double helix, the direction of the nucleotides in one strand is opposite to their direction in the other strand: the strands are antiparallel. The asymmetric ends of DNA strands are said to have a directionality of five prime end (5′), and three prime end (3′), with the 5′ end having a terminal phosphate group and the 3′ end a terminal hydroxyl group. One major difference between DNA and RNA is the sugar, with the 2-deoxyribose in DNA being replaced by the alternative pentose sugar ribose in RNA.

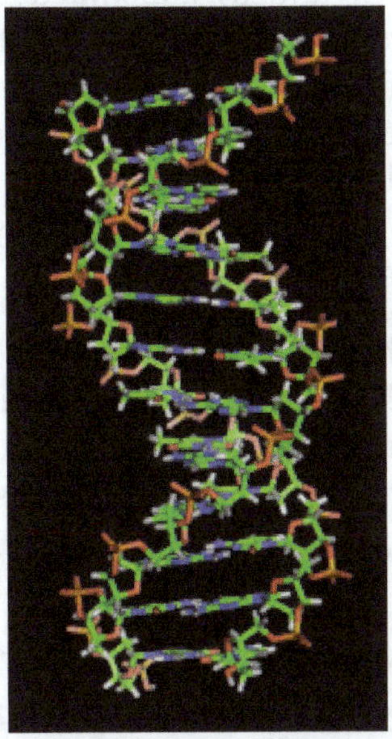

A section of DNA. The bases lie horizontally between the two spiraling strands.

The DNA double helix is stabilized primarily by two forces: hydrogen bonds between nucleotides and base-stacking interactions among aromatic nucleobases. In the cytosol of the cell, the conjugated pi bonds of nucleotide bases align perpendicular to the axis of the DNA molecule, minimizing their interaction with the solvation shell. The four bases found in DNA are adenine (A), cytosine (C), guanine (G) and thymine (T). These four bases are attached to the sugar-phosphate to form the complete nucleotide, as shown for adenosine monophosphate. Adenine pairs with thymine and guanine pairs with cytosine, forming A-T and G-C base pairs.

Nucleobase Classification

The nucleobases are classified into two types: the purines, A and G, which are fused five- and six-membered heterocyclic compounds, and the pyrimidines, the six-membered rings C and T. A fifth pyrimidine nucleobase, uracil (U), usually takes the place of thymine in RNA and differs from thymine by lacking a methyl group on its ring. In addition to RNA and DNA, many artificial nucleic acid analogues have been created to study the properties of nucleic acids, or for use in biotechnology.

Non-canonical Bases

Uracil is not usually found in DNA, occurring only as a breakdown product of cytosine. However, in several bacteriophages, such as *Bacillus subtilis* phages PBS1 and PBS2 and *Yersinia* phage piR1-37, thymine has been replaced by uracil. Another phage—Staphylococcal phage S6—has been identified with a genome where thymine has been replaced by uracil.

Uracil is also found in the DNA of *Plasmodium falciparum* It is present is relatively small amounts (7–10 uracil residues per million bases).

5-hydroxymethyldeoxyuridine,(hm5dU) is also known to replace thymidine in several genomes including the *Bacillus phages* SPO1, φe, SP8, H1, 2C and SP82. Another modified uracil—5-dihydroxypentauracil—has also been described.

Base J (beta-d-glucopyranosyloxymethyluracil), a modified form of uracil, is also found in several organisms: the flagellates *Diplonema* and *Euglena*, and all the kinetoplastid genera. Biosynthesis of J occurs in two steps: in the first step, a specific thymidine in DNA is converted into hydroxymethyldeoxyuridine; in the second, HOMedU is glycosylated to form J. Proteins that bind specifically to this base have been identified. These proteins appear to be distant relatives of the Tet1 oncogene that is involved in the pathogenesis of acute myeloid leukemia. J appears to act as a termination signal for RNA polymerase II.

In 1976, the S-2La bacteriophage, which infects species of the genus *Synechocystis*, was found to have all the adenosine bases within its genome replaced by 2,6-diaminopurine. In 2016 deoxyarchaeosine was found to be present in the genomes of several bacteria and the *Escherichia* phage 9g. It seems likely that this base may be present in other viruses also.

Modified bases also occur in DNA. The first of these recognised was 5-methylcytosine, which was found in the genome of *Mycobacterium tuberculosis* in 1925. The complete replacement of cytosine by 5-glycosylhydroxymethylcytosine in T even phages (T2, T4 and T6) was observed in 1953. In the genomes of Xanthomonas oryzae bacteriophage Xp12 and halovirus FH the full complement of cystosine has been replaced by 5-methylcytosine. 6N-methyladenine was discovered to be present in DNA in 1955. N6-carbamoyl-methyladenine was described in 1975. 7-Methylguanine was described in 1976. N4-methylcytosine in DNA was described in 1983. In 1985 5-hydroxycytosine was found in the genomes of the Rhizobium phages RL38JI and N17. α-putrescinylthymine occurs in both the genomes of the *Delftia* phage ΦW-14 and the *Bacillus* phage SP10. α-glutamylthymidine is found in the Bacillus phage SP01 and 5-dihydroxypentyluracil is found in the Bacillus phage SP15.

The reason for the presence of these non canonical bases in DNA is not known. It seems likely that at least part of the reason for their presence in bacterial viruses (phages) is to avoid the restriction enzymes present in bacteria. This enzyme system acts at least in part as a molecular immune system protecting bacteria from infection by viruses.

This does not appear to be the entire story. Four modifications to the cytosine residues in human DNA have been reported. These modifications are the addition of methyl (CH_3^-), hydroxymethyl (CH_2OH^-), formyl (CHO^-) and carboxyl ($COOH^-$) groups. These modifications are thought to have regulatory functions.

Uracil is found in the centromeric regions of at least two human chromosomes (chromosome 6 and chromosome 11).

Listing of Non-canonical bases Found in DNA

Seventeen non canonical bases are known to occur in DNA. Most of these are modifications of the canonical bases plus uracil.

- Modified Adenosine

- N6-carbamoyl-methyladenine
- N6-methyadenine

- Modified Guanine
  - 7-Methylguanine
- Modified Cytosine
  - N4-Methylcytosine
  - 5-Carboxylcytosine
  - 5-Formylcytosine
  - 5-Glycosylhydroxymethylcytosine
  - 5-Hydroxycytosine
  - 5-Methylcytosine
- Modified Thymidine
  - α-Glutamythymidine
  - α-Putrescinylthymine
- Uracil and modifications
  - Base J
  - Uracil
  - 5-Dihydroxypentauracil
  - 5-Hydroxymethyldeoxyuracil
- Others
  - Deoxyarchaeosine
  - 2,6-Diaminopurine

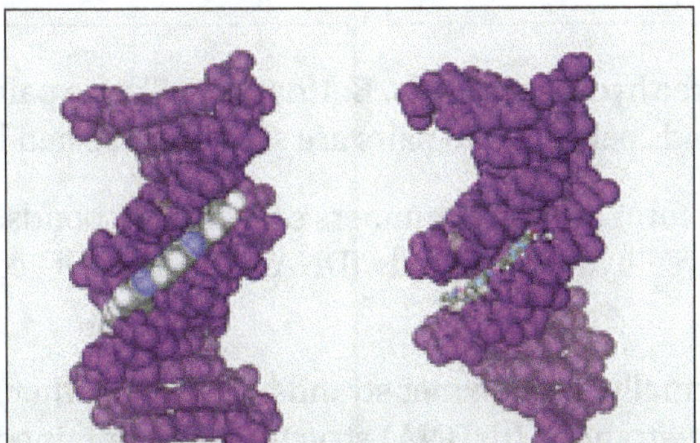

DNA major and minor grooves. The latter is a binding site for the Hoechst stain dye 33258.

Grooves

Twin helical strands form the DNA backbone. Another double helix may be found tracing the spaces, or grooves, between the strands. These voids are adjacent to the base pairs and may provide a binding site. As the strands are not symmetrically located with respect to each other, the grooves are unequally sized. One groove, the major groove, is 22 angstroms (Å) wide and the other, the minor groove, is 12 Å wide. The width of the major groove means that the edges of the bases are more accessible in the major groove than in the minor groove. As a result, proteins such as transcription factors that can bind to specific sequences in double-stranded DNA usually make contact with the sides of the bases exposed in the major groove. This situation varies in unusual conformations of DNA within the cell, but the major and minor grooves are always named to reflect the differences in size that would be seen if the DNA is twisted back into the ordinary B form.

Base Pairing

In a DNA double helix, each type of nucleobase on one strand bonds with just one type of nucleobase on the other strand. This is called complementary base pairing. Here, purines form hydrogen bonds to pyrimidines, with adenine bonding only to thymine in two hydrogen bonds, and cytosine bonding only to guanine in three hydrogen bonds. This arrangement of two nucleotides binding together across the double helix is called a Watson-Crick base pair. Another type of base pairing is Hoogsteen base pairing where two hydrogen bonds form between guanine and cytosine. As hydrogen bonds are not covalent, they can be broken and rejoined relatively easily. The two strands of DNA in a double helix can thus be pulled apart like a zipper, either by a mechanical force or high temperature. As a result of this base pair complementarity, all the information in the double-stranded sequence of a DNA helix is duplicated on each strand, which is vital in DNA replication. This reversible and specific interaction between complementary base pairs is critical for all the functions of DNA in organisms.

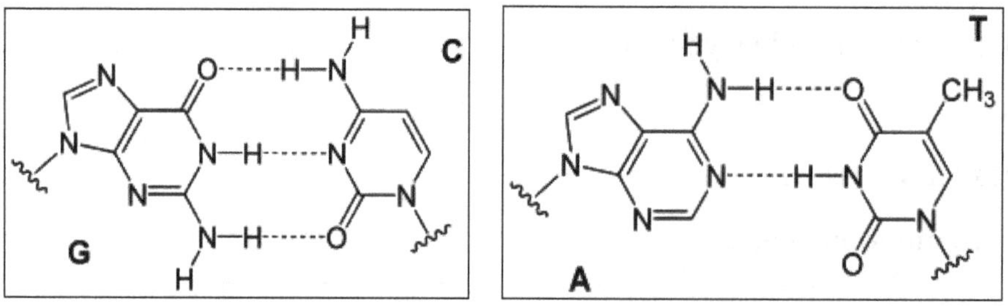

Top, a GC base pair with three hydrogen bonds. Bottom, an AT base pair with two hydrogen bonds. Non-covalent hydrogen bonds between the pairs are shown as dashed lines.

The two types of base pairs form different numbers of hydrogen bonds, AT forming two hydrogen bonds, and GC forming three hydrogen bonds. DNA with high GC-content is more stable than DNA with low GC-content.

Most DNA molecules are actually two polymer strands, bound together in a helical fashion by non-covalent bonds; this double-stranded (dsDNA) structure is maintained largely by the intrastrand base stacking interactions, which are strongest for G,C stacks. The two strands can come apart—a process known as melting—to form two single-stranded DNA (ssDNA) molecules. Melting occurs at high temperature, low salt and high pH (low pH also melts DNA, but since DNA is unstable due to acid depurination, low pH is rarely used).

The stability of the dsDNA form depends not only on the GC-content (% G,C basepairs) but also on sequence (since stacking is sequence specific) and also length (longer molecules are more stable). The stability can be measured in various ways; a common way is the "melting temperature", which is the temperature at which 50% of the ds molecules are converted to ss molecules; melting temperature is dependent on ionic strength and the concentration of DNA. As a result, it is both the percentage of GC base pairs and the overall length of a DNA double helix that determines the strength of the association between the two strands of DNA. Long DNA helices with a high GC-content have stronger-interacting strands, while short helices with high AT content have weaker-interacting strands. In biology, parts of the DNA double helix that need to separate easily, such as the TATAAT Pribnow box in some promoters, tend to have a high AT content, making the strands easier to pull apart.

In the laboratory, the strength of this interaction can be measured by finding the temperature necessary to break the hydrogen bonds, their melting temperature (also called T_m value). When all the base pairs in a DNA double helix melt, the strands separate and exist in solution as two entirely independent molecules. These single-stranded DNA molecules have no single common shape, but some conformations are more stable than others.

Sense and Antisense

A DNA sequence is called a "sense" sequence if it is the same as that of a messenger RNA copy that is translated into protein. The sequence on the opposite strand is called the "antisense" sequence. Both sense and antisense sequences can exist on different parts of the same strand of DNA (i.e. both strands can contain both sense and antisense sequences). In both prokaryotes and eukaryotes, antisense RNA sequences are produced, but the functions of these RNAs are not entirely clear. One proposal is that antisense RNAs are involved in regulating gene expression through RNA-RNA base pairing.

A few DNA sequences in prokaryotes and eukaryotes, and more in plasmids and viruses, blur the distinction between sense and antisense strands by having overlapping genes. In these cases, some DNA sequences do double duty, encoding one protein when read along one strand, and a second protein when read in the opposite direction along the other strand. In bacteria, this overlap may be involved in the regulation of gene transcription, while in viruses, overlapping genes increase the amount of information that can be encoded within the small viral genome.

Supercoiling

DNA can be twisted like a rope in a process called DNA supercoiling. With DNA in its "relaxed" state, a strand usually circles the axis of the double helix once every 10.4 base pairs, but if the DNA is twisted the strands become more tightly or more loosely wound. If the DNA is twisted in the direction of the helix, this is positive supercoiling, and the bases are held more tightly together. If they are twisted in the opposite direction, this is negative supercoiling, and the bases come apart more easily. In nature, most DNA has slight negative supercoiling that is introduced by enzymes called topoisomerases. These enzymes are also needed to relieve the twisting stresses introduced into DNA strands during processes such as transcription and DNA replication.

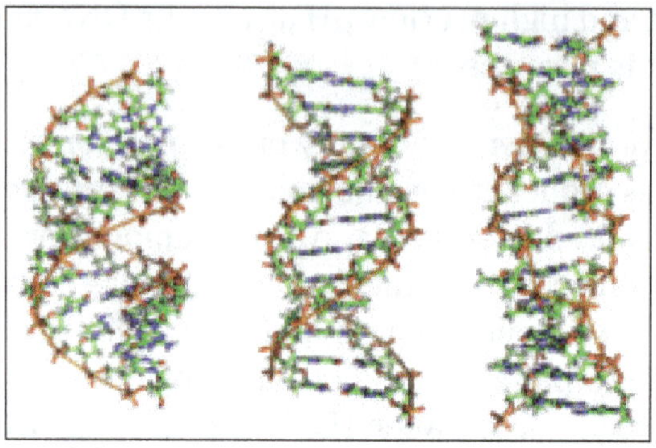

From left to right, the structures of A, B and Z DNA.

Alternative DNA Structures

DNA exists in many possible conformations that include A-DNA, B-DNA, and Z-DNA forms, although, only B-DNA and Z-DNA have been directly observed in functional organisms. The conformation that DNA adopts depends on the hydration level, DNA sequence, the amount and direction of supercoiling, chemical modifications of the bases, the type and concentration of metal ions, and the presence of polyamines in solution.

The first published reports of A-DNA X-ray diffraction patterns—and also B-DNA—used analyses based on Patterson transforms that provided only a limited amount of structural information for oriented fibers of DNA. An alternative analysis was then proposed by Wilkins *et al.*, in 1953, for the *in vivo* B-DNA X-ray diffraction-scattering patterns of highly hydrated DNA fibers in terms of squares of Bessel functions. In the same journal, James Watson and Francis Crick presented their molecular modeling analysis of the DNA X-ray diffraction patterns to suggest that the structure was a double-helix.

Although the *B-DNA form* is most common under the conditions found in cells, it is not a well-defined conformation but a family of related DNA conformations that occur at the high hydration levels present in cells. Their corresponding X-ray diffraction and scattering patterns are characteristic of molecular paracrystals with a significant degree of disorder.

Compared to B-DNA, the A-DNA form is a wider right-handed spiral, with a shallow, wide minor groove and a narrower, deeper major groove. The A form occurs under non-physiological conditions in partly dehydrated samples of DNA, while in the cell it may be produced in hybrid pairings of DNA and RNA strands, and in enzyme-DNA complexes. Segments of DNA where the bases have been chemically modified by methylation may undergo a larger change in conformation and adopt the Z form. Here, the strands turn about the helical axis in a left-handed spiral, the opposite of the more common B form. These unusual structures can be recognized by specific Z-DNA binding proteins and may be involved in the regulation of transcription.

Alternative DNA Chemistry

For many years, exobiologists have proposed the existence of a shadow biosphere, a postulated microbial biosphere of Earth that uses radically different biochemical and molecular processes than currently known life. One of the proposals was the existence of lifeforms that use arsenic instead of phosphorus in DNA. A report in 2010 of the possibility in the bacterium GFAJ-1, was announced, though the research was disputed, and evidence suggests the bacterium actively prevents the incorporation of arsenic into the DNA backbone and other biomolecules.

Quadruplex Structures

At the ends of the linear chromosomes are specialized regions of DNA called telomeres. The main function of these regions is to allow the cell to replicate chromosome ends using the enzyme telomerase, as the enzymes that normally replicate DNA cannot copy the extreme 3′ ends of chromosomes. These specialized chromosome caps also help protect the DNA ends, and stop the DNA repair systems in the cell from treating them as damage to be corrected. In human cells, telomeres are usually lengths of single-stranded DNA containing several thousand repeats of a simple TTAGGG sequence.

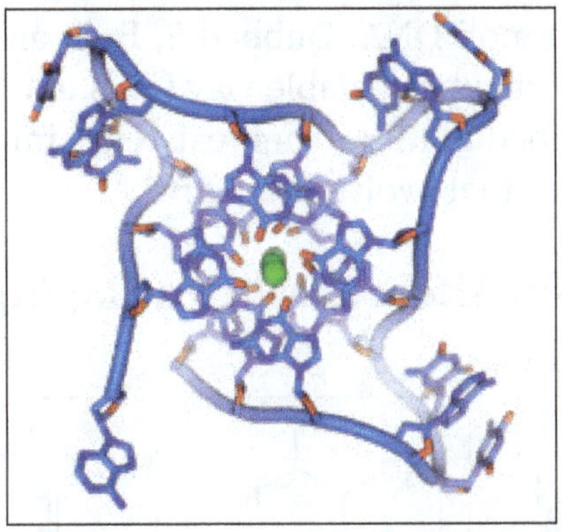

DNA quadruplex formed by telomere repeats. The looped conformation of the DNA backbone is very different from the typical DNA helix. The green spheres in the center represent potassium ions.

These guanine-rich sequences may stabilize chromosome ends by forming structures of stacked sets of four-base units, rather than the usual base pairs found in other DNA molecules. Here, four guanine bases form a flat plate and these flat four-base units then stack on top of each other, to form a stable G-quadruplex structure. These structures are stabilized by hydrogen bonding between the edges of the bases and chelation of a metal ion in the centre of each four-base unit. Other structures can also be formed, with the central set of four bases coming from either a single strand folded around the bases, or several different parallel strands, each contributing one base to the central structure.

In addition to these stacked structures, telomeres also form large loop structures called telomere loops, or T-loops. Here, the single-stranded DNA curls around in a long circle stabilized by telomere-binding proteins. At the very end of the T-loop, the single-stranded telomere DNA is held onto a region of double-stranded DNA by the telomere strand disrupting the double-helical DNA and base pairing to one of the two strands. This triple-stranded structure is called a displacement loop or D-loop.

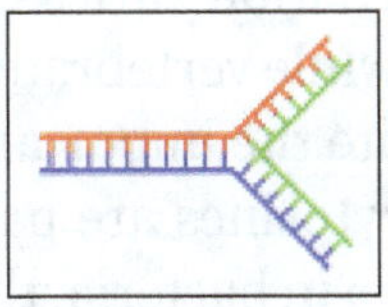

Single branch

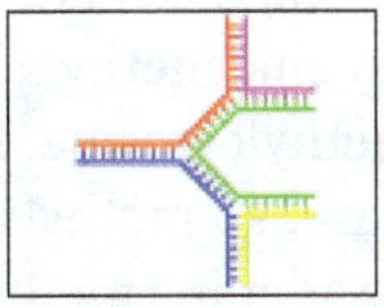

Multiple branches

Branched DNA can form networks containing multiple branches.

Branched DNA

In DNA, fraying occurs when non-complementary regions exist at the end of an otherwise complementary double-strand of DNA. However, branched DNA can occur if a third strand of DNA is introduced and contains adjoining regions able to hybridize with the frayed regions of the pre-existing double-strand. Although the simplest example of branched DNA involves only three strands of DNA, complexes involving additional strands and multiple branches are also possible. Branched DNA can be used in nanotechnology to construct geometric shapes.

Artificial Bases

Several artificial nucleobases have been synthesized, and successfully incorporated in the eight-base DNA analogue named Hachimoji DNA. Dubbed S, B, P, and Z, these artificial bases are capable of bonding with each other in a predictable way (S–B and P–Z), maintain the double helix structure of DNA, and be transcribed to RNA. Their existence implies that there is nothing special about the four natural nucleobases that evolved on Earth.

Chemical Modifications and Altered DNA Packaging

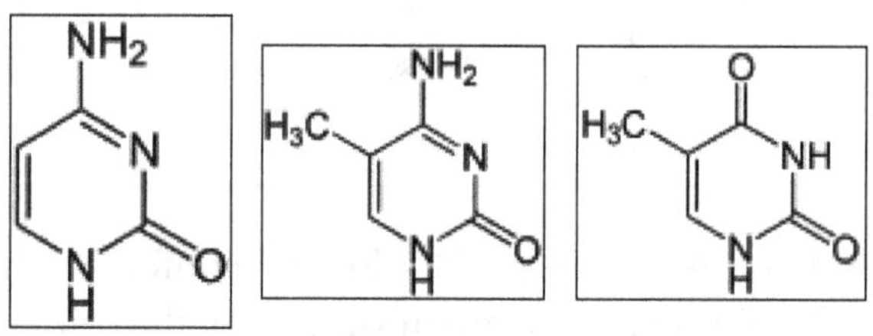

cytosine 5-methylcytosine thymine

Structure of cytosine with and without the 5-methyl group. Deamination converts 5-methylcytosine into thymine.

Base Modifications and DNA Packaging

The expression of genes is influenced by how the DNA is packaged in chromosomes, in a structure called chromatin. Base modifications can be involved in packaging, with regions that have low or no gene expression usually containing high levels of methylation of cytosine bases. DNA packaging and its influence on gene expression can also occur by covalent modifications of the histone protein core around which DNA is wrapped in the chromatin structure or else by remodeling carried out by chromatin remodeling complexes. There is, further, crosstalk between DNA methylation and histone modification, so they can coordinately affect chromatin and gene expression.

For one example, cytosine methylation produces 5-methylcytosine, which is important for X-inactivation of chromosomes. The average level of methylation varies between organisms—the worm *Caenorhabditis elegans* lacks cytosine methylation, while vertebrates have higher levels, with up to 1% of their DNA containing 5-methylcytosine. Despite the importance of 5-methylcytosine, it can deaminate to leave a thymine base, so methylated cytosines are particularly prone to mutations. Other base modifications include adenine methylation in bacteria, the presence of 5-hydroxymethylcytosine in the brain, and the glycosylation of uracil to produce the "J-base" in kinetoplastids.

Damage

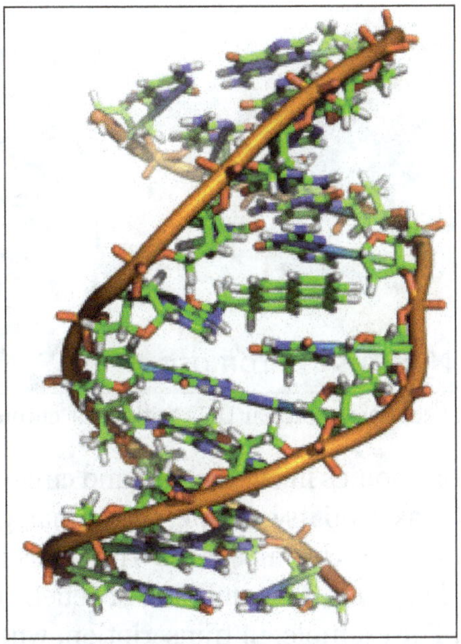

A covalent adduct between a metabolically activated form of benzo[*a*]pyrene, the major mutagen in tobacco smoke, and DNA.

DNA can be damaged by many sorts of mutagens, which change the DNA sequence. Mutagens include oxidizing agents, alkylating agents and also high-energy electromagnetic radiation such as ultraviolet light and X-rays. The type of DNA damage produced depends on the type of mutagen. For example, UV light can damage DNA by producing thymine dimers, which are cross-links between pyrimidine bases. On the other hand, oxidants such as free radicals or hydrogen peroxide produce multiple forms of damage, including base modifications, particularly of guanosine, and double-strand breaks. A typical human cell contains about 150,000 bases that have suffered oxidative damage. Of these oxidative lesions, the most dangerous are double-strand breaks, as these are difficult to repair and can produce point mutations, insertions, deletions from the DNA sequence, and chromosomal translocations. These mutations can cause cancer. Because of inherent limits in the DNA repair mechanisms, if humans lived long enough, they would all eventually develop cancer. DNA damages that are naturally occurring, due to normal cellular processes that produce reactive oxygen species, the hydrolytic activities of cellular water, etc., also occur frequently. Although most of these damages are repaired, in any cell some DNA damage may remain despite the action of repair processes. These remaining DNA damages accumulate with age in mammalian postmitotic tissues. This accumulation appears to be an important underlying cause of aging.

Many mutagens fit into the space between two adjacent base pairs, this is called *intercalation*. Most intercalators are aromatic and planar molecules; examples include ethidium bromide, acridines, daunomycin, and doxorubicin. For an intercalator to fit between base pairs, the bases must separate, distorting the DNA strands by unwinding of the double helix. This inhibits both transcription and DNA replication, causing toxicity and mutations. As a result, DNA intercalators may be carcinogens, and in the case of thalidomide, a teratogen. Others such as benzo[*a*]pyrene diol epoxide and aflatoxin form DNA adducts that induce errors in replication. Nevertheless, due to their ability to inhibit DNA transcription and replication, other similar toxins are also used in chemotherapy to inhibit rapidly growing cancer cells.

Biological Functions

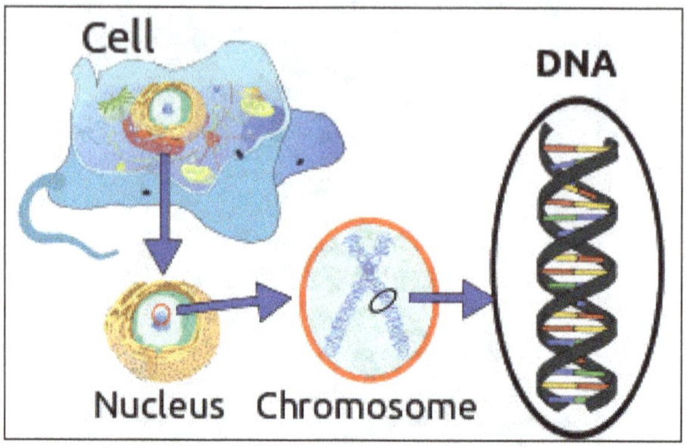

Location of eukaryote nuclear DNA within the chromosomes.

DNA usually occurs as linear chromosomes in eukaryotes, and circular chromosomes in prokaryotes. The set of chromosomes in a cell makes up its genome; the human genome has approximately 3 billion base pairs of DNA arranged into 46 chromosomes. The information carried by DNA is held in the sequence of pieces of DNA called genes. Transmission of genetic information in genes is achieved via complementary base pairing. For example, in transcription, when a cell uses the information in a gene, the DNA sequence is copied into a complementary RNA sequence through the attraction between the DNA and the correct RNA nucleotides. Usually, this RNA copy is then used to make a matching protein sequence in a process called translation, which depends on the same interaction between RNA nucleotides. In alternative fashion, a cell may simply copy its genetic information in a process called DNA replication.

Genes and Genomes

Genomic DNA is tightly and orderly packed in the process called DNA condensation, to fit the small available volumes of the cell. In eukaryotes, DNA is located in the cell nucleus, with small amounts in mitochondria and chloroplasts. In prokaryotes, the DNA is held within an irregularly shaped body in the cytoplasm called the nucleoid. The genetic information in a genome is held within genes, and the complete set of this information in an organism is called its genotype. A gene is a unit of heredity and is a region of DNA that influences a particular characteristic in an organism. Genes contain an open reading frame that can be transcribed, and regulatory sequences such as promoters and enhancers, which control transcription of the open reading frame.

In many species, only a small fraction of the total sequence of the genome encodes protein. For example, only about 1.5% of the human genome consists of protein-coding exons, with over 50% of human DNA consisting of non-coding repetitive sequences. The reasons for the presence of so much noncoding DNA in eukaryotic genomes and the extraordinary differences in genome size, or *C-value*, among species, represent a long-standing puzzle known as the "C-value enigma". However, some DNA sequences that do not code protein may still encode functional non-coding RNA molecules, which are involved in the regulation of gene expression.

Some noncoding DNA sequences play structural roles in chromosomes. Telomeres and centromeres typically contain few genes but are important for the function and stability of chromosomes. An abundant form of noncoding DNA in humans are pseudogenes, which are copies of genes that have been disabled by mutation. These sequences are usually just molecular fossils, although they can occasionally serve as raw genetic material for the creation of new genes through the process of gene duplication and divergence.

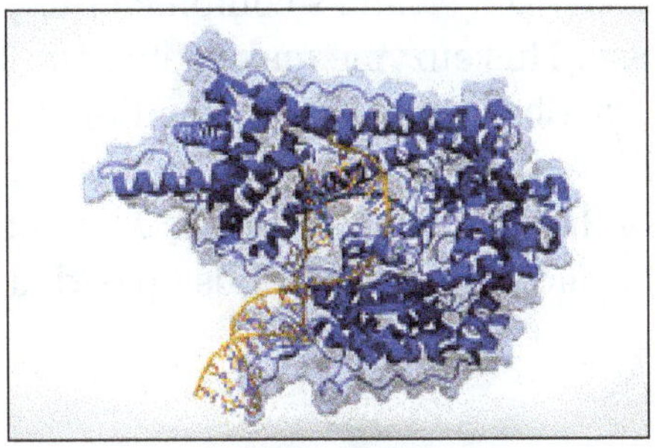

T7 RNA polymerase (blue) producing an mRNA (green) from a DNA template (orange).

Transcription and Translation

A gene is a sequence of DNA that contains genetic information and can influence the phenotype of an organism. Within a gene, the sequence of bases along a DNA strand defines a messenger RNA sequence, which then defines one or more protein sequences. The relationship between the nucleotide sequences of genes and the amino-acid sequences of proteins is determined by the rules of translation, known collectively as the genetic code. The genetic code consists of three-letter 'words' called *codons* formed from a sequence of three nucleotides (e.g. ACT, CAG, TTT).

In transcription, the codons of a gene are copied into messenger RNA by RNA polymerase. This RNA copy is then decoded by a ribosome that reads the RNA sequence by base-pairing the messenger RNA to transfer RNA, which carries amino acids. Since there are 4 bases in 3-letter combinations, there are 64 possible codons (4^3 combinations). These encode the twenty standard amino acids, giving most amino acids more than one possible codon. There are also three 'stop' or 'nonsense' codons signifying the end of the coding region; these are the TAA, TGA, and TAG codons.

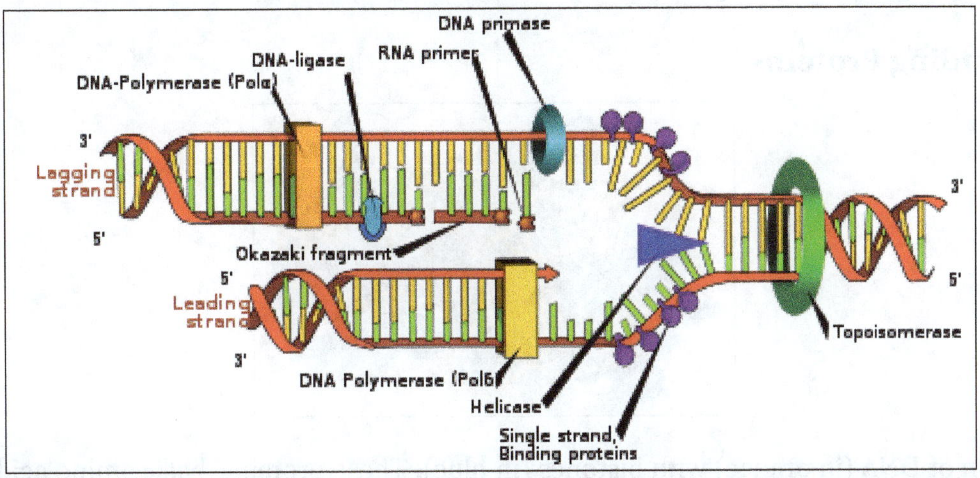

DNA replication. The double helix is unwound by a helicase and topoisomerase. Next, one DNA polymerase produces the leading strand copy. Another DNA polymerase binds to the lagging strand. This enzyme makes discontinuous segments (called Okazaki fragments) before DNA ligase joins them together.

Cell division is essential for an organism to grow, but, when a cell divides, it must replicate the DNA in its genome so that the two daughter cells have the same genetic information as their parent. The double-stranded structure of DNA provides a simple mechanism for DNA replication. Here, the two strands are separated and then each strand's complementary DNA sequence is recreated by an enzyme called DNA polymerase. This enzyme makes the complementary strand by finding the correct base through complementary base pairing and bonding it onto the original strand. As DNA polymerases can only extend a DNA strand in a 5′ to 3′ direction, different mechanisms are used to copy the antiparallel strands of the double helix. In this way, the base on the old strand dictates which base appears on the new strand, and the cell ends up with a perfect copy of its DNA.

Extracellular Nucleic Acids

Naked extracellular DNA (eDNA), most of it released by cell death, is nearly ubiquitous in the environment. Its concentration in soil may be as high as 2 μg/L, and its concentration in natural aquatic environments may be as high at 88 μg/L. Various possible functions have been proposed for eDNA: it may be involved in horizontal gene transfer; it may provide nutrients; and it may act as a buffer to recruit or titrate ions or antibiotics. Extracellular DNA acts as a functional extracellular matrix component in the biofilms of several bacterial species. It may act as a recognition factor to regulate the attachment and dispersal of specific cell types in the biofilm; it may contribute to biofilm formation; and it may contribute to the biofilm's physical strength and resistance to biological stress.

Cell-free fetal DNA is found in the blood of the mother, and can be sequenced to determine a great deal of information about the developing fetus.

Interactions with Proteins

All the functions of DNA depend on interactions with proteins. These protein interactions can be non-specific, or the protein can bind specifically to a single DNA sequence. Enzymes can also bind to DNA and of these, the polymerases that copy the DNA base sequence in transcription and DNA replication are particularly important.

DNA-binding Proteins

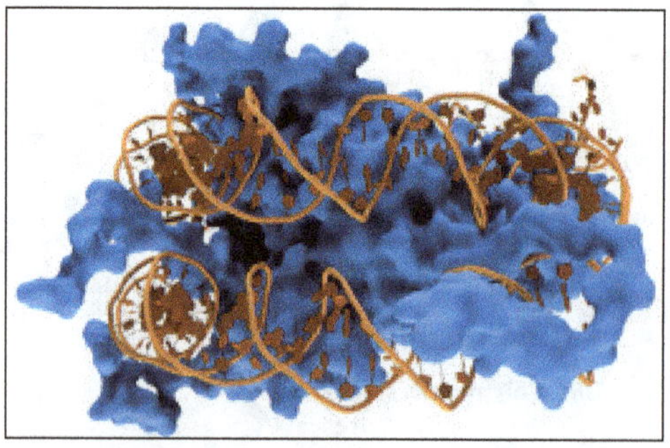

Interaction of DNA (in orange) with histones (in blue). These proteins' basic amino acids bind to the acidic phosphate groups on DNA.

Structural proteins that bind DNA are well-understood examples of non-specific DNA-protein interactions. Within chromosomes, DNA is held in complexes with structural proteins. These proteins organize the DNA into a compact structure called chromatin. In eukaryotes, this structure involves DNA binding to a complex of small basic proteins called histones, while in prokaryotes multiple types of proteins are involved. The histones form a disk-shaped complex called a nucleosome, which contains two complete turns of double-stranded DNA wrapped around its surface. These non-specific interactions are formed through basic residues in the histones, making ionic bonds to the acidic sugar-phosphate backbone of the DNA, and are thus largely independent of the base sequence. Chemical modifications of these basic amino acid residues include methylation, phosphorylation, and acetylation. These chemical changes alter the strength of the interaction between the DNA and the histones, making the DNA more or less accessible to transcription factors and changing the rate of transcription. Other non-specific DNA-binding proteins in chromatin include the high-mobility group proteins, which bind to bent or distorted DNA. These proteins are important in bending arrays of nucleosomes and arranging them into the larger structures that make up chromosomes.

A distinct group of DNA-binding proteins is the DNA-binding proteins that specifically bind single-stranded DNA. In humans, replication protein A is the best-understood member of this family and is used in processes where the double helix is separated, including DNA replication, recombination, and DNA repair. These binding proteins seem to stabilize single-stranded DNA and protect it from forming stem-loops or being degraded by nucleases.

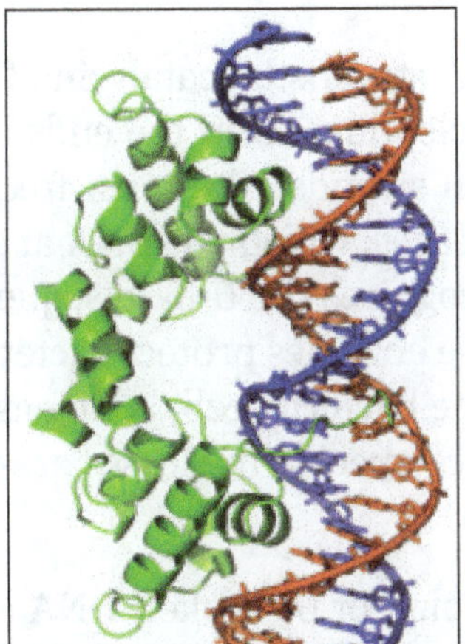

The lambda repressor helix-turn-helix transcription factor bound to its DNA target.

In contrast, other proteins have evolved to bind to particular DNA sequences. The most intensively studied of these are the various transcription factors, which are proteins that regulate transcription. Each transcription factor binds to one particular set of DNA sequences and activates or inhibits the transcription of genes that have these sequences close to their promoters. The transcription factors do this in two ways. Firstly, they can bind the RNA polymerase responsible for transcription, either directly or through other mediator proteins; this locates the polymerase at the promoter and allows it to begin transcription. Alternatively, transcription factors can bind enzymes that modify the histones at the promoter. This changes the accessibility of the DNA template to the polymerase.

As these DNA targets can occur throughout an organism's genome, changes in the activity of one type of transcription factor can affect thousands of genes. Consequently, these proteins are often the targets of the signal transduction processes that control responses to environmental changes or cellular differentiation and development. The specificity of these transcription factors' interactions with DNA come from the proteins making multiple contacts to the edges of the DNA bases, allowing them to "read" the DNA sequence. Most of these base-interactions are made in the major groove, where the bases are most accessible.

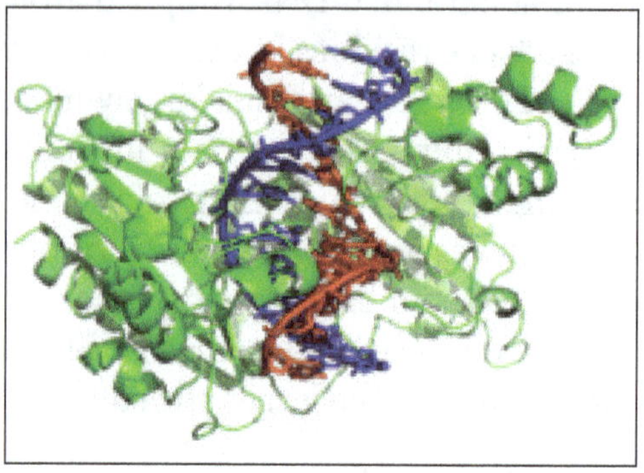

The restriction enzyme EcoRV (green) in a complex with its substrate DNA.

DNA-modifying Enzymes

Nucleases and Ligases

Nucleases are enzymes that cut DNA strands by catalyzing the hydrolysis of the phosphodiester bonds. Nucleases that hydrolyse nucleotides from the ends of DNA strands are called exonucleases, while endonucleases cut within strands. The most frequently used nucleases in molecular biology are the restriction endonucleases, which cut DNA at specific sequences. For instance, the EcoRV enzyme shown to the left recognizes the 6-base sequence 5′-GATATC-3′ and makes a cut at the horizontal line. In nature, these enzymes protect bacteria against phage infection by digesting the phage DNA when it enters the bacterial cell, acting as part of the restriction modification system. In technology, these sequence-specific nucleases are used in molecular cloning and DNA fingerprinting.

Enzymes called DNA ligases can rejoin cut or broken DNA strands. Ligases are particularly important in lagging strand DNA replication, as they join together the short segments of DNA produced at the replication fork into a complete copy of the DNA template. They are also used in DNA repair and genetic recombination.

Topoisomerases and Helicases

Topoisomerases are enzymes with both nuclease and ligase activity. These proteins change the amount of supercoiling in DNA. Some of these enzymes work by cutting the DNA helix and allowing one section to rotate, thereby reducing its level of supercoiling; the enzyme then seals the DNA break. Other types of these enzymes are capable of cutting one DNA helix and then passing a second strand of DNA through this break, before rejoining the helix. Topoisomerases are required for many processes involving DNA, such as DNA replication and transcription.

Helicases are proteins that are a type of molecular motor. They use the chemical energy in nucleoside triphosphates, predominantly adenosine triphosphate (ATP), to break hydrogen bonds between bases and unwind the DNA double helix into single strands. These enzymes are essential for most processes where enzymes need to access the DNA bases.

Polymerases

Polymerases are enzymes that synthesize polynucleotide chains from nucleoside triphosphates. The sequence of their products is created based on existing polynucleotide chains—which are called *templates*. These enzymes function by repeatedly adding a nucleotide to the 3′ hydroxyl group at the end of the growing polynucleotide chain. As a consequence, all polymerases work in a 5′ to 3′ direction. In the active site of these enzymes, the incoming nucleoside triphosphate base-pairs to the template: this allows polymerases to accurately synthesize the complementary strand of their template. Polymerases are classified according to the type of template that they use.

In DNA replication, DNA-dependent DNA polymerases make copies of DNA polynucleotide chains. To preserve biological information, it is essential that the sequence of bases in each copy are precisely complementary to the sequence of bases in the template strand. Many DNA polymerases have a proofreading activity. Here, the polymerase recognizes the occasional mistakes in the synthesis reaction by the lack of base pairing between the mismatched nucleotides. If a mismatch is detected, a 3′ to 5′ exonuclease activity is activated and the incorrect base removed. In most organisms, DNA polymerases function in a large complex called the replisome that contains multiple accessory subunits, such as the DNA clamp or helicases.

RNA-dependent DNA polymerases are a specialized class of polymerases that copy the sequence of an RNA strand into DNA. They include reverse transcriptase, which is a viral enzyme involved in the infection of cells by retroviruses, and telomerase, which is required for the replication of telomeres. For example, HIV reverse transcriptase is an enzyme for AIDS virus replication. Telomerase is an unusual polymerase because it contains its own RNA template as part of its structure. It synthesizes telomeres at the ends of chromosomes. Telomeres prevent fusion of the ends of neighboring chromosomes and protect chromosome ends from damage.

Transcription is carried out by a DNA-dependent RNA polymerase that copies the sequence of a DNA strand into RNA. To begin transcribing a gene, the RNA polymerase binds to a sequence of DNA called a promoter and separates the DNA strands. It then copies the gene sequence into a messenger RNA transcript until it reaches a region of DNA called the terminator, where it halts and detaches from the DNA. As with human DNA-dependent DNA polymerases, RNA polymerase II, the enzyme that transcribes most of the genes in the human genome, operates as part of a large protein complex with multiple regulatory and accessory subunits.

Genetic Recombination

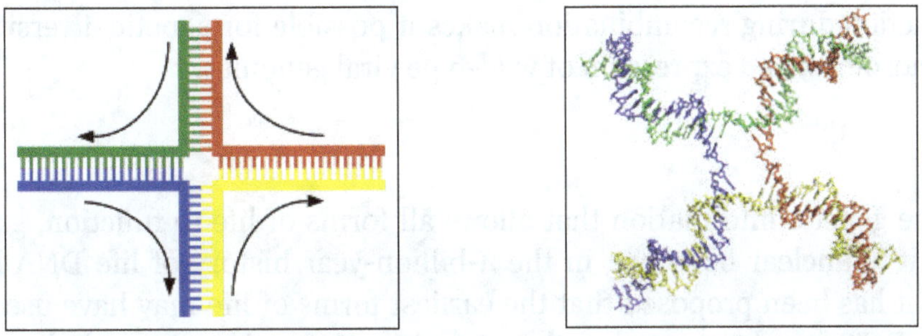

Structure of the Holliday junction intermediate in genetic recombination.
The four separate DNA strands are coloured red, blue, green and yellow.

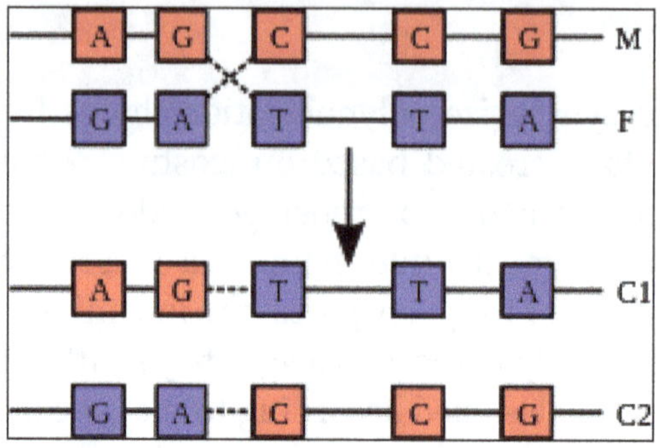

Recombination involves the breaking and rejoining of two chromosomes (M and F) to produce two rearranged chromosomes (C1 and C2).

A DNA helix usually does not interact with other segments of DNA, and in human cells, the different chromosomes even occupy separate areas in the nucleus called "chromosome territories". This physical separation of different chromosomes is important for the ability of DNA to function as a stable repository for information, as one of the few times chromosomes interact is in chromosomal crossover which occurs during sexual reproduction, when genetic recombination occurs. Chromosomal crossover is when two DNA helices break, swap a section and then rejoin.

Recombination allows chromosomes to exchange genetic information and produces new combinations of genes, which increases the efficiency of natural selection and can be important in the rapid evolution of new proteins. Genetic recombination can also be involved in DNA repair, particularly in the cell's response to double-strand breaks.

The most common form of chromosomal crossover is homologous recombination, where the two chromosomes involved share very similar sequences. Non-homologous recombination can be damaging to cells, as it can produce chromosomal translocations and genetic abnormalities. The recombination reaction is catalyzed by enzymes known as recombinases, such as RAD51. The first step in recombination is a double-stranded break caused by either an endonuclease or damage to the DNA. A series of steps catalyzed in part by the recombinase then leads to joining of the two helices by at least one Holliday junction, in which a segment of a single strand in each helix is annealed to the complementary strand in the other helix. The Holliday junction is a tetrahedral junction structure that can be moved along the pair of chromosomes, swapping one strand for another. The recombination reaction is then halted by cleavage of the junction and re-ligation of the released DNA. Only strands of like polarity exchange DNA during recombination. There are two types of cleavage: east-west cleavage and north-south cleavage. The north-south cleavage nicks both strands of DNA, while the east-west cleavage has one strand of DNA intact. The formation of a Holliday junction during recombination makes it possible for genetic diversity, genes to exchange on chromosomes, and expression of wild-type viral genomes.

Evolution

DNA contains the genetic information that allows all forms of life to function, grow and reproduce. However, it is unclear how long in the 4-billion-year history of life DNA has performed this function, as it has been proposed that the earliest forms of life may have used RNA as their genetic material. RNA may have acted as the central part of early cell metabolism as it can both transmit genetic information and carry out catalysis as part of ribozymes. This ancient RNA world where nucleic acid would have been used for both catalysis and genetics may have influenced the evolution of the current genetic code based on four nucleotide bases. This would occur, since the number of different bases in such an organism is a trade-off between a small number of bases increasing replication accuracy and a large number of bases increasing the catalytic efficiency of ribozymes. However, there is no direct evidence of ancient genetic systems, as recovery of DNA from most fossils is impossible because DNA survives in the environment for less than one million years, and slowly degrades into short fragments in solution. Claims for older DNA have been made, most notably a report of the isolation of a viable bacterium from a salt crystal 250 million years old, but these claims are controversial.

Building blocks of DNA (adenine, guanine, and related organic molecules) may have been formed extraterrestrially in outer space. Complex DNA and RNA organic compounds of life, including uracil, cytosine, and thymine, have also been formed in the laboratory under conditions mimicking those found in outer space, using starting chemicals, such as pyrimidine, found in meteorites. Pyrimidine, like polycyclic aromatic hydrocarbons (PAHs), the most carbon-rich chemical found in the universe, may have been formed in red giants or in interstellar cosmic dust and gas clouds.

Uses in Technology

Genetic Engineering

Methods have been developed to purify DNA from organisms, such as phenol-chloroform extraction, and to manipulate it in the laboratory, such as restriction digests and the polymerase chain reaction. Modern biology and biochemistry make intensive use of these techniques in recombinant DNA technology. Recombinant DNA is a man-made DNA sequence that has been assembled from other DNA sequences. They can be transformed into organisms in the form of plasmids or in the appropriate format, by using a viral vector. The genetically modified organisms produced can be used to produce products such as recombinant proteins, used in medical research, or be grown in agriculture.

DNA Profiling

Forensic scientists can use DNA in blood, semen, skin, saliva or hair found at a crime scene to identify a matching DNA of an individual, such as a perpetrator. This process is formally termed DNA profiling, also called *DNA fingerprinting*. In DNA profiling, the lengths of variable sections of repetitive DNA, such as short tandem repeats and minisatellites, are compared between people. This method is usually an extremely reliable technique for identifying a matching DNA. However, identification can be complicated if the scene is contaminated with DNA from several people. DNA profiling was developed in 1984 by British geneticist Sir Alec Jeffreys, and first used in forensic science to convict Colin Pitchfork in the 1988 Enderby murders case.

The development of forensic science and the ability to now obtain genetic matching on minute samples of blood, skin, saliva, or hair has led to re-examining many cases. Evidence can now be uncovered that was scientifically impossible at the time of the original examination. Combined with the removal of the double jeopardy law in some places, this can allow cases to be reopened where prior trials have failed to produce sufficient evidence to convince a jury. People charged with serious crimes may be required to provide a sample of DNA for matching purposes. The most obvious defense to DNA matches obtained forensically is to claim that cross-contamination of evidence has occurred. This has resulted in meticulous strict handling procedures with new cases of serious crime.

DNA profiling is also used successfully to positively identify victims of mass casualty incidents, bodies or body parts in serious accidents, and individual victims in mass war graves, via matching to family members.

DNA profiling is also used in DNA paternity testing to determine if someone is the biological parent or grandparent of a child with the probability of parentage is typically 99.99% when the alleged parent is biologically related to the child. Normal DNA sequencing methods happen after birth, but there are new methods to test paternity while a mother is still pregnant.

DNA Enzymes or Catalytic DNA

Deoxyribozymes, also called DNAzymes or catalytic DNA, were first discovered in 1994. They are mostly single stranded DNA sequences isolated from a large pool of random DNA sequences through a combinatorial approach called in vitro selection or systematic evolution of ligands by exponential enrichment (SELEX). DNAzymes catalyze variety of chemical reactions including RNA-DNA cleavage, RNA-DNA ligation, amino acids phosphorylation-dephosphorylation, carbon-carbon bond formation, and etc. DNAzymes can enhance catalytic rate of chemical reactions up to 100,000,000,000-fold over the uncatalyzed reaction. The most extensively studied class of DNAzymes is RNA-cleaving types which have been used to detect different metal ions and designing therapeutic agents. Several metal-specific DNAzymes have been reported including the GR-5 DNAzyme (lead-specific), the CA1-3 DNAzymes (copper-specific), the 39E DNAzyme (uranyl-specific) and the NaA43 DNAzyme (sodium-specific). The NaA43 DNAzyme, which is reported to be more than 10,000-fold selective for sodium over other metal ions, was used to make a real-time sodium sensor in cells.

Bioinformatics

Bioinformatics involves the development of techniques to store, data mine, search and manipulate biological data, including DNA nucleic acid sequence data. These have led to widely applied advances in computer science, especially string searching algorithms, machine learning, and database theory. String searching or matching algorithms, which find an occurrence of a sequence of letters inside a larger sequence of letters, were developed to search for specific sequences of nucleotides. The DNA sequence may be aligned with other DNA sequences to identify homologous sequences and locate the specific mutations that make them distinct. These techniques, especially multiple sequence alignment, are used in studying phylogenetic relationships and protein function. Data sets representing entire genomes' worth of DNA sequences, such as those produced by the Human Genome Project, are difficult to use without the annotations that identify the locations of genes and regulatory elements on each chromosome. Regions of DNA sequence that have the characteristic patterns associated with protein- or RNA-coding genes can be identified by gene finding algorithms, which allow researchers to predict the presence of particular gene products and their possible functions in an organism even before they have been isolated experimentally. Entire genomes may also be compared, which can shed light on the evolutionary history of particular organism and permit the examination of complex evolutionary events.

DNA Nanotechnology

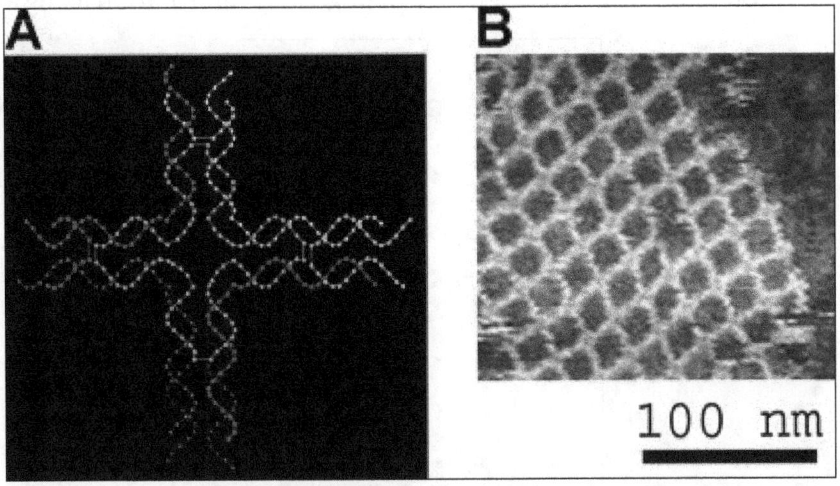

The DNA structure at left (schematic shown) will self-assemble into the structure visualized by atomic force microscopy at right. DNA nanotechnology is the field that seeks to design nanoscale structures using the molecular recognition properties of DNA molecules.

DNA nanotechnology uses the unique molecular recognition properties of DNA and other nucleic acids to create self-assembling branched DNA complexes with useful properties. DNA is thus used as a structural material rather than as a carrier of biological information. This has led to the creation of two-dimensional periodic lattices (both tile-based and using the *DNA origami* method) and three-dimensional structures in the shapes of polyhedra. Nanomechanical devices and algorithmic self-assembly have also been demonstrated, and these DNA structures have been used to template the arrangement of other molecules such as gold nanoparticles and streptavidin proteins.

Because DNA collects mutations over time, which are then inherited, it contains historical information, and, by comparing DNA sequences, geneticists can infer the evolutionary history of organisms, their phylogeny. This field of phylogenetics is a powerful tool in evolutionary biology. If DNA sequences within a species are compared, population geneticists can learn the history of particular populations. This can be used in studies ranging from ecological genetics to anthropology.

Information Storage

DNA as a storage device for information has enormous potential since it has much higher storage density compared to electronic devices. However high costs, extremely slow read and write times (memory latency), and insufficient reliability has prevented its practical use.

RNA

Ribonucleic acid (RNA) is a polymeric molecule essential in various biological roles in coding, decoding, regulation and expression of genes. RNA and DNA are nucleic acids, and, along with lipids, proteins and carbohydrates, constitute the four major macromolecules essential for all known forms of life. Like DNA, RNA is assembled as a chain of nucleotides, but unlike DNA it is more often found in nature as a single-strand folded onto itself, rather than a paired double-strand. Cellular organisms use messenger RNA (*mRNA*) to convey genetic information (using the nitrogenous bases of guanine, uracil, adenine, and cytosine, denoted by the letters G, U, A, and C) that directs synthesis of specific proteins. Many viruses encode their genetic information using an RNA genome.

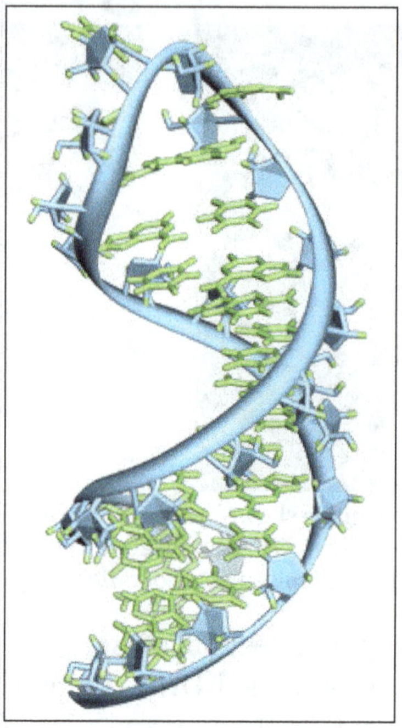

A hairpin loop from a pre-mRNA. Highlighted are the nucleobases (green) and the ribose-phosphate backbone (blue). This is a single strand of RNA that folds back upon itself.

Some RNA molecules play an active role within cells by catalyzing biological reactions, controlling gene expression, or sensing and communicating responses to cellular signals. One of these active processes is protein synthesis, a universal function in which RNA molecules direct the synthesis of proteins on ribosomes. This process uses transfer RNA (*tRNA*) molecules to deliver amino acids to the ribosome, where ribosomal RNA (*rRNA*) then links amino acids together to form coded proteins.

Comparison with DNA

Three-dimensional representation of the 50S ribosomal subunit. Ribosomal RNA is in ochre, proteins in blue. The active site is a small segment of rRNA, indicated in red.

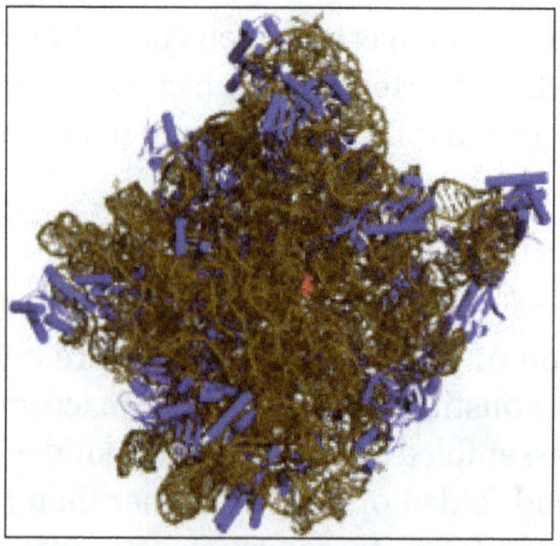

The chemical structure of RNA is very similar to that of DNA, but differs in three primary ways:

- Unlike double-stranded DNA, RNA is a single-stranded molecule in many of its biological roles and consists of much shorter chains of nucleotides. However, a single RNA molecule can, by complementary base pairing, form intrastrand double helixes, as in tRNA.
- While the sugar-phosphate "backbone" of DNA contains *deoxyribose*, RNA contains *ribose* instead. Ribose has a hydroxyl group attached to the pentose ring in the 2' position, whereas deoxyribose does not. The hydroxyl groups in the ribose backbone make RNA more chemically labile than DNA by lowering the activation energy of hydrolysis.
- The complementary base to adenine in DNA is thymine, whereas in RNA, it is uracil, which is an unmethylated form of thymine.

Like DNA, most biologically active RNAs, including mRNA, tRNA, rRNA, snRNAs, and other non-coding RNAs, contain self-complementary sequences that allow parts of the RNA to fold and pair with itself to form double helices. Analysis of these RNAs has revealed that they are highly structured. Unlike DNA, their structures do not consist of long double helices, but rather collections of short helices packed together into structures akin to proteins. In this fashion, RNAs can achieve chemical catalysis (like enzymes). For instance, determination of the structure of the ribosome—an RNA-protein complex that catalyzes peptide bond formation—revealed that its active site is composed entirely of RNA.

Structure

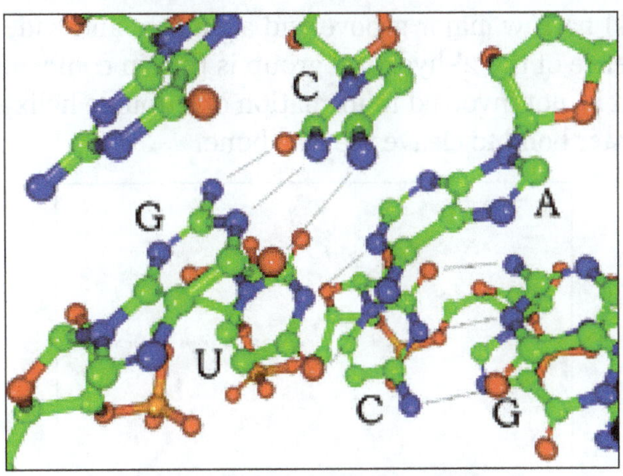

Watson-Crick base pairs in a siRNA (hydrogen atoms are not shown).

Each nucleotide in RNA contains a ribose sugar, with carbons numbered 1' through 5'. A base is attached to the 1' position, in general, adenine (A), cytosine (C), guanine (G), or uracil (U). Adenine and guanine are purines, cytosine and uracil are pyrimidines. A phosphate group is attached to the 3' position of one ribose and the 5' position of the next. The phosphate groups have a negative charge each, making RNA a charged molecule (polyanion). The bases form hydrogen bonds between cytosine and guanine, between adenine and uracil and between guanine and uracil. However, other interactions are possible, such as a group of adenine bases binding to each other in a bulge, or the GNRA tetraloop that has a guanine–adenine base-pair.

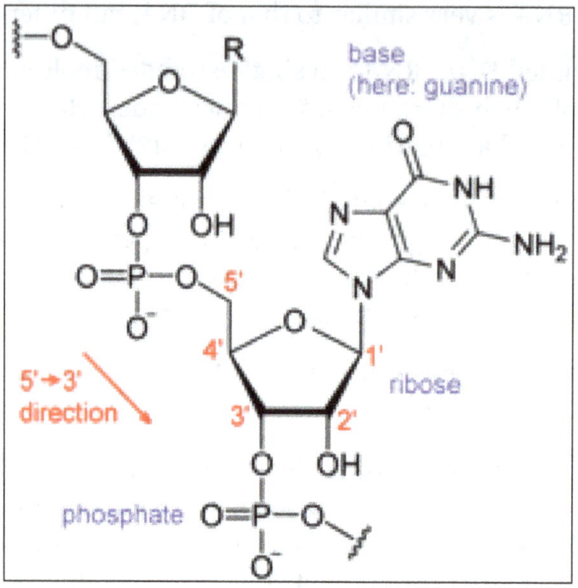

Chemical Structure of RNA

An important structural component of RNA that distinguishes it from DNA is the presence of a hydroxyl group at the 2' position of the ribose sugar. The presence of this functional group causes the helix to mostly take the A-form geometry, although in single strand dinucleotide contexts, RNA can rarely also adopt the B-form most commonly observed in DNA. The A-form geometry results in a very deep and narrow major groove and a shallow and wide minor groove. A second consequence of the presence of the 2'-hydroxyl group is that in conformationally flexible regions of an RNA molecule (that is, not involved in formation of a double helix), it can chemically attack the adjacent phosphodiester bond to cleave the backbone.

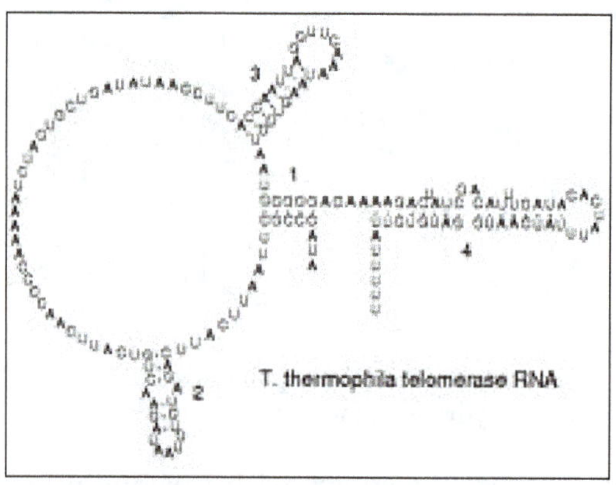

Secondary structure of a telomerase RNA.

RNA is transcribed with only four bases (adenine, cytosine, guanine and uracil), but these bases and attached sugars can be modified in numerous ways as the RNAs mature. Pseudouridine (Ψ), in which the linkage between uracil and ribose is changed from a C–N bond to a C–C bond, and ribothymidine (T) are found in various places (the most notable ones being in the TΨC loop of tRNA). Another notable modified base is hypoxanthine, a deaminated adenine base whose nucleoside is called inosine (I). Inosine plays a key role in the wobble hypothesis of the genetic code.

There are more than 100 other naturally occurring modified nucleosides. The greatest structural diversity of modifications can be found in tRNA, while pseudouridine and nucleosides with 2'-O-methylribose often present in rRNA are the most common. The specific roles of many of these modifications in RNA are not fully understood. However, it is notable that, in ribosomal RNA, many of the post-transcriptional modifications occur in highly functional regions, such as the peptidyl transferase center and the subunit interface, implying that they are important for normal function.

The functional form of single-stranded RNA molecules, just like proteins, frequently requires a specific tertiary structure. The scaffold for this structure is provided by secondary structural elements that are hydrogen bonds within the molecule. This leads to several recognizable "domains" of secondary structure like hairpin loops, bulges, and internal loops. Since RNA is charged, metal ions such as Mg^{2+} are needed to stabilise many secondary and tertiary structures.

The naturally occurring enantiomer of RNA is D-RNA composed of D-ribonucleotides. All chirality centers are located in the D-ribose. By the use of L-ribose or rather L-ribonucleotides, L-RNA can be synthesized. L-RNA is much more stable against degradation by RNase.

Like other structured biopolymers such as proteins, one can define topology of a folded RNA molecule. This is often done based on arrangement of intra-chain contacts within a folded RNA, termed as circuit topology.

Synthesis

Synthesis of RNA is usually catalyzed by an enzyme—RNA polymerase—using DNA as a template, a process known as transcription. Initiation of transcription begins with the binding of the enzyme to a promoter sequence in the DNA (usually found "upstream" of a gene). The DNA double helix is unwound by the helicase activity of the enzyme. The enzyme then progresses along the template strand in the 3' to 5' direction, synthesizing a complementary RNA molecule with elongation occurring in the 5' to 3' direction. The DNA sequence also dictates where termination of RNA synthesis will occur.

Primary transcript RNAs are often modified by enzymes after transcription. For example, a poly(A) tail and a 5' cap are added to eukaryotic pre-mRNA and introns are removed by the spliceosome.

There are also a number of RNA-dependent RNA polymerases that use RNA as their template for synthesis of a new strand of RNA. For instance, a number of RNA viruses (such as poliovirus) use this type of enzyme to replicate their genetic material. Also, RNA-dependent RNA polymerase is part of the RNA interference pathway in many organisms.

Types of RNA

Messenger RNA (mRNA) is the RNA that carries information from DNA to the ribosome, the sites of protein synthesis (translation) in the cell. The coding sequence of the mRNA determines the amino acid sequence in the protein that is produced. However, many RNAs do not code for protein (about 97% of the transcriptional output is non-protein-coding in eukaryotes).

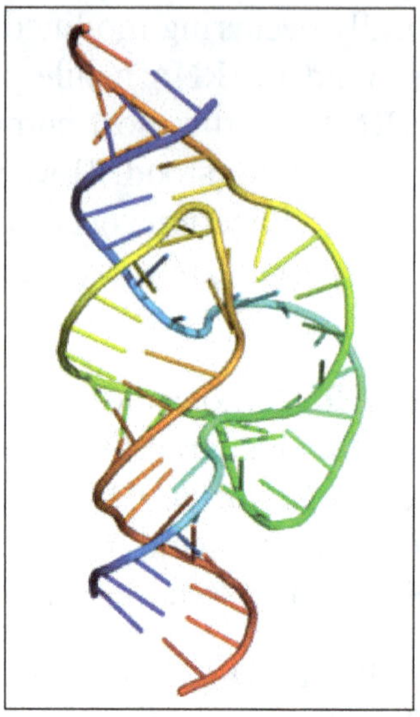

Structure of a hammerhead ribozyme, a ribozyme that cuts RNA.

These so-called non-coding RNAs ("ncRNA") can be encoded by their own genes (RNA genes), but can also derive from mRNA introns. The most prominent examples of non-coding RNAs are transfer RNA (tRNA) and ribosomal RNA (rRNA), both of which are involved in the process of translation. There are also non-coding RNAs involved in gene regulation, RNA processing and other roles. Certain RNAs are able to catalyse chemical reactions such as cutting and ligating other RNA molecules, and the catalysis of peptide bond formation in the ribosome; these are known as ribozymes.

Length

According to the length of RNA chain, RNA includes small RNA and long RNA. Usually, small RNAs are shorter than 200 nt in length, and long RNAs are greater than 200 nt long. Long RNAs, also called large RNAs, mainly include long non-coding RNA (lncRNA) and mRNA. Small RNAs mainly include 5.8S ribosomal RNA (rRNA), 5S rRNA, transfer RNA (tRNA), microRNA (miRNA), small interfering RNA (siRNA), small nucleolar RNA (snoRNAs), Piwi-interacting RNA (piRNA), tRNA-derived small RNA (tsRNA) and small rDNA-derived RNA (srRNA).

Translation

Messenger RNA (mRNA) carries information about a protein sequence to the ribosomes, the protein synthesis factories in the cell. It is coded so that every three nucleotides (a codon) corresponds to one amino acid. In eukaryotic cells, once precursor mRNA (pre-mRNA) has been transcribed from DNA, it is processed to mature mRNA. This removes its introns—non-coding sections of the pre-mRNA. The mRNA is then exported from the nucleus to the cytoplasm, where it is bound to ribosomes and translated into its corresponding protein form with the help of tRNA. In prokaryotic cells, which do not have nucleus and cytoplasm compartments, mRNA can bind to ribosomes while it is being transcribed from DNA. After a certain amount of time, the message degrades into its component nucleotides with the assistance of ribonucleases.

Transfer RNA (tRNA) is a small RNA chain of about 80 nucleotides that transfers a specific amino acid to a growing polypeptide chain at the ribosomal site of protein synthesis during translation.

It has sites for amino acid attachment and an anticodon region for codon recognition that binds to a specific sequence on the messenger RNA chain through hydrogen bonding.

Ribosomal RNA (rRNA) is the catalytic component of the ribosomes. Eukaryotic ribosomes contain four different rRNA molecules: 18S, 5.8S, 28S and 5S rRNA. Three of the rRNA molecules are synthesized in the nucleolus, and one is synthesized elsewhere. In the cytoplasm, ribosomal RNA and protein combine to form a nucleoprotein called a ribosome. The ribosome binds mRNA and carries out protein synthesis. Several ribosomes may be attached to a single mRNA at any time. Nearly all the RNA found in a typical eukaryotic cell is rRNA.

Transfer-messenger RNA (tmRNA) is found in many bacteria and plastids. It tags proteins encoded by mRNAs that lack stop codons for degradation and prevents the ribosome from stalling.

Regulatory RNA

The earliest known regulators of gene expression were proteins known as repressors and activators, regulators with specific short binding sites within enhancer regions near the genes to be regulated. More recently, RNAs have been found to regulate genes as well. There are several kinds of RNA-dependent processes in eukaryotes regulating the expression of genes at various points, such as RNAi repressing genes post-transcriptionally, long non-coding RNAs shutting down blocks of chromatin epigenetically, and enhancer RNAs inducing increased gene expression. In addition to these mechanisms in eukaryotes, both bacteria and archaea have been found to use regulatory RNAs extensively. Bacterial small RNA and the CRISPR system are examples of such prokaryotic regulatory RNA systems. Fire and Mello were awarded the 2006 Nobel Prize in Physiology or Medicine for discovering microRNAs (miRNAs), specific short RNA molecules that can base-pair with mRNAs.

RNA Interference by miRNAs

Post-transcriptional expression levels of many genes can be controlled by RNA interference, in which miRNAs, specific short RNA molecules, pair with meRNA regions and target them for degradation. This antisense-based process involves steps that first process the RNA so that it can base-pair with a region of its target mRNAs. Once the base pairing occurs, other proteins direct the mRNA to be destroyed by nucleases. Fire and Mello were awarded the 2006 Nobel Prize in Physiology or Medicine for this discovery.

Long Non-coding RNAs

Next to be linked to regulation were Xist and other long noncoding RNAs associated with X chromosome inactivation. Their roles, at first mysterious, were shown by Jeannie T. Lee and others to be the silencing of blocks of chromatin via recruitment of Polycomb complex so that messenger RNA could not be transcribed from them. Additional lncRNAs, currently defined as RNAs of more than 200 base pairs that do not appear to have coding potential, have been found associated with regulation of stem cell pluripotency and cell division.

Enhancer RNAs

The third major group of regulatory RNAs is called enhancer RNAs. It is not clear at present whether they are a unique category of RNAs of various lengths or constitute a distinct subset of lncRNAs.

In any case, they are transcribed from enhancers, which are known regulatory sites in the DNA near genes they regulate. They up-regulate the transcription of the gene(s) under control of the enhancer from which they are transcribed.

Regulatory RNA in Prokaryotes

At first, regulatory RNA was thought to be a eukaryotic phenomenon, a part of the explanation for why so much more transcription in higher organisms was seen than had been predicted. But as soon as researchers began to look for possible RNA regulators in bacteria, they turned up there as well. Currently, the ubiquitous nature of systems of RNA regulation of genes has been discussed as support for the RNA World theory. Bacterial small RNAs generally act via antisense pairing with mRNA to down-regulate its translation, either by affecting stability or affecting cis-binding ability. Riboswitches have also been discovered. They are cis-acting regulatory RNA sequences acting allosterically. They change shape when they bind metabolites so that they gain or lose the ability to bind chromatin to regulate expression of genes.

Archaea also have systems of regulatory RNA. The CRISPR system, recently being used to edit DNA *in situ*, acts via regulatory RNAs in archaea and bacteria to provide protection against virus invaders.

RNA Processing

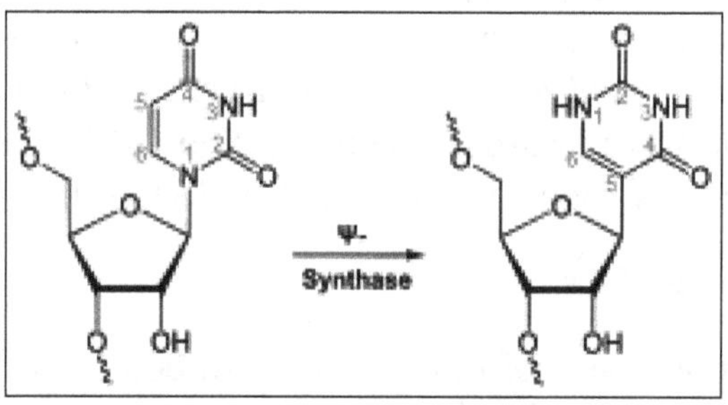

Uridine to pseudouridine is a common RNA modification.

Many RNAs are involved in modifying other RNAs. Introns are spliced out of pre-mRNA by spliceosomes, which contain several small nuclear RNAs (snRNA), or the introns can be ribozymes that are spliced by themselves. RNA can also be altered by having its nucleotides modified to nucleotides other than A, C, G and U. In eukaryotes, modifications of RNA nucleotides are in general directed by small nucleolar RNAs (snoRNA; 60–300 nt), found in the nucleolus and cajal bodies. snoRNAs associate with enzymes and guide them to a spot on an RNA by basepairing to that RNA. These enzymes then perform the nucleotide modification. rRNAs and tRNAs are extensively modified, but snRNAs and mRNAs can also be the target of base modification. RNA can also be methylated.

RNA Genomes

Like DNA, RNA can carry genetic information. RNA viruses have genomes composed of RNA that encodes a number of proteins. The viral genome is replicated by some of those proteins, while other proteins protect the genome as the virus particle moves to a new host cell. Viroids are another group of pathogens, but they consist only of RNA, do not encode any protein and are replicated by a host plant cell's polymerase.

Reverse Transcription

Reverse transcribing viruses replicate their genomes by reverse transcribing DNA copies from their RNA; these DNA copies are then transcribed to new RNA. Retrotransposons also spread by copying DNA and RNA from one another, and telomerase contains an RNA that is used as template for building the ends of eukaryotic chromosomes.

Double-stranded RNA

Double-stranded RNA (dsRNA) is RNA with two complementary strands, similar to the DNA found in all cells, but with the replacement of thymine by uracil. dsRNA forms the genetic material of some viruses (double-stranded RNA viruses). Double-stranded RNA, such as viral RNA or siRNA, can trigger RNA interference in eukaryotes, as well as interferon response in vertebrates.

Circular RNA

In the late 1970s, it was shown that there is a single stranded covalently closed, i.e. circular form of RNA expressed throughout the animal and plant kingdom. circRNAs are thought to arise via a "back-splice" reaction where the spliceosome joins a downstream donor to an upstream acceptor splice site. So far the function of circRNAs is largely unknown, although for few examples a microRNA sponging activity has been demonstrated.

References

- Biomolecules-in-living-organisms: byjus.com, Retrieved 25 May, 2019
- Introduction-to-proteins-and-amino-acids, proteins-and-amino-acids: khanacademy.org, Retrieved 05 August, 2019
- Carbohydrate: britannica.com, Retrieved 26 January, 2019
- Lipid: britannica.com, Retrieved 28 April, 2019
- Nucleic-acids, wmopen-biology1: courses.lumenlearning.com, Retrieved 08 July, 2019
- Protein: britannica.com, Retrieved 16 June, 2019

3

Enzymes and Enzyme Activity

The biological catalysts that accelerate a chemical reaction by lowering the activation energy are known as enzymes. The activities of enzymes can be understood through protein dynamics, enzyme kinetics, enzyme inhibition and enzyme catalysis. The following chapter elucidates these varied activities and mechanisms associated with enzymes.

Enzyme

Enzyme is a substance that acts as a catalyst in living organisms, regulating the rate at which chemical reactions proceed without itself being altered in the process.

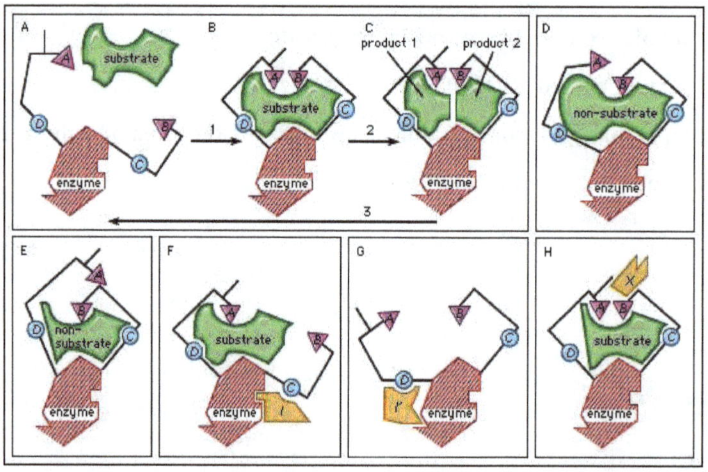

In the induced-fit theory of enzyme-substrate binding, a substrate approaches the surface of an enzyme (step 1 in box A, B, C) and causes a change in the enzyme shape that results in the correct alignment of the catalytic groups (triangles *A* and *B*; circles *C* and *D* represent substrate-binding groups on the enzyme that are essential for catalytic activity). The catalytic groups react with the substrate to form products (step 2). The products then separate from the enzyme, freeing it to repeat the sequence (step 3). Boxes D and E represent examples of molecules that are too large or too small for proper catalytic alignment. Boxes F and G demonstrate binding of an inhibitor molecule (*I* and *I′*) to an allosteric site, thereby preventing interaction of the enzyme with the substrate. Box H illustrates binding of an allosteric activator (*X*), a nonsubstrate molecule capable of reacting with the enzyme.

The biological processes that occur within all living organisms are chemical reactions, and most are regulated by enzymes. Without enzymes, many of these reactions would not take place at a perceptible rate. Enzymes catalyze all aspects of cell metabolism. This includes the digestion of food, in which large nutrient molecules (such as proteins, carbohydrates, and fats) are broken down into smaller molecules; the conservation and transformation of chemical energy; and the construction of cellular macromolecules from smaller precursors. Many inherited human diseases, such as albinism and phenylketonuria, result from a deficiency of a particular enzyme.

Enzymes also have valuable industrial and medical applications. The fermenting of wine, leavening of bread, curdling of cheese, and brewing of beer have been practiced from earliest times, but not until the 19th century were these reactions understood to be the result of the catalytic activity of enzymes. Since then, enzymes have assumed an increasing importance in industrial processes that involve organic chemical reactions. The uses of enzymes in medicine include killing disease-causing microorganisms, promoting wound healing, and diagnosing certain diseases.

Enzyme (cheese making): Rennet, which contains the protease enzyme chymosin, is added to milk during cheese making.

Chemical Nature

All enzymes were once thought to be proteins, but since the 1980s the catalytic ability of certain nucleic acids, called ribozymes (or catalytic RNAs), has been demonstrated, refuting this axiom. Because so little is yet known about the enzymatic functioning of RNA, this discussion will focus primarily on protein enzymes.

A large protein enzyme molecule is composed of one or more amino acid chains called polypeptide chains. The amino acid sequence determines the characteristic folding patterns of the protein's structure, which is essential to enzyme specificity. If the enzyme is subjected to changes, such as fluctuations in temperature or pH, the protein structure may lose its integrity (denature) and its enzymatic ability. Denaturation is sometimes, but not always, reversible.

Bound to some enzymes is an additional chemical component called a cofactor, which is a direct participant in the catalytic event and thus is required for enzymatic activity. A cofactor may be either a coenzyme—an organic molecule, such as a vitamin—or an inorganic metal ion; some enzymes require both. A cofactor may be either tightly or loosely bound to the enzyme. If tightly connected, the cofactor is referred to as a prosthetic group.

Nomenclature

An enzyme will interact with only one type of substance or group of substances, called the substrate, to catalyze a certain kind of reaction. Because of this specificity, enzymes often have been named by adding the suffix "-ase" to the substrate's name (as in urease, which catalyzes the breakdown of urea). Not all enzymes have been named in this manner, however, and to ease the confusion surrounding enzyme nomenclature, a classification system has been developed based on the type of reaction the enzyme catalyzes. There are six principal categories and their reactions: (1) oxidoreductases, which are involved in electron transfer; (2) transferases, which transfer a chemical group from one substance to another; (3) hydrolases, which cleave the substrate by uptake of a water molecule (hydrolysis); (4) lyases, which form double bonds by adding or removing a chemical group; (5) isomerases, which transfer a group within a molecule to form an isomer; and (6) ligases, or synthetases, which couple the formation of various chemical bonds to the breakdown of a pyrophosphate bond in adenosine triphosphate or a similar nucleotide.

Mechanism of Enzyme Action

In most chemical reactions, an energy barrier exists that must be overcome for the reaction to occur. This barrier prevents complex molecules such as proteins and nucleic acids from spontaneously degrading, and so is necessary for the preservation of life. When metabolic changes are required in a cell, however, certain of these complex molecules must be broken down, and this energy barrier must be surmounted. Heat could provide the additional needed energy (called activation energy), but the rise in temperature would kill the cell. The alternativeis to lower the activation energy level through the use of a catalyst. This is the role that enzymes play. They react with the substrate to form an intermediate complex—a "transition state"—that requires less energy for the reaction to proceed. The unstable intermediate compound quickly breaks down to form reaction products, and the unchanged enzyme is free to react with other substrate molecules.

Only a certain region of the enzyme, called the active site, binds to the substrate. The active site is a groove or pocket formed by the folding pattern of the protein. This three-dimensional structure, together with the chemical and electrical properties of the amino acids and cofactors within the active site, permits only a particular substrate to bind to the site, thus determining the enzyme's specificity.

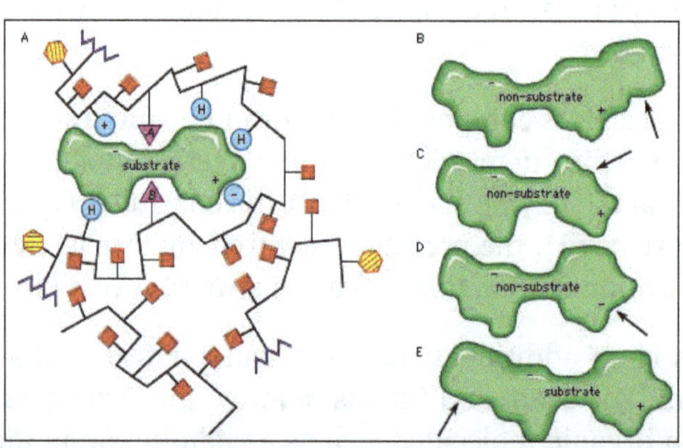

Enzyme (active site): The active site of an enzyme is a groove or pocket that binds a specific substrate.

Enzyme synthesis and activity also are influenced by genetic control and distribution in a cell. Some enzymes are not produced by certain cells, and others are formed only when required. Enzymes are not always found uniformly within a cell; often they are compartmentalized in the nucleus, on the cell membrane, or in subcellular structures. The rates of enzyme synthesis and activity are further influenced by hormones, neurosecretions, and other chemicals that affect the cell's internal environment.

Factors affecting Enzyme Activity

Because enzymes are not consumed in the reactions they catalyze and can be used over and over again, only a very small quantity of an enzyme is needed to catalyze a reaction. A typical enzyme molecule can convert 1,000 substrate molecules per second. The rate of an enzymatic reaction increases with increased substrate concentration, reaching maximum velocity when all active sites of the enzyme molecules are engaged. The enzyme is then said to be saturated, the rate of the reaction being determined by the speed at which the active sites can convert substrate to product.

Enzyme activity can be inhibited in various ways. Competitive inhibition occurs when molecules very similar to the substrate molecules bind to the active site and prevent binding of the actual substrate. Penicillin, for example, is a competitive inhibitor that blocks the active site of an enzyme that many bacteria use to construct their cell walls.

Noncompetitive inhibition occurs when an inhibitor binds to the enzyme at a location other than the active site. In some cases of noncompetitive inhibition, the inhibitor is thought to bind to the enzyme in such a way as to physically block the normal active site. In other instances, the binding of the inhibitor is believed to change the shape of the enzyme molecule, thereby deforming its active site and preventing it from reacting with its substrate. This latter type of noncompetitive inhibition is called allosteric inhibition; the place where the inhibitor binds to the enzyme is called the allosteric site. Frequently, an end-product of a metabolic pathway serves as an allosteric inhibitor on an earlier enzyme of the pathway. This inhibition of an enzyme by a product of its pathway is a form of negative feedback.

Allosteric control can involve stimulation of enzyme action as well as inhibition. An activator molecule can be bound to an allosteric site and induce a reaction at the active site by changing its shape to fit a substrate that could not induce the change by itself. Common activators include hormones and the products of earlier enzymatic reactions. Allosteric stimulation and inhibition allow production of energy and materials by the cell when they are needed and inhibit production when the supply is adequate.

Protein Dynamic

Proteins are generally thought to adopt unique structures determined by their amino acid sequences, as outlined by Anfinsen's dogma. However, proteins are not strictly static objects, but rather populate ensembles of (sometimes similar) conformations. Transitions between these states occur on a variety of length scales (tenths of Å to nm) and time scales (ns to s), and have been linked to functionally relevant phenomena such as allosteric signaling and enzyme catalysis.

The study of protein dynamics is most directly concerned with the transitions between these states, but can also involve the nature and equilibrium populations of the states themselves. These two perspectives—kinetics and thermodynamics, respectively—can be conceptually synthesized in an "energy landscape" paradigm: highly populated states and the kinetics of transitions between them can be described by the depths of energy wells and the heights of energy barriers, respectively.

Local Flexibility: Atoms and Residues

Portions of protein structures often deviate from the equilibrium state. Some such excursions are harmonic, such as stochastic fluctuations of chemical bonds and bond angles. Others are anharmonic, such as sidechains that jump between separate discrete energy minima, or rotamers.

Evidence for local flexibility is often obtained from NMR spectroscopy. Flexible and potentially disordered regions of a protein can be detected using the random coil index. Flexibility in folded proteins can be identified by analyzing the spin relaxation of individual atoms in the protein. Flexibility can also be observed in very high-resolution electron density maps produced by X-ray crystallography, particularly when diffraction data is collected at room temperature instead of the traditional cryogenic temperature (typically near 100 K). Information on the frequency distribution and dynamics of local protein flexibility can be obtained using Raman and optical Kerr-effect spectroscopy in the terahertz frequency domain.

Regional Flexibility: Intra-domain Multi-residue Coupling

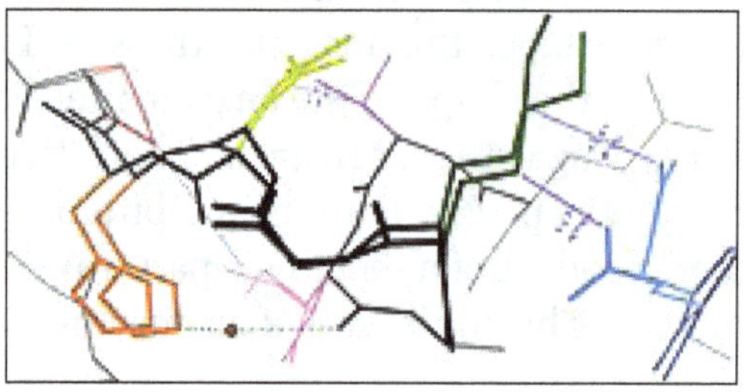

A network of alternative conformations in catalase (Protein Data Bank code: 1gwe) with diverse properties. Multiple phenomena define the network: van der Waals interactions (blue dots and line segments) between sidechains, a hydrogen bond (dotted green line) through a partial-occupancy water (brown), coupling through the locally mobile backbone (black), and perhaps electrostatic forces between the Lys (green) and nearby polar residues (blue: Glu, yellow: Asp, purple: Ser). This particular network is distal from the active site and is therefore putatively not critical for function.

Many residues are in close spatial proximity in protein structures. This is true for most residues that are contiguous in the primary sequence, but also for many that are distal in sequence yet are brought into contact in the final folded structure. Because of this proximity, these residues's energy landscapes become coupled based on various biophysical phenomena such as hydrogen bonds, ionic bonds, and van der Waals interactions. Transitions between states for such sets of residues therefore become correlated.

This is perhaps most obvious for surface-exposed loops, which often shift collectively to adopt different conformations in different crystal structures. However, coupled conformational heterogeneity is also sometimes evident in secondary structure. For example, consecutive residues and residues offset by 4 in the primary sequence often interact in α helices. Also, residues offset by 2 in the primary sequence point their sidechains toward the same face of β sheets and are close enough to interact sterically, as are residues on adjacent strands of the same β sheet.

An "ensemble" of 44 crystal structures of hen egg white lysozyme from the Protein Data Bank, showing that different crystallization conditions lead to different conformations for various surface-exposed loops and termini (red arrows).

When these coupled residues form pathways linking functionally important parts of a protein, they may participate in allosteric signaling. For example, when a molecule of oxygen binds to one subunit of the hemoglobin tetramer, that information is allosterically propagated to the other three subunits, thereby enhancing their affinity for oxygen. In this case, the coupled flexibility in hemoglobin allows for cooperative oxygen binding, which is physiologically useful because it allows rapid oxygen loading in lung tissue and rapid oxygen unloading in oxygen-deprived tissues (e.g. muscle).

Global Flexibility: Multiple Domains

The presence of multiple domains in proteins gives rise to a great deal of flexibility and mobility, leading to protein domain dynamics. Domain motions can be inferred by comparing different structures of a protein (as in Database of Molecular Motions), or they can be directly observed using spectra measured by neutron spin echo spectroscopy. They can also be suggested by sampling in extensive molecular dynamics trajectories and principal component analysis. Domain motions are important for:

- ABC transporters.
- Catalysis.
- Cellular locomotion and motor proteins.
- Formation of protein complexes.
- Ion channels.
- Mechanoreceptors and mechanotransduction.
- Regulatory activity.
- Transport of metabolites across cell membranes.

One of the largest observed domain motions is the 'swivelling' mechanism in pyruvate phosphate dikinase. The phosphoinositide domain swivels between two states in order to bring a phosphate group from the active site of the nucleotide binding domain to that of the phosphoenolpyruvate/ pyruvate domain. The phosphate group is moved over a distance of 45 Å involving a domain motion of about 100 degrees around a single residue. In enzymes, the closure of one domain onto another captures a substrate by an induced fit, allowing the reaction to take place in a controlled way. A detailed analysis by Gerstein led to the classification of two basic types of domain motion; hinge and shear. Only a relatively small portion of the chain, namely the inter-domain linker and side chains undergo significant conformational changes upon domain rearrangement.

Hinges by Secondary Structures

A study by Hayward found that the termini of α-helices and β-sheets form hinges in a large number of cases. Many hinges were found to involve two secondary structure elements acting like hinges of a door, allowing an opening and closing motion to occur. This can arise when two neighbouring strands within a β-sheet situated in one domain, diverge apart as they join the other domain. The two resulting termini then form the bending regions between the two domains. α-helices that preserve their hydrogen bonding network when bent are found to behave as mechanical hinges, storing `elastic energy' that drives the closure of domains for rapid capture of a substrate.

Helical to Extended Conformation

The interconversion of helical and extended conformations at the site of a domain boundary is not uncommon. In calmodulin, torsion angles change for five residues in the middle of a domain linking α-helix. The helix is split into two, almost perpendicular, smaller helices separated by four residues of an extended strand.

Shear Motions

Shear motions involve a small sliding movement of domain interfaces, controlled by the amino acid side chains within the interface. Proteins displaying shear motions often have a layered architecture: stacking of secondary structures. The interdomain linker has merely the role of keeping the domains in close proximity.

Domain Motion and Functional Dynamics in Enzymes

The analysis of the internal dynamics of structurally different, but functionally similar enzymes has highlighted a common relationship between the positioning of the active site and the two principal protein sub-domains. In fact, for several members of the hydrolase superfamily, the catalytic site is located close to the interface separating the two principal quasi-rigid domains. Such positioning appears instrumental for maintaining the precise geometry of the active site, while allowing for an appreciable functionally oriented modulation of the flanking regions resulting from the relative motion of the two sub-domains.

Implications for Macromolecular Evolution

Evidence suggests that protein dynamics are important for function, e.g. enzyme catalysis in DHFR, yet they are also posited to facilitate the acquisition of new functions by molecular evolution.

This argument suggests that proteins have evolved to have stable, mostly unique folded structures, but the unavoidable residual flexibility leads to some degree of functional promiscuity, which can be amplified/harnessed/diverted by subsequent mutations.

However, there is growing awareness that intrinsically unstructured proteins are quite prevalent in eukaryotic genomes, casting further doubt on the simplest interpretation of Anfinsen's dogma: "sequence determines structure (singular)". In effect, the new paradigm is characterized by the addition of two caveats: "sequence and cellular environment determine structural ensemble".

Enzyme Kinetics

Kinetic information is useful for examining possible mechanisms for the reaction. This is true for all types of reactions; kinetic principles are used to understand both catalyzed and non-catalyzed reactions.

For enzymes, kinetic information is useful for understanding how metabolism is regulated and how it will occur under different conditions. For enzymes, kinetic information is useful for understanding pathological states. Diseases and disorders often involve alterations in enzymes or enzyme activities. Understanding the way that enzymes work is critical for understanding how drugs work, because many drugs function by interacting with enzymes. In addition, the more you know about an enzymatic reaction, the more information you will have for designing new drugs.

Finally, essentially all of biochemistry is based on enzymes. It is nearly impossible to understand biochemistry without understanding enzymes, and it is impossible to understand enzymes without understanding the kinetic principles of the reactions they mediate.

Thermodynamic considerations and transition state theory Consider a simple system involving the conversion of one molecule, S into another molecule, P. The molecules S and P have different energies.

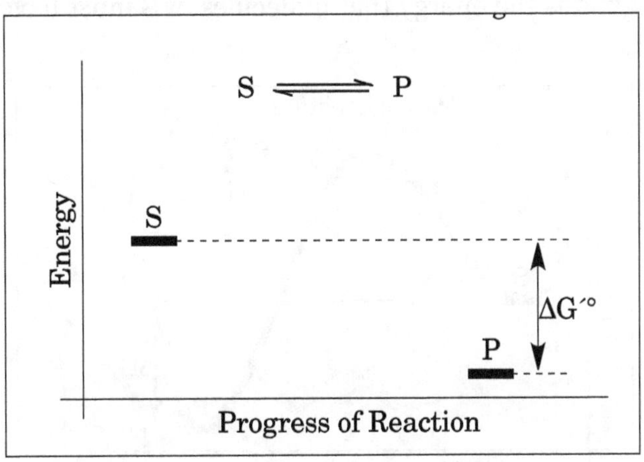

For this system, as for any other, $\Delta G'^{\circ} = -RT \ln K_{eq}$.

$\Delta G'^{\circ}$ is a measure of the relative energies of the two molecules, and thus gives the direction for the reaction when the concentrations of the two molecules are equal. For reactions involving multiple substrates and products, $\Delta G'^{\circ}$ gives the direction in which the reaction will proceed under standard conditions (i.e. when the concentration of all participating species is one molar, except protons and water: protons and water are exceptions, because $\Delta G'^{\circ}$ refers to pH 7.0, and includes the high concentration of water in the term as a constant).

Thermodynamics determines which direction is preferred; however, standard thermodynamics yields no information about the rate of the reaction. Because the tools and concepts of thermodynamics are very powerful, they were extended to allow an understanding of why reactions occur with different rates.

If a maximum energy state exists intermediate between S and P, and if this state is assumed to be in equilibrium with S and P, it is possible to apply the concepts of thermodynamics to the reaction process. This transient high energy intermediate is usually called a transition state, and will be abbreviated X‡ (the "‡" symbol is typically used to designate the transition state). The extension of thermodynamics to consider rates of reaction is called transition state theory to emphasize the importance of this concept.

$$S \rightleftharpoons X^{\ddagger} \rightleftharpoons P$$

The equilibrium constant for the process of S going to the transition state is:

$$K_{eq}^{\ddagger} = \frac{[X^{\ddagger}]}{[S]}$$

Note that "concentration of S" is usually abbreviated as "[S]". This allows the calculation of the ΔG for the transition state.

$$\Delta G^{\ddagger} = -RT \ln K_{eq}^{\ddagger}$$

The $\Delta G^{\ddagger}$ is the free energy difference between S and the transition state. Note that there is also a separate $\Delta G^{\ddagger}$ (in this case, a larger one) between P and the transition state. $\Delta G^{\ddagger}$ is often referred to as the activation energy; it is the energy that molecules of S must have in order to form molecules of P.

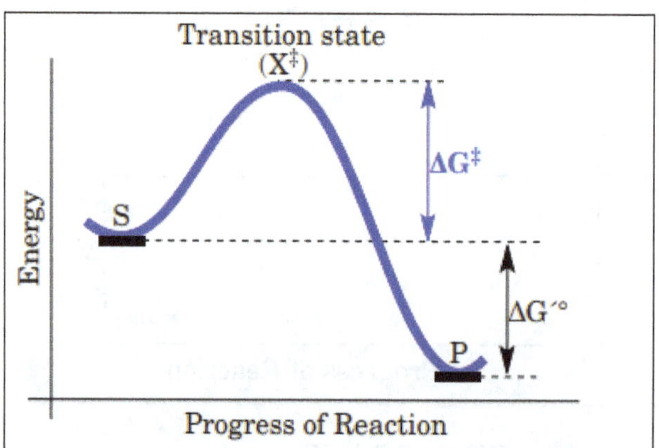

The rate of a chemical reaction depends on the $\Delta G^{\ddagger}$ for the reaction. Using the principles of transition state theory, J.H. van't Hoff and later Svante Arrhenius derived an equation for the reaction rate constant:

$$k = Ae^{\frac{-\Delta G^{\ddagger}}{RT}}$$

In this, the Arrhenius equation, k is the rate constant for the reaction, and A is the Arrhenius constant for the reaction. The Arrhenius constant for a reaction is a measurable quantity. Transition state theory predicts that $A = k_B T/h$, where k_B is the Boltzmann constant (1.380649×10^{-23} J/K), T is absolute temperature, and h is Planck's constant (6.6261×10^{-34} J•sec); for real reactions the measured value of A tends to vary from this theoretical value.

The Arrhenius equation states that, if the $\Delta G^{\ddagger}$ decreases, the rate of the reaction will increase (note that it also states that the rate of the reaction will increase with increasing temperature; changing temperatures is rarely possible for biological processes, but control of temperature can be important for in vitro experiments).

Based on transition state theory, the rate of a reaction is dependent on the energy difference between the initial state and the highest energy transient state along the reaction pathway. Assuming that this is true, rate enhancements by enzymes must be mediated at least in part by a decrease in the energy of the highest energy transition state, although in some reactions, an enzyme may function in part by destabilizing the state that immediately precedes the transition state.

A decrease in the energy of the highest energy transition state can be accomplished in one of two main ways. Some enzymes use one of these ways, and some use both. 1) The enzyme stabilizes the transition state. The same transition state that would normally be present in the reaction pathway is also present in the enzymecatalyzed pathway, but in the enzyme-catalyzed pathway, this state has a lower energy. 2) The enzyme allows a different pathway for the process. Without the enzyme, the reaction might proceed by some pathway. The enzyme allows a different series of reactions to occur that would otherwise be either of much higher energy or impossible. In the enzyme-catalyzed process, the highest energy transition state (which does not exist in the non-catalyzed process) has a lower energy than the energy of the transition state for the non-catalyzed process.

Let us consider the first of these possibilities. Consider a reaction in which S is converted to P, either as a simple chemical reaction or as an enzyme-catalyzed process. Instead of merely stating S→P, for transition state theory we need to add an additional term:

$$S \underset{}{\overset{k_{uncat}}{\rightleftharpoons}} X^{\ddagger} \rightleftharpoons P$$

$$E + S \rightleftharpoons ES \underset{}{\overset{k_{cat}}{\rightleftharpoons}} EX^{\ddagger} \rightleftharpoons EP \rightleftharpoons E + P$$

In each case, the reaction pathway passes through an identical transition state, X‡ (in the uncatalyzed pathway, this species is not bound to the enzyme, but it is the same species in both cases).

In the reaction diagram, we can show the binding of the substrate and dissociation of the product from the enzyme explicitly. Note that the ES complex is lower in energy than the free S, which is why this complex forms spontaneously.

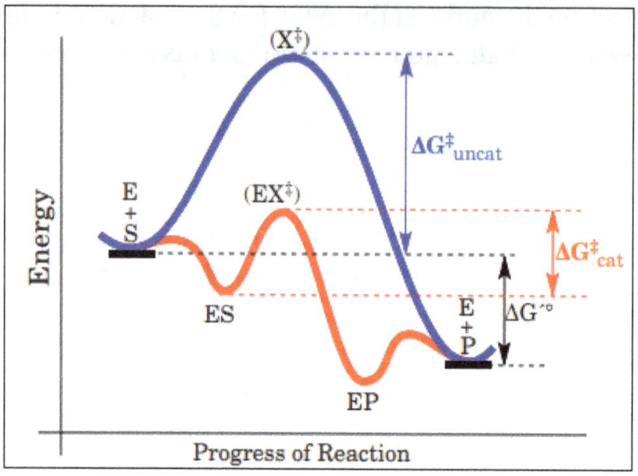

Using the principles of transition state theory, we can derive an equation for the rate enhancement mediated by the enzyme. Recall that the rate constant is related to the difference in energy of between the initial state and the transition state.

$$\text{Arrhenius equation}: \; k = Ae^{\frac{-\Delta G\ddagger}{RT}}$$

If we divide the rate constant for the catalyzed reaction by that of the uncatalyzed reaction:

$$\frac{k_{cat}}{k_{uncat}} = \frac{Ae^{\frac{-\Delta G^{\ddagger}_{cat}}{RT}}}{Ae^{\frac{-\Delta G^{\ddagger}_{uncat}}{RT}}}$$

which simplifies to:

$$\frac{k_{cat}}{k_{uncat}} = e^{\frac{\Delta G^{\ddagger}_{uncat} - \Delta G^{\ddagger}_{cat}}{RT}}$$

Recall that K_{eq} and $\Delta G°$ are related. Rearranging the standard equation for $\Delta G°$ to solve for K_{eq}, we obtain:

$$K_{eq} = e^{\frac{-\Delta G°}{RT}}$$

Using this, we see that:

$$\frac{k_{cat}}{k_{uncat}} = e^{\frac{\Delta G^{\ddagger}_{uncat} - \Delta G^{\ddagger}_{cat}}{RT}} = \frac{K^{\ddagger}_{eq}}{K_{eq}}$$

Thus, the ratio of the rate constant for the catalyzed reaction to that of the uncatalyzed reaction is equal to the ratio of the equilibrium constant for the binding of the enzyme to the transition state to that for the binding to the substrate.

$$E + S \underset{}{\overset{K_{eq}}{\rightleftharpoons}} ES \qquad E + X^{\ddagger} \underset{}{\overset{K^{\ddagger}_{eq}}{\rightleftharpoons}} EX^{\ddagger}$$

This deserves a few comments. The first is that while neither the $X^\ddagger$ nor the $EX^\ddagger$ are present in significant concentrations due to the transient nature of these species, it is, at least in principle, possible for the $X^\ddagger$ to bind to the free enzyme, or for the $EX^\ddagger$ complex to dissociate. The second is more important: the interaction between the enzyme and the transition state exhibits much higher affinity than the interaction between the enzyme and the substrate. The magnitude of this difference in affinity is proportional to the magnitude of the increase in rate constant.

The function of a catalyst is to lower the energy of the transition state for the reaction: in other words, enzymes change $\Delta G^\ddagger$. The most common method for allowing this is for the enzyme to bind the transition state more tightly than the substrate. The additional energy obtained from this binding stabilizes the transition state and therefore accelerates the reaction. Note that enzymes do not change the energy of S and P (because the free S and P are not bound to the enzyme). As a result, enzymes do not alter the $\Delta G'^\circ$ for the reaction, but instead only alter the rate of the reaction, usually by altering $\Delta G^\ddagger$.

Chemical Kinetics

Zero[th]-order kinetics Some reactions occur at rates that are independent of reactant concentration. In these cases, the rate of the reaction (often called the velocity of the reaction) is constant, regardless of the concentration of the participating compounds.

Consider a simple reaction: $S \rightleftharpoons P$

For this reaction the velocity of the reaction can be expressed as:

$$v = \frac{d[p]}{dt} = \frac{-d[S]}{dt} = k$$

These equations state that at any given instant, the velocity is equal to the increase in concentration of P divided by the time interval, and that the velocity is also equal to the decrease in concentration of S divided by the time interval (note the minus sign in the $-d[S]/dt$ expression). In this case, because the rate is independent of [S] and [P], the velocity = k, where k is the rate constant for the reaction. These (somewhat unusual) reactions are called zero[th]-order reactions.

First-order kinetics Again, consider a simple reaction, but now a first-order reaction in which the rate depends on the reactant concentration:

$$S \rightleftharpoons P, \qquad v = \frac{d[p]}{dt} = \frac{-d[S]}{dt}$$

If you assume that the reverse reaction does not occur (which is a valid assumption if the initial concentration of P is zero), then at any given time, $v = k[S]$. For zero[th]-order and first-order reactions, k is the rate constant for the reaction; k is a measure of how rapidly the reaction will occur at any concentration of S, and has the units of time^{-1} (usually seconds^{-1} or minutes^{-1}). In a first-order reaction, the rate depends on the concentration of a single species (in this case, the reactant S).

The equation $v = k[S]$ states that velocity is a linear function of S concentration. In the graph (below) the zero[th]-order reaction has a constant velocity regardless of the concentration of S, while the first-order reaction velocity increases linearly with increasing S concentration.

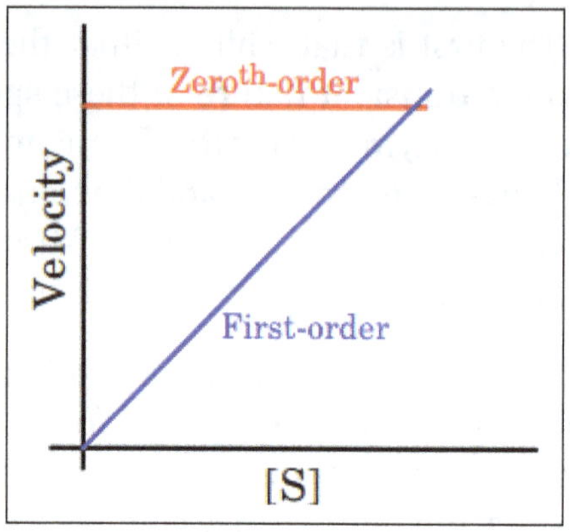

Another way to look at the course of these types of reactions is to consider the concentration of S as a function of time. For a zeroth-order reaction:

$$[S]_t = [S]_0 - kt$$

while for a first-order reaction:

$$[S]_t = [S]_0 e^{kt}$$

where $[S]_0$ is the concentration of S at the beginning of the reaction, and $[S]_t$ is the concentration of S at time t.

Plotting these equations reveals that S concentration decreases linearly with time for a zeroth-order reaction, and that S concentration decreases exponentially with time for a first-order reaction. These observations make sense: for a first-order reaction, the rate of the reaction decreases as S is used up, because the rate depends on the S concentration.

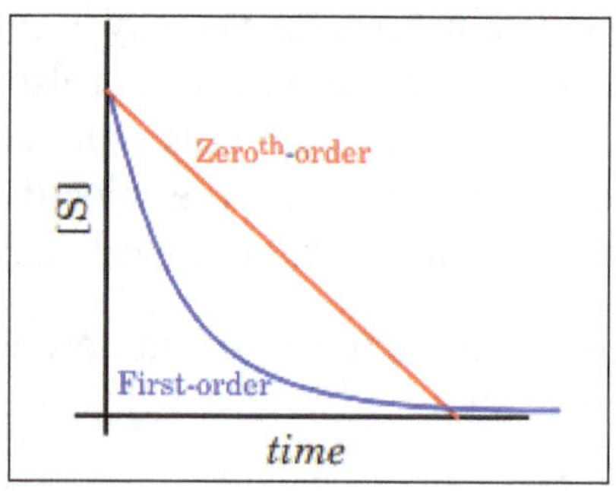

The following, slightly simplified, derivation of the equation for the first-order reaction is included for completeness.

The rate equation: $\frac{-d[S]}{dt} = k[S]$ rearranges to: $\frac{d[S]}{[S]} = -\,kdt$

Integrating both sides: $\int_{[S]_0}^{[S]} \frac{d[S]}{[S]} = \int_0^t -k\,dt$ gives: $\ln\,[S] - \ln[S]_0 = -kt$

Raising both sides to the e power and rearranging gives: $[S]_t = [S]_0 = -kt$

Second-order Kinetics

More complicated reactions can also occur: $S + R \rightleftharpoons P + Q$

For these reactions:

$$v = \frac{d[\mathrm{P}]}{dt} = \frac{d[\mathrm{Q}]}{dt} = \frac{-d[\mathrm{S}]}{dt} = \frac{-d[\mathrm{R}]}{dt}$$

and

$$v = k[\mathrm{S}][\mathrm{R}]$$

Reactions of this type are second-order, and k is a second-order rate constant, because the rate of the reaction depends on the product of [S] and [R]. If the reaction involved the collision of two molecules of S, the velocity equation would be:

$$v = k[\mathrm{S}][\mathrm{S}] = k[\mathrm{S}]^2$$

The order of the reaction comes from the exponent that describes the number of reactants. Second-order rate constants have units of $M^{-1} \cdot sec^{-1}$.

Real chemical reactions rarely have more than two molecules interacting at one time, because the simultaneous collision of more than two molecules is unlikely.

There are a few examples of trimolecular reactions; in most reactions that appear to involve more than two reactants, two reactants form an intermediate, which then reacts with the other compound to form the final product.

Pseudo-first-order kinetics Studying second-order reactions is usually more difficult than studying first-order reactions. One way around this difficulty is to create "pseudo-first-order conditions". These are conditions in which the concentration of one compound is very high. If the concentration of R is very high compared to S, then the concentration of R will essentially be constant during the reaction. This allows the equation to be rewritten:

$$v = k[\mathrm{S}][\mathrm{R}] = k_{\mathrm{pseudo}}[\mathrm{S}]$$

The term for concentration of R did not disappear; because the concentration of R is approximately a constant, it was merely incorporated into the k_{pseudo} first-order rate constant.

Rate-limiting steps Many reaction pathways involve multiple steps. In most cases, one step will be appreciably slower than the others. This step is the rate-limiting step; it is the step upon which the rate of the overall reaction depends. Analysis for rate-limiting steps is important for understanding all types of reactions. (In biochemistry, analysis for rate-limiting steps in metabolic pathways is especially important, because these steps tend to be the ones that act as regulated control points for the pathway.)

Michaelis-Menten kinetics Initially, it was assumed that a simple enzyme catalyzed reaction, such as S→P would be a first-order reaction:

$$\mathrm{S} \underset{k_{-1}}{\overset{k_1}{\rightleftharpoons}} \mathrm{P}$$

However, when the reaction data were analyzed, instead of a linear plot of velocity versus [S], the experiment yielded a hyperbolic curve.

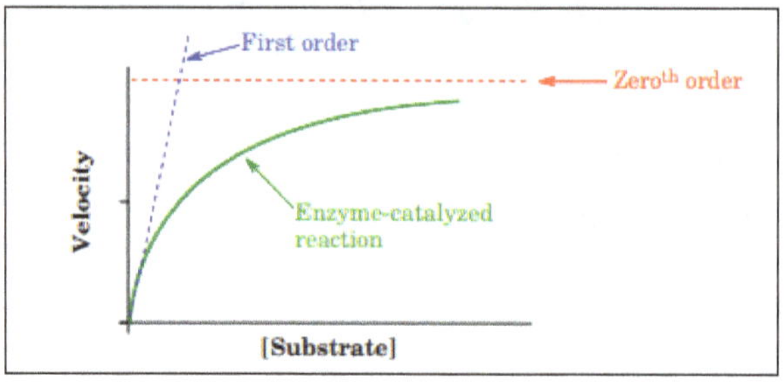

Analysis of the data suggested that the reaction was first-order at very low concentrations of S (*i.e.* $v \approx k[\mathrm{S}]$, where k is a rate constant appropriate for the conditions of the enzymatic reaction), and that as $[\mathrm{S}]$ increased, the reaction became progressively closer to zeroth-order (*i.e.* $v \approx k[\mathrm{E}] = k_{cat}[\mathrm{E}]$, where k_{cat} is the catalytic rate constant for the enzyme, and $[\mathrm{E}]$ is the enzyme concentration).

To explain this phenomenon, it was necessary to hypothesize the formation of a complex between the enzyme and S. This hypothesis means that, even for the simple conversion of S to P, the reaction was actually somewhat more complicated than the expected first-order process:

$$\mathrm{E} + [\mathrm{S}] \underset{k_{-1}}{\overset{k_1}{\rightleftharpoons}} [\mathrm{ES}] \xrightarrow{k_2} \mathrm{E} + \mathrm{P}$$

The velocity for the reaction is then:

$$v = \frac{d[\mathrm{P}]}{dt} = k_2[\mathrm{ES}]$$

Note that this appears to be a first-order rate equation; however, the concentration of ES is not a linear function of $[\mathrm{S}]$. Instead, the change in $[\mathrm{ES}]$ concentration as a function of time is more complex:

$$\frac{d[\mathrm{ES}]}{dt} = k_1[\mathrm{E}]_{free}[\mathrm{S}] - k_{-1} - k_2[\mathrm{ES}]$$

Note that the above equation consists of the rate of formation of the ES complex $\left(k_1[\mathrm{E}]_{free}[\mathrm{S}]\right)$ minus the rates at which the ES complex disappears (both $k_{-1}[\mathrm{ES}]$ and $k_2[\mathrm{ES}]$ describe rates of ES complex disappearance). This equation is rather difficult to use unless some simplifying assumptions are made. This is true for several reasons; the most obvious reason is that neither $[\mathrm{ES}]$ nor $[\mathrm{E}]$ *free* (the concentration of enzyme unbound to S) can be calculated (or measured) accurately, because neither one is an independent variable.

Steady State Assumption

Most enzyme kinetics derivations depend on several assumptions:

Assumption (1): $\dfrac{d[\mathrm{E}]}{dt} = 0$

The total amount of enzyme does not change during measurement of the reaction. This is usually a good assumption, because enzymes are catalysts that are not consumed during the reaction.

Assumption (2): $k_2 \ll k_1$ and $k_2 \ll k_{-1}$

The rate constant for the reaction, k_2, is much smaller than the rate constants for ES formation, k_1 and for ES dissociation, k_{-1}. In other words, the rate-limiting step for the reaction is the actual catalytic step, while the ES complex rapidly reaches an equilibrium concentration.

Assumption (3): $E]_{total} \ll [S]$

The total concentration of enzyme is much less than the total concentration of S. Consequently, $[S]$ remains approximately constant during the reaction. This assumption is very similar to that used in pseudo-first-order kinetics experiments, but it is a somewhat flawed assumption, because if $[S]$ were not changing, no product would be formed.

Assumption (4) $[P]_{t-0} = 0$

The initial concentration of P is zero. This is implicit in the rate equation given above, because there is no term for formation of ES from P. In experimental systems, this assumption is usually valid, because the product can be omitted from the reaction vessel.

Leonor Michaelis and Maud Menten, and later G.E. Briggs and James B.S. Haldane made these assumptions in slightly different ways (the Briggs-Haldane assumptions probably more closely model reality, but because both methods result in the same equation, the majority of textbooks refer to Michaelis and Menten as the originators of the concepts). Using these assumptions, it is possible to derive equations that describe the behavior of enzyme-catalyzed reactions.

If the ES complex forms rapidly, and if the concentration of S does not change significantly during the reaction, then, for much of the reaction, the concentration of the ES complex will be constant. Mathematically, this concept is expressed as:

$$\frac{d[\mathrm{ES}]}{dt} = k_1[\mathrm{E}]_{free}[\mathrm{S}] - k_{-1}[\mathrm{ES}] - k_2[\mathrm{ES}] = 0$$

The equation above describes steady state conditions, and is critical to enzyme kinetics. Note that the steady state assumption is an approximation; it is rarely absolutely correct, unless $k_2 = 0$ or in situations where the substrate is replenished and the product removed during the reaction. The term "steady state" refers to the fact that, while the concentration of ES is constant, it is continuously being both formed and disrupted. Thus, substrate is flowing, via the enzyme, in a steady stream to form product.

Assuming that the amount of enzyme does not change during the reaction, the total enzyme concentration $[\mathrm{E}]_{total} = [\mathrm{E}]_{free} + [\mathrm{ES}]$. Using this observation and the steady state equation, Michaelis and Menten derived an equation for velocity as a function of $[S]$. This important equation is known as the Michaelis-Menten equation.

$$v = \frac{\mathrm{V}_{max}[\mathrm{S}]}{\mathrm{K}_m + [\mathrm{S}]}$$

Note the derivation of the Michaelis-Menten equation is given in nearly all biochemistry textbooks, and it is unnecessary to reproduce it here.

The Michaelis-Menten equation is applicable to most enzymes, and is critically important to understanding enzyme action in biological systems. At least for simple systems, the Michaelis-Menten equation describes the way that the reaction velocity depends on the substrate concentration.

The parameters K_m and V_{max} cannot be determined from a single measurement; instead, they must be determined by measuring velocity at a variety of $[S]$.

The meaning of V_{max} and K_m

V_{max} is the velocity observed when all of the enzyme present is fully saturated with substrate; in other words, when $[ES]=[E]_{total}$. This is only completely true if the concentration of S is infinitely high (which is obviously impossible in the real world).

For the simple reaction we have been discussing, $V_{max}=k_2[E]_{total}$. For more complex reactions, $V_{max}=k_{cat}[E]_{total}$, where k_{cat} is the rate constant for the slowest step of the reaction. Note that the V_{max} is not an intrinsic property of the enzyme, because it is dependent on the enzyme concentration; the actual intrinsic property is the k_{cat}.

The Michaelis-Menten equation, and the definition of V_{max} have one major consequence for biological systems: the velocity is directly proportional to the enzyme concentration. This means that one simple method for increasing the velocity is to synthesize more enzyme molecules. Increased enzyme concentrations result in higher velocities at any substrate concentration.

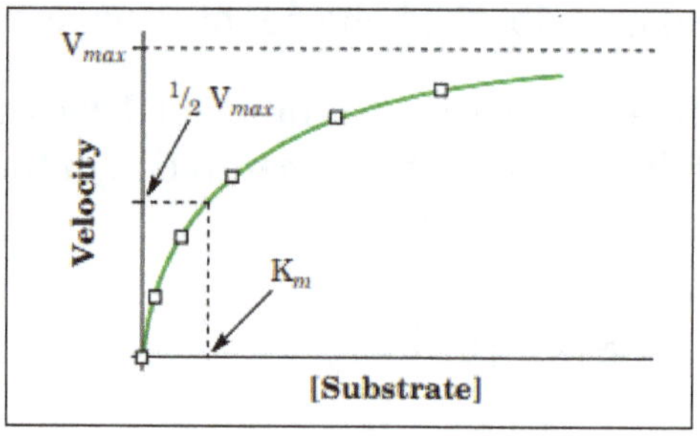

In contrast to V_{max}, the parameter K_m is an intrinsic parameter of the enzyme. When properly performed, measurements of Km yield constant results, regardless of enzyme concentration. Note that if $[S]=K_m$, the Michaelis-Menten equation reduces to $v = 1/2\ V_{max}$. Therefore, K_m is a measure of the ability of the substrate to interact with the enzyme. Altering K_m (either by having multiple isozymes with different K_m values, or by having an enzyme with a K_m that can be regulated), also allows alteration in the velocity of a reaction.

Another, related, way of looking at K_m is to compare it to K_d, the equilibrium dissociation constant for formation of ES complex. The dissociation constant is a measure of affinity, with higher values indicating lower affinity.

$$K_m=\frac{k_{-1}+k_2}{k_1}, \qquad K_d=\frac{k_{-1}}{k_1}$$

If $k_2 = 0$, then $K_m = K_d$. Because, for most enzymes, k_2 is relatively small compared to k_{-1}, the K_m value is often close to the K_d value.

Note that k_1 is a second-order rate constant, and has units of $M^{-1} \bullet sec^{-1}$. The other rate constants are first-order, and have units of sec^{-1}. This means that Km has units of M; in other words, both K_m and K_d are expressed in concentration units.

Contemplation of the Michaelis-Menten equation suggests that a low K_m means a high affinity, and therefore, for a given substrate concentration, a high velocity. In contrast, a high K_m means low affinity, and therefore low velocity at any $[S]$.

Uses of K_m

K_m can act as a measure of several useful properties of enzymes.

1) Measurement of K_m is used to determine the substrate preferences of an enzyme. If more than one endogenous compound can act as a substrate for an enzyme, the substrate with the lowest K_m is probably the preferred physiological substrate.

2) Measurement of K_m is used to distinguish isozymes. Isozymes often have different affinities for the same substrate.

3) Measurement of K_m is used to check for abnormalities in an enzyme: An altered K_m reflects some change in the way the enzyme binds the substrate. K_m is therefore sensitive to modifications to the enzyme; measurement of Km can often reveal extremely useful information regarding mutations or other changes in the structure of an enzyme.

Determining V_{max} and K_m

A cursory examination of a velocity versus $[S]$ plot (such as the one above) may suggest that the graph could be used to determine K_m and V_{max}. In practice, for any type of plot, accurately determining the values from a curve is difficult; for enzyme kinetics, it is especially difficult, because achieving $v = V_{max}$ is impossible, and because nothing about the curve states that: "the K_m is right here!"

To avoid this problem, several scientists derived linear forms of the MichaelisMenten equation. Hans Lineweaver and Dean Burk were the first to do one of these derivations, and developed the double reciprocal plot (also called the LineweaverBurk plot) in 1934. This plot has its deficiencies, but it is still useful, and all biochemists must be able to understand the information it presents.

The equations below reveal that a plot of 1/v versus $1/[S]$ has a slope of K_m/V_{max} and a y-intercept of $1/V_{max}$. Further examination shows that x-intercept = $-1/K_m$.

$$\frac{1}{v} = \frac{K_m}{V_{max}} \frac{1}{[S]} + \frac{1}{V_{max}}$$

$$y = m\ x + b$$

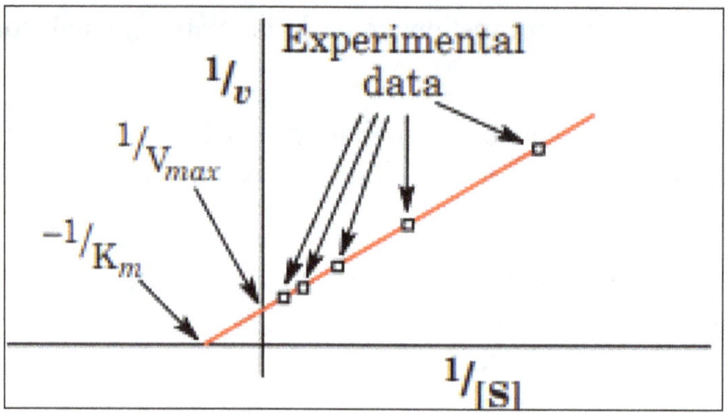

The Lineweaver-Burk plot is very useful for illustrative purposes, but tends to yield aberrant values of K_m and V_{max} unless the v versus [S] data fit the hyperbolic curve closely; in other words, for real-world data containing experimental errors, the Lineweaver-Burk plot is frequently inaccurate.

When the Lineweaver-Burk plot was invented, computers effectively did not exist, and linearization of the Michaelis-Menten equation allowed for a simple analysis of the data. These linearization routines remain popular because most calculators and all graphing software packages are capable of calculating a least-squares linear regression analysis exactly. In contrast, most non-linear equations cannot be solved exactly. However, current computer technology allows for rapid iteration and optimization of the K_m and V_{max} parameters that best fit a set of data. The Excel Solver package and many other optimization routines are available to allow analysis of the non-linear Michaelis-Menten equation. For analysis of actual experimental v versus [S] data, non-linear regression methods that fit the data to the unmodified Michaelis-Menten equation are far more likely to yield satisfactory values for K_m and V_{max}.

The collision of molecules is limited by diffusion. Diffusion limits2 for proteins in aqueous solution are about $10^9\ M^{-1}\bullet sec^{-1}$. This is a measure of the maximum possible rate for any reaction involving proteins in aqueous solution. Note that the diffusion limit is a second-order rate constant; it has a concentration term because collision between molecules is concentration dependent.

$\boldsymbol{k_{cat}}$ is a term used for the rate constant for the overall reaction. In complex reactions with several steps, where the maximal catalytic rate depends on several rate constants, k_{cat} is the rate constant for the rate-limiting step.

$\boldsymbol{k_{cat}}$ = turnover number = the number of product molecules formed by one enzyme molecule in one second (or, for slow enzymes, in one minute).

$\boldsymbol{k_{cat}/K_m}$ is a measure of the catalytic efficiency of an enzyme; in effect, it takes into account both substrate binding and conversion to product. k_{cat}/k_m cannot be faster than the diffusion limit (E and S must collide in order to react).

Some enzymes operate at efficiencies approaching the diffusion limit.

Enzyme[3]	K_m (M)	k_{cat} (sec^{-1})	k_{cat}/K_m ($M^{-1}\bullet sec^{-1}$)
Acetylcholinesterase	1×10^{-4}	1.4×10^4	1.5×10^8
Carbonic anhydrase	1.2×10^{-2}	1.0×10^6	8.3×10^7

Catalase	2.5 x 10^{-2}	1.0 x 10^{7}	4.0 x 10^{8}
Superoxide dismutase	3.6 x 10^{-4}	1.0 x 10^{6}	1.8 x 10^{9}
Ferredoxin reductase	1.0 x 10^{-7}	1.0 x 10^{1}	1.0 x 10^{8}

Specific activity is a term used when the molar enzyme concentration is not known. (If the moles of enzyme present is unknown, it is impossible to calculate k_{cat}.) Specific activity is similar to turnover number, but is usually given in terms of velocity per milligram of total protein.

Multisubstrate enzymes So far we have been looking at the kinetics of the enzyme reaction with one substrate. However, many enzyme reactions involve two substrates, and some involve even more. Kinetic equations for multi-substrate reactions have been derived that are analogous to the Michaelis-Menten treatment for single substrate reactions.

By keeping the concentration of one substrate constant, and varying the concentration of the other substrate, K_m and V_{max} values can be obtained for multisubstrate enzymes. These values will only approximate the true K_m and V_{max} values because the concentration of the constant substrate used will not fully saturate the enzyme. Because the values are approximate, the values obtained in this manner are usually referred to as "apparent" K_m and V_{max} values.

Mechanisms of multi-substrate reactions It was initially assumed that two substrate reactions would involve a ternary complex: $[ES_1S_2]$. However, this is not necessarily true.

Consider the reaction catalyzed by sucrose phosphorylase:

$$\text{glucose-1-phosphate} + \text{fructose} \rightleftharpoons \text{sucrose} + \text{phosphate}$$

In the absence of fructose, and in the presence of ^{32}P-labeled inorganic phosphate (note that ^{32}P is radioactive, and is therefore readily detected) the following reaction was observed:

$$\text{glucose-1-phosphate} + {}^{32}\text{P-phosphate} \rightleftharpoons \text{glucose-1-}{}^{32}\text{P-phosphate} + \text{phosphate}$$

In other words, the enzyme catalyzed the exchange between phosphate covalently bound to the glucose and the phosphate in solution. This reaction was possible because the enzyme reacted with glucose (and fructose) in separate steps:

Step 1:

$$\text{glucose-1-phosphate} + \text{Enzyme} \rightleftharpoons \text{glucose-Enzyme} + \text{phosphate}$$

where "glucose-Enzyme" is a covalent reaction intermediate between glucose and sucrose phosphorylase.

Step 2:

$$\text{glucose-Enzyme} + \text{fructose} \rightleftharpoons \text{sucrose} + \text{Enzyme}$$

There are two general types of 2-substrate reactions: Double displacement and Sequential.

Double displacement reactions (also called "ping-pong" reactions) include the reaction mechanism shown in the sucrose phosphorylase example. One substrate binds the enzyme, is modified, and is released. The other substrate has to wait until the first product is released before binding to the enzyme.

In contrast, in sequential reactions all substrates bind to the enzyme before catalysis occurs.

There are two types of sequential enzymes, ordered and random. For ordered sequential enzymes, the reactants combine in a specific order and products are released in a specific order.

For example, $substrate_1$ binds, then $substrate_2$ binds. The reaction occurs, and then $product_2$ is released followed by the release of $product_1$.

For random sequence enzymes, there are no restrictions to the order of either substrate binding or product release.

Enzyme Inhibition

Enzymes are required for most, if not all, of the processes required for life. Enzymes catalyse a reaction by reducing the activation energy needed for the reaction to occur. However, enzymes need to be tightly regulated to ensure that levels of the product do not rise to undesired levels. This is accomplished by enzyme inhibition.

Types of Inhibition

Reversible and irreversible inhibitors are chemicals which bind to an enzyme to suppress its activity. One method to accomplish this is to almost permanently bind to an enzyme. These types of inhibitors are called irreversible. However, other chemicals can transiently bind to an enzyme. These are called reversible. Reversible inhibitors either bind to an active site (competitive inhibitors), or to another site on the enzyme (non-competitive inhibitors).

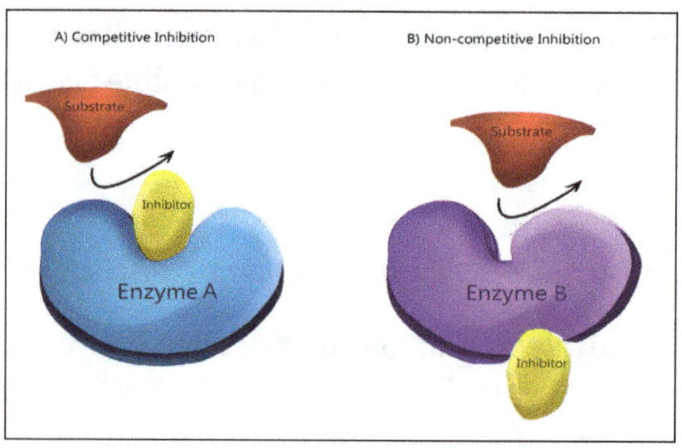

Diagram to show the effect of competitive and non-competitive enzyme inhibitors.

Competitive Inhibitors

Competitive inhibitors compete with the substrate at the active site, and therefore increase Km (the Michaelis-Menten constant). However, Vmax is unchanged because, with enough substrate concentration, the reaction can still complete. The graph plot of enzyme activity against substrate concentration would be shifted to the right due to the increase of the Km, whilst the Lineweaver-Burke plot would be steeper when compared with no inhibitor.

Non-competitive Inhibitors

Non-competitive inhibitors bind to another location on the enzyme and as such decrease VMAX. However, KM is unchanged. This is demonstrated by a lower maximum on a graph plotting enzyme activity against substrate concentration and a higher y-intercept on a Lineweaver-Burke plot when compared with no inhibitor.

	Competitive	Non-Competitive
Where does it act?	Active Site	Alternative site
Change of Km	Increase	Unchanged
Change of Vmax	Unchanged	Decrease

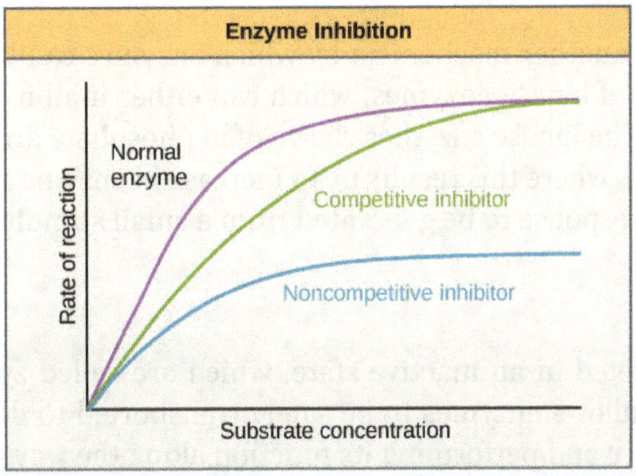

Diagram to show the effect of enzyme inhibitors on the rate of reaction and how it varies with substrate concentration.

Allosteric Inhibition

Allosteric enzymes display a sigmoidal curve in contrast to the hyperbolic curve displayed by Michaelis-Menten Enzymes. This is because most allosteric enzymes contain multiple sub-units which can affect each other when the substrate binds to the enzyme. Inhibition can affect either $K_{0.5}$, which is the substrate concentration for half-saturation, Vmax or both. This results in a shift of the curve to the right, and in the case of reducing Vmax, shifts the curve down.

Allosteric enzymes have two states: a low affinity state dubbed the "T" state and the high affinity "R" state. Inhibitors work by preferentially binding to the T state of an allosteric enzyme, causing the enzyme to maintain this low affinity state.

This is extremely useful to limit the amount of an enzyme's product, as the product can then go on to inhibit the same type of enzyme to ensure the amount of product is not excessive. This is known as feedback inhibition. For example, ATP allosterically inhibits pyruvate kinase to prevent increased formation of pyruvate, so less ATP is eventually formed. Additionally, phosphofructokinase is allosterically inhibited by citrate, an intermediate of the Kreb's cycle. This means that glycolysis will be limited when there is high ATP generation from the Kreb's cycle.

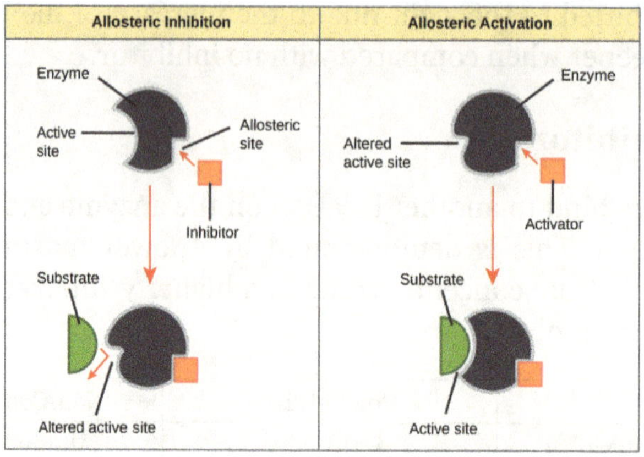

Diagram to show the mechanism of both allosteric inhibition and activation.

Phosphorylation

Phosphorylation provides another mechanism by which enzymes can be inhibited. This typically occurs through the action of kinase enzymes, which can either inhibit or activate an enzyme depending on the situation. The kinase enzymes cleave off a phosphate group from ATP and binds it to the enzyme. In situations where this results in an increase in enzyme activity it creates a cascade reaction, allowing a large response to be generated from a small stimulus.

Zymogens

Enzymes can also be secreted in an inactive state, which are called zymogens. Zymogens are a useful mechanism which allows enzymes to be safely transported to different locations, without the enzyme becoming active and performing its function along the way. They remain inactive due to an addition of amino acids in the protein. Therefore, to activate a zymogen, another enzyme must cleave off these additional amino acids. For example, chymotrypsinogen is synthesised by the pancreas, yet is inactivated and cannot function. Upon travelling to the intestines, another enzyme (trypsin) cleaves off the additional amino acids to produce the activated form, chymotripsin.

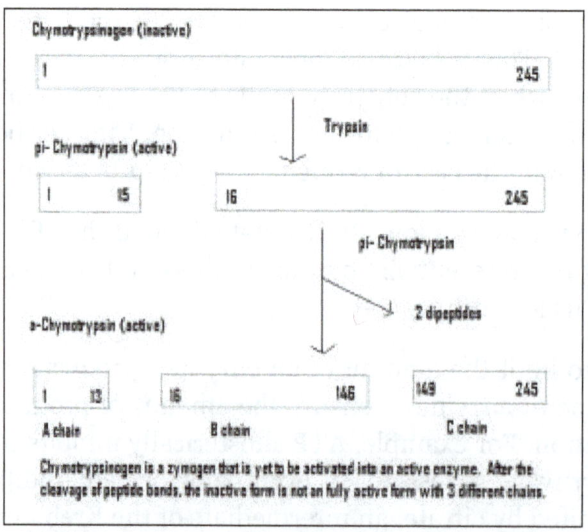

Diagram to show the activation of chymotrypsinogen within the gastrointestinal system and the enzymes involved at each stage.

Enzyme Catalysis

Enzyme catalysis is the increase in the rate of a process by a biological molecule, an “enzyme”. Most enzymes are proteins, and most such processes are chemical reactions. Within the enzyme, generally catalysis occurs at a localized site, called the active site.

Most enzymes are made predominantly of proteins, either a single protein chain or many such chains in a multi-subunit complex. Enzymes often also incorporate non-protein components, such as metal ions or specialized organic molecules known as cofactor (e.g. adenosine triphosphate). Many cofactors are vitamins, and their role as vitamins is directly linked to their use in the catalysis of biological process within metabolism. Catalysis of biochemical reactions in the cell is vital since many but not all metabolically essential reactions have very low rates when uncatalysed. One driver of protein evolution is the optimization of such catalytic activities, although only the most crucial enzymes operate near catalytic efficiency limits, and many enzymes are far from optimal. Important factors in enzyme catalysis include general acid and base catalysis, orbital steering, entropic restriction, orientation effects (i.e. lock and key catalysis), as well as motional effects involving protein dynamics

Mechanisms of enzyme catalysis vary, but are all similar in principle to other types of chemical catalysis in that the crucial factor is a reduction of energy barrier(s) separating the reactants from the products. The reduction of activation energy (E_a) increases the fraction of reactant molecules that can overcome this barrier and form the product. An important principle is that that since they only reduce energy barriers between products and reactants, enzymes always catalyze reactions in both directions, and cannot drive a reaction forward or affect the equilibrium position - only the speed with which is it achieved. As with other catalysts, the enzyme is not consumed or changed by the reaction (as a substrate is) but is recycled such that a single enzyme performs many rounds of catalysis.

Induced Fit

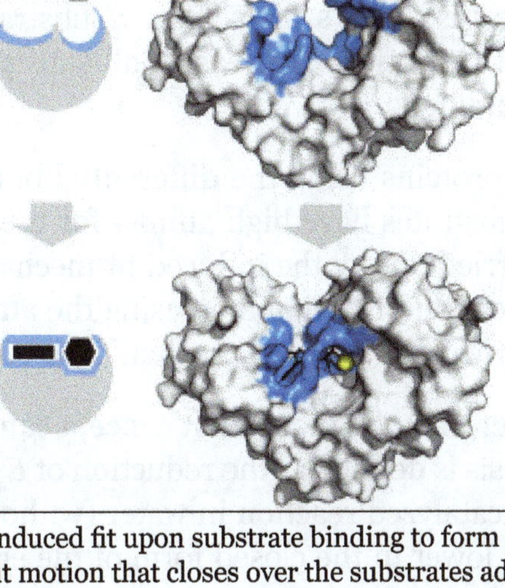

Enzyme changes shape by induced fit upon substrate binding to form enzyme-substrate complex. Hexokinase has a large induced fit motion that closes over the substrates adenosine triphosphate and xylose. Binding sites in blue, substrates in black and Mg^{2+} cofactor in yellow.

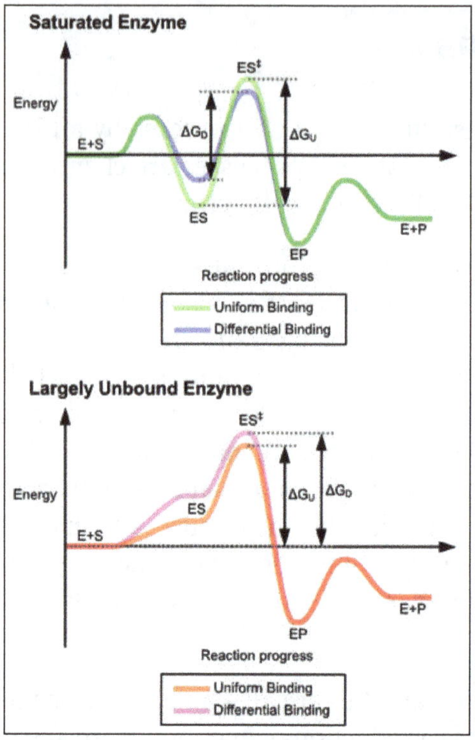

The different mechanisms of substrate binding.

The favored model for the enzyme-substrate interaction is the induced fit model. This model proposes that the initial interaction between enzyme and substrate is relatively weak, but that these weak interactions rapidly induce conformational changes in the enzyme that strengthen binding.

The advantages of the induced fit mechanism arise due to the stabilizing effect of strong enzyme binding. There are two different mechanisms of substrate binding: uniform binding, which has strong substrate binding, and differential binding, which has strong transition state binding. The stabilizing effect of uniform binding increases both substrate and transition state binding affinity, while differential binding increases only transition state binding affinity. Both are used by enzymes and have been evolutionarily chosen to minimize the activation energy of the reaction. Enzymes that are saturated, that is, have a high affinity substrate binding, require differential binding to reduce the energy of activation, whereas small substrate unbound enzymes may use either differential or uniform binding.

These effects have led to most proteins using the differential binding mechanism to reduce the energy of activation, so most substrates have high affinity for the enzyme while in the transition state. Differential binding is carried out by the induced fit mechanism - the substrate first binds weakly, then the enzyme changes conformation increasing the affinity to the transition state and stabilizing it, so reducing the activation energy to reach it.

It is important to clarify, however, that the induced fit concept cannot be used to rationalize catalysis. That is, the chemical catalysis is defined as the reduction of $E_a^{\ddagger}$ (when the system is already in the $ES^{\ddagger}$) relative to $E_a^{\ddagger}$ in the uncatalyzed reaction in water (without the enzyme). The induced fit only suggests that the barrier is lower in the closed form of the enzyme but does not tell us what the reason for the barrier reduction is.

Induced fit may be beneficial to the fidelity of molecular recognition in the presence of competition and noise via the conformational proofreading mechanism .

Mechanisms of an Alternative Reaction Route

These conformational changes also bring catalytic residues in the active site close to the chemical bonds in the substrate that will be altered in the reaction. After binding takes place, one or more mechanisms of catalysis lowers the energy of the reaction's transition state, by providing an alternative chemical pathway for the reaction. There are six possible mechanisms of "over the barrier" catalysis as well as a "through the barrier" mechanism:

Proximity and Orientation

Enzyme-substrate interactions align the reactive chemical groups and hold them close together in an optimal geometry, which increases the rate of the reaction. This reduces the entropy of the reactants and thus makes addition or transfer reactions less unfavorable, since a reduction in the overall entropy when two reactants become a single product. However this is a general effect and is seen in non-addition or transfer reactions where it occurs due to an increase in the "effective concentration" of the reagents. This is understood when considering how increases in concentration leads to increases in reaction rate: essentially when the reactants are more concentrated, they collide more often and so react more often. In enzyme catalysis, the binding of the reagents to the enzyme restricts the conformational space of the reactants, holding them in the 'proper orientation' and close to each other, so that the collide more frequently, and with the correct geometry, to facilitate the desired reaction. The "effective concentration" is the concentration the reactant would have to be, free in solution, to experiences the same collisional frequency. Often such theoretical effective concentrations are unphysical and impossible to realize in reality - which is a testament to the great catalytic power of many enzymes, with massive rate increases over the uncatalyzed state.

For example: Similar reactions will occur far faster if the reaction is intramolecular.

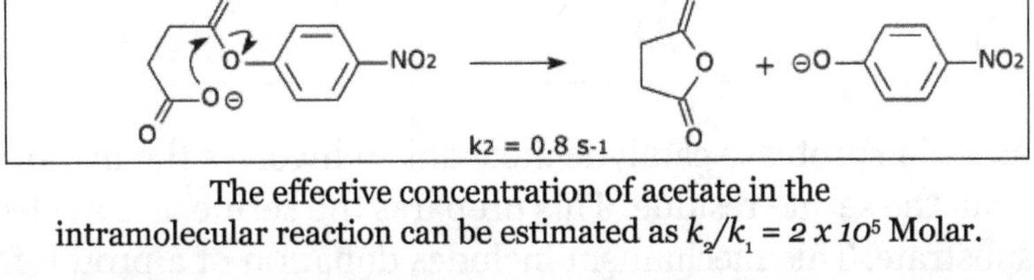

The effective concentration of acetate in the intramolecular reaction can be estimated as $k_2/k_1 = 2 \times 10^5$ Molar.

However, the situation might be more complex, since modern computational studies have established that traditional examples of proximity effects cannot be related directly to enzyme entropic effects. Also, the original entropic proposal has been found to largely overestimate the contribution of orientation entropy to catalysis.

Proton Donors or Acceptors

Proton donors and acceptors, i.e. acids and base may donate and accept protons in order to stabilize developing charges in the transition state. This is related to the overall principle of catalysis, that of reducing energy barriers, since in general transition stated are high energy states, and by stabilizing them this high energy is reduced, lowering the barrier. A key feature of enzyme catalysis over many non-biological catalysis, is that both acid and base catalysis can be combined in the same reaction. In many abiotic systems, acids (large $[H^+]$) or bases (large concentration H^+ sinks, or species with electron pairs) can increase the rate of the reaction; but of course the environamen can only have one overall pH (measure of acidity or basicity (alkalinity)). However since enzymes are large molecules, they can position both acid groups and basic groups in their active site to interact with their substrates, an meploy both modes independed of the bulk pH.

Often general acid of base catalysis is employed to activate nucleophile and/or electrophile groups, or to stabilizing leaving groups. Many amino acids with acidic or basic groups are this employed in the active site, such as the glutamic and aspartic acid, histidine, cystine, tyrosine, lysine and arginine, as well as serine and threonine . In addition the peptide backbone, with carbonyl and amide N groups is often employed. Cystine and Histidine are very commonly involved , since they both have a pKa close to neutral pH and can therefore both accept and donate protons.

Many reaction mechanisms involving acid/base catalysis assume a substantially altered pKa. This alteration of pKa is possible through the local environment of the residue.

Conditions	Acids	Bases
Hydrophobic environment	Increase pKa	Decrease pKa
Adjacent residues of like charge	Increase pKa	Decrease pKa
Salt bridge (and hydrogen bond) formation	Decrease pKa	Increase pKa

pKa can also be influenced significantly by the surrounding environment, to the extent that residues which are basic in solution may act as proton donors, and vice versa.

For example: Catalytic triad of a serine protease.

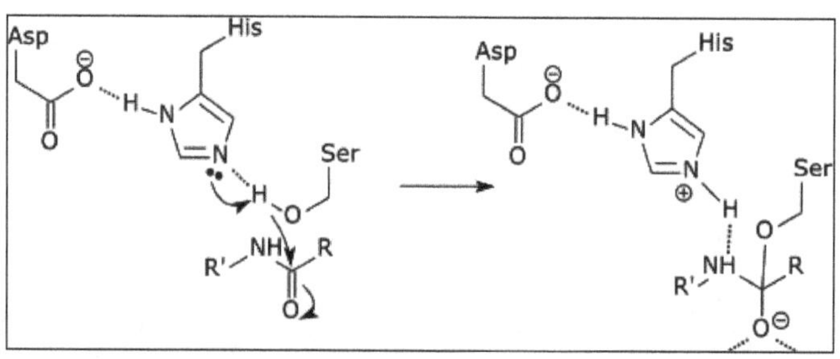

The initial step of the serine protease catalytic mechanism involves the histidine of the active site accepting a proton from the serine residue. This prepares the serine as a nucleophile to attack the amide bond of the substrate. This mechanism includes donation of a proton from serine (a base, pKa 14) to histidine (an acid, pKa 6), made possible due to the local environment of the bases.

It is important to clarify that the modification of the pKa's is a pure part of the electrostatic mechanism. Furthermore, the catalytic effect of the above example is mainly associated with the reduction of the pKa of the oxyanion and the increase in the pKa of the histidine, while the proton transfer from the serine to the histidine is not catalyzed significantly, since it is not the rate determining barrier.

Electrostatic Catalysis

Stabilization of charged transition states can also be by residues in the active site forming ionic bonds (or partial ionic charge interactions) with the intermediate. These bonds can either come from acidic or basic side chains found on amino acids such as lysine, arginine, aspartic acid or glutamic acid or come from metal cofactors such as zinc. Metal ions are particularly effective and can reduce the pKa of water enough to make it an effective nucleophile.

Systematic computer simulation studies established that electrostatic effects give, by far, the largest contribution to catalysis. In particular, it has been found that enzyme provides an environment which is more polar than water, and that the ionic transition states are stabilized by fixed dipoles. This is very different from transition state stabilization in water, where the water molecules must pay with "reorganization energy". In order to stabilize ionic and charged states. Thus, the catalysis is associated with the fact that the enzyme polar groups are preorganized

The magnitude of the electrostatic field exerted by an enzyme's active site has been shown to be highly correlated with the enzyme's catalytic rate enhancement

Binding of substrate usually excludes water from the active site, thereby lowering the local dielectric constant to that of an organic solvent. This strengthens the electrostatic interactions between the charged/polar substrates and the active sites. In addition, studies have shown that the charge distributions about the active sites are arranged so as to stabilize the transition states of the catalyzed reactions. In several enzymes, these charge distributions apparently serve to guide polar substrates toward their binding sites so that the rates of these enzymatic reactions are greater than their apparent diffusion-controlled limits.

For example: Carboxypeptidase Catalytic Mechanism.

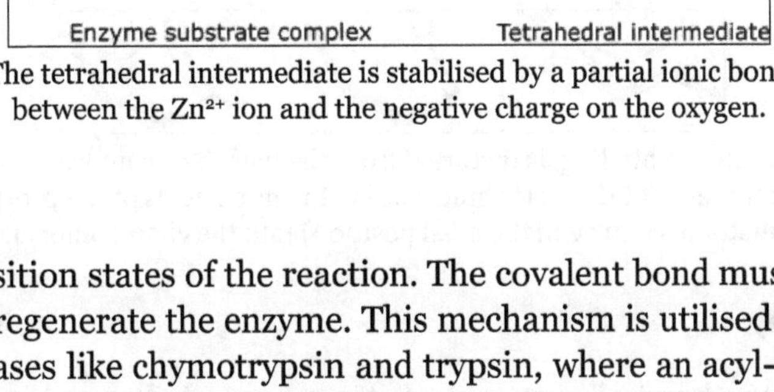

The tetrahedral intermediate is stabilised by a partial ionic bond between the Zn^{2+} ion and the negative charge on the oxygen.

Covalent Catalysis

Covalent catalysis transition states of the reaction. The covalent bond must, at a later stage in the reaction, be broken to regenerate the enzyme. This mechanism is utilised by the catalytic triad of enzymes such as proteases like chymotrypsin and trypsin, where an acyl-enzyme intermediate is formed. An alternative mechanism is schiff base formation using the free amine from a lysine residue, as seen in the enzyme aldolase during glycolysis.

Some enzymes utilize non-amino acid cofactors such as pyridoxal phosphate (PLP) or thiamine pyrophosphate (TPP) to form covalent intermediates with reactant molecules. Such covalent intermediates function to reduce the energy of later transition states, similar to how covalent intermediates formed with active site amino acid residues allow stabilization, but the capabilities of cofactors allow enzymes to carryout reactions that amino acid side residues alone could not. Enzymes utilizing such cofactors include the PLP-dependent enzyme aspartate transaminase and the TPP-dependent enzyme pyruvate dehydrogenase.

Rather than lowering the activation energy for a reaction pathway, covalent catalysis provides an alternative pathway for the reaction (via to the covalent intermediate) and so is distinct from true catalysis. For example, the energetics of the covalent bond to the serine molecule in chymotrypsin should be compared to the well-understood covalent bond to the nucleophile in the uncatalyzed solution reaction. A true proposal of a covalent catalysis (where the barrier is lower than the corresponding barrier in solution) would require, for example, a partial covalent bond to the transition state by an enzyme group (e.g., a very strong hydrogen bond), and such effects do not contribute significantly to catalysis.

Metal Ion Catalysis

A metal ion in the active site participates in catalysis by coordinating charge stabilization and shielding. Because of a metal's positive charge, only negative charges can be stabilized through metal ions. However, metal ions are advantageous in biological catalysis because they are not affected by changes in pH. Metal ions can also act to ionize water by acting as a Lewis acid. Metal ions may also be agents of oxidation and reduction.

Bond Strain

This is the principal effect of induced fit binding, where the affinity of the enzyme to the transition state is greater than to the substrate itself. This induces structural rearrangements which strain substrate bonds into a position closer to the conformation of the transition state, so lowering the energy difference between the substrate and transition state and helping catalyze the reaction.

However, the strain effect is, in fact, a ground state destabilization effect, rather than transition state stabilization effect. Furthermore, enzymes are very flexible and they cannot apply large strain effect.

In addition to bond strain in the substrate, bond strain may also be induced within the enzyme itself to activate residues in the active site.

For example: Substrate, bound substrate, and transition state conformations of lysozyme.

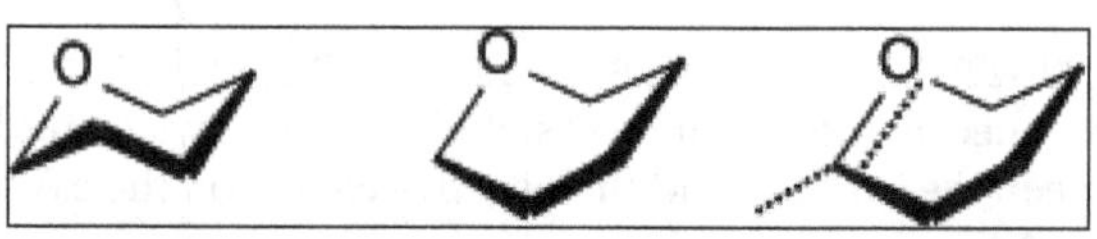

The substrate, on binding, is distorted from the half chair conformation of the hexose ring (because of the steric hindrance with amino acids of the protein forcing the equatorial c6 to be in the axial position) into the chair conformation.

Quantum Tunneling

These traditional "over the barrier" mechanisms have been challenged in some cases by models and observations of "through the barrier" mechanisms (quantum tunneling). Some enzymes operate with kinetics which are faster than what would be predicted by the classical $\Delta G^{\ddagger}$. In "through the barrier" models, a proton or an electron can tunnel through activation barriers. Quantum tunneling for protons has been observed in tryptamine oxidation by aromatic amine dehydrogenase.

Quantum tunneling does not appear to provide a major catalytic advantage, since the tunneling contributions are similar in the catalyzed and the uncatalyzed reactions in solution. However, the tunneling contribution (typically enhancing rate constants by a factor of ~1000 compared to the rate of reaction for the classical 'over the barrier' route) is likely crucial to the viability of biological organisms. This emphasizes the general importance of tunneling reactions in biology.

In 1971-1972 the first quantum-mechanical model of enzyme catalysis was formulated.

Active Enzyme

The binding energy of the enzyme-substrate complex cannot be considered as an external energy which is necessary for the substrate activation. The enzyme of high energy content may firstly transfer some specific energetic group X_1 from catalytic site of the enzyme to the final place of the first bound reactant, then another group X_2 from the second bound reactant (or from the second group of the single reactant) must be transferred to active site to finish substrate conversion to product and enzyme regeneration.

We can present the whole enzymatic reaction as a two coupling reactions:

$$S_1 + EX_1 \rightarrow S_1EX_1 \rightarrow P_1 + EP_2$$

$$S_2 + EP_2 \rightarrow S_2EP_2 \rightarrow P_2 + EX_2$$

It may be seen from reaction ($S_1 + EX_1 \rightarrow S_1EX_1 \rightarrow P_1 + EP_2$) that the group X_1 of the active enzyme appears in the product due to possibility of the exchange reaction inside enzyme to avoid both electrostatic inhibition and repulsion of atoms. So we represent the active enzyme as a powerful reactant of the enzymatic reaction. The reaction ($S_2 + EP_2 \rightarrow S_2EP_2 \rightarrow P_2 + EX_2$) shows incomplete conversion of the substrate because its group X_2 remains inside enzyme. This approach as idea had formerly proposed relying on the hypothetical extremely high enzymatic conversions (catalytically perfect enzyme).

The crucial point for the verification of the present approach is that the catalyst must be a complex of the enzyme with the transfer group of the reaction. This chemical aspect is supported by the well-studied mechanisms of the several enzymatic reactions. Consider the reaction of peptide bond hydrolysis catalyzed by a pure protein α-chymotrypsin (an enzyme acting without a cofactor), which is a well-studied member of the serine proteases family, see.

We present the experimental results for this reaction as two chemical steps:

$$S_1 + EH \rightarrow P_1 + EP_2$$

$$EP_2 + H\text{-}O\text{-}H \rightarrow EH + P_2$$

where S_1 is a polypeptide, P_1 and P_2 are products. The first chemical step ($S_1 + EH \rightarrow P_1 + EP_2$) includes the formation of a covalent acyl-enzyme intermediate. The second step ($EP_2 + H\text{-}O\text{-}H \rightarrow EH + P_2$) is the deacylation step. It is important to note that the group H+, initially found on the enzyme, but not in water, appears in the product before the step of hydrolysis, therefore it may be considered as an additional group of the enzymatic reaction.

Thus, the reaction ($S_1 + EH \rightarrow P_1 + EP_2$) shows that the enzyme acts as a powerful reactant of the reaction. According to the proposed concept, the H transport from the enzyme promotes the first reactant conversion, breakdown of the first initial chemical bond (between groups P_1 and P_2). The step of hydrolysis leads to a breakdown of the second chemical bond and regeneration of the enzyme.

The proposed chemical mechanism does not depend on the concentration of the substrates or products in the medium. However, a shift in their concentration mainly causes free energy changes in the first and final steps of the reactions ($S_1 + EX_1 \rightarrow S_1EX_1 \rightarrow P_1 + EP_2$) and ($S_2 + EP_2 \rightarrow S_2EP_2 \rightarrow P_2 + EX_2$) due to the changes in the free energy content of every molecule, whether S or P, in water solution. This approach is in accordance with the following mechanism of muscle contraction. The final step of ATP hydrolysis in skeletal muscle is the product release caused by the association of myosin heads with actin. The closing of the actin-binding cleft during the association reaction is structurally coupled with the opening of the nucleotide-binding pocket on the myosin active site.

Notably, the final steps of ATP hydrolysis include the fast release of phosphate and the slow release of ADP. The release of a phosphate anion from bound ADP anion into water solution may be considered as an exergonic reaction because the phosphate anion has low molecular mass.

Thus, we arrive at the conclusion that the primary release of the inorganic phosphate $H_2PO_4^-$ leads to transformation of a significant part of the free energy of ATP hydrolysis into the kinetic energy of the solvated phosphate, producing active streaming. This assumption of a local mechano-chemical transduction is in accord with Tirosh's mechanism of muscle contraction, where the muscle force derives from an integrated action of active streaming created by ATP hydrolysis.

Examples of Catalytic Mechanisms

In reality, most enzyme mechanisms involve a combination of several different types of catalysis.

Triose Phosphate Isomerase

Triose phosphate isomerase catalyses the reversible interconvertion of the two triose phosphates isomers dihydroxyacetone phosphate and D-glyceraldehyde 3-phosphate.

Trypsin

Trypsin is a serine protease that cleaves protein substrates after lysine or arginine residues using a catalytic triad to perform covalent catalysis, and an oxyanion hole to stabilise charge-buildup on the transition states.

Aldolase

Aldolase catalyses the breakdown of fructose 1,6-bisphosphate (F-1,6-BP) into glyceraldehyde 3-phosphate and dihydroxyacetone phosphate (DHAP).

References

- Enzyme: britannica.com, Retrieved 05 August, 2019
- Enzyme-inhibition, enzyme-activity: teachmephysiology.com, Retrieved 25 June, 2019
- Enzyme-kinetics: rose-hulman.edu, Retrieved 28 April, 2019
- Bu Z, Callaway DJ (2011). Proteins move! Protein dynamics and long-range allostery in cell signaling. Advances in Protein Chemistry and Structural Biology. 83. pp. 163–221. doi:10.1016/B978-0-12-381262-9.00005-7. ISBN 9780123812629. PMID 21570668

4

Metabolism and Metabolic Pathway

The set of chemical reactions in organisms that are essential for life and provide energy for vital processes is referred to as metabolism. It includes the conversion of food to energy, proteins, lipids, carbohydrates and nucleic acids. All these different processes of metabolism and metabolic pathway have been carefully analyzed in this chapter.

Metabolism

Metabolism is the sum of the chemical reactions that take place within each cell of a living organism and that provide energy for vital processes and for synthesizing new organic material.

Living organisms are unique in that they can extract energy from their environments and use it to carry out activities such as movement, growth and development, and reproduction. But how do living organisms—or, their cells—extract energy from their environments, and how do cells use this energy to synthesize and assemble the components from which the cells are made?

The answers to these questions lie in the enzyme-mediated chemical reactions that take place in living matter (metabolism). Hundreds of coordinated, multistep reactions, fueled by energy obtained from nutrients and/or solar energy, ultimately convert readily available materials into the molecules required for growth and maintenance.

The Unity of Life

At the cellular level of organization, the main chemical processes of all living matter are similar, if not identical. This is true for animals, plants, fungi, or bacteria; where variations occur (such as, for example, in the secretion of antibodies by some molds), the variant processes are but variations on common themes. Thus, all living matter is made up of large molecules called proteins, which provide support and coordinated movement, as well as storage and transport of small molecules, and, as catalysts, enable chemical reactions to take place rapidly and specifically under mild temperature, relatively low concentration, and neutral conditions (i.e., neither acidic nor basic). Proteins are assembled from some 20 amino acids, and, just as the 26 letters of the alphabet can be assembled in specific ways to form words of various lengths and meanings, so may tens or even hundreds of the 20 amino-acid "letters" be joined to form specific proteins. Moreover, those portions of protein molecules involved in performing similar functions in different organisms often comprise the same sequences of amino acids.

There is the same unity among cells of all types in the manner in which living organisms preserve their individuality and transmit it to their offspring. For example, hereditary information is encoded in a specific sequence of bases that make up the DNA(deoxyribonucleic acid) molecule in the nucleus of each cell. Only four bases are used in synthesizing DNA: adenine, guanine, cytosine, and thymine. Just as the Morse Code consists of three simple signals—a dash, a dot, and a space—the precise arrangement of which suffices to convey coded messages, so the precise arrangement of the bases in DNA contains and conveys the information for the synthesis and assembly of cell components. Some primitive life-forms, however, use RNA (ribonucleic acid; a nucleic acid differing from DNA in containing the sugar ribose instead of the sugar deoxyribose and the base uracil instead of the base thymine) in place of DNA as a primary carrier of genetic information. The replication of the genetic material in these organisms must, however, pass through a DNA phase. With minor exceptions, the genetic code used by all living organisms is the same.

The chemical reactions that take place in living cells are similar as well. Green plants use the energy of sunlight to convert water (H_2O) and carbon dioxide (CO_2) to carbohydrates (sugars and starches), other organic (carbon-containing) compounds, and molecular oxygen (O_2). The process of photosynthesis requires energy, in the form of sunlight, to split one water molecule into one-half of an oxygen molecule (O_2; the oxidizing agent) and two hydrogen atoms (H; the reducing agent), each of which dissociates to one hydrogen ion (H^+) and one electron. Through a series of oxidation-reduction reactions, electrons (denoted e^-) are transferred from a donating molecule (oxidation), in this case water, to an accepting molecule (reduction) by a series of chemical reactions; this "reducing power" may be coupled ultimately to the reduction of carbon dioxide to the level of carbohydrate. In effect, carbon dioxide accepts and bonds with hydrogen, forming carbohydrates ($Cn[H_2O]n$).

Living organisms that require oxygen reverse this process: they consume carbohydrates and other organic materials, using oxygen synthesized by plants to form water, carbon dioxide, and energy. The process that removes hydrogen atoms (containing electrons) from the carbohydrates and passes them to the oxygen is an energy-yielding series of reactions.

In plants, all but two of the steps in the process that converts carbon dioxide to carbohydrates are the same as those steps that synthesize sugars from simpler starting materials in animals, fungi, and bacteria. Similarly, the series of reactions that take a given starting material and synthesize certain molecules that will be used in other synthetic pathways are similar, or identical, among all cell types. From a metabolic point of view, the cellular processes that take place in a lion are only marginally different from those that take place in a dandelion.

Biological Energy Exchanges

The energy changes associated with physicochemical processes are the province of thermodynamics, a subdiscipline of physics. The first two laws of thermodynamics state, in essence, that energy can be neither created nor destroyed and that the effect of physical and chemical changes is to increase the disorder, or randomness (i.e., entropy), of the universe. Although it might be supposed that biological processes—through which organisms grow in a highly ordered and complex manner, maintain order and complexity throughout their life, and pass on the instructions for order to succeeding generations—are in contravention of these laws, this is not so. Living organisms neither consume nor create energy: they can only transform it from one form to another. From the environment they absorb energy in a form useful to them; to the environment they return an equivalent amount of energy in a biologically less useful form. The useful energy, or free energy, may be defined as energy capable of doing work under isothermal conditions (conditions in which no temperature differential exists); free energy is associated with any chemical change. Energy less useful than free energy is returned to the environment, usually as heat. Heat cannot perform work in biological systems because all parts of cells have essentially the same temperature and pressure.

The Carrier of Chemical Energy

At any given time, a neutral molecule of water dissociates into a hydrogen ion (H^+) and a hydroxide ion (OH^-), and the ions are continually re-forming into the neutral molecule. Under normal conditions (neutrality), the concentration of hydrogen ions (acidic ions) is equal to that of the hydroxide ions (basic ions); each are at a concentration of 10^{-7} mole per litre, which is described as a pH of 7.

All cells either are bounded by membranes or contain organelles that have membranes. These membranes do not permit water or the ions derived from water to pass into or out of the cells or organelles. In green plants, sunlight is absorbed by chlorophyll and other pigments in the chloroplasts of the cells, called photosystem II. As shown previously, when a water molecule is split by light energy, one-half of an oxygen molecule and two hydrogen atoms (which dissociate to two electrons and two hydrogen ions, H^+) are formed. When excited by sunlight, chlorophyll loses one electron to an electron carrier molecule but quickly recovers it from a hydrogen atom of the split water molecule, which sends H^+ into solution in the process. Two oxygen atoms come together to form a molecule of oxygen gas (O_2). The free electrons are passed to photosystem I, but, in doing so, an excess concentration of positively charged hydrogen ions (H^+) appears on one side of the membrane in the chloroplast, whereas an excess of negatively charged hydroxide ions (OH^-) builds up on the other side. The free energy released as H^+ ions move through a specific "pore" in the membrane, to equalize the concentrations of ions, is sufficient to make some biological processes work, such as the uptake of certain nutrients by bacteria and the rotation of the whiplike protein-based propellers that enable such bacteria to move. Equally important, however, is that this gradient across the membrane powers the formation of adenosine triphosphate (ATP) from inorganic phosphate (HPO_4^{2-}, abbreviated P*i*) and adenosine diphosphate (ADP). ATP is the major carrier of biologically utilizable energy in all forms of living matter. The interrelationships of energy-yielding and energy-requiring metabolic reactions may be considered largely as processes that couple the formation of ATP with its breakdown.

Synthesis of ATP by green plants is similar to the synthesis of ATP that takes place in the mitochondria of animal, plant, and fungus cells, and in the plasma membranes of bacteria that use oxygen (or other inorganic electron acceptors, such as nitrate) to accept electrons from the removal of hydrogen atoms from a molecule of food. Through these processes most of the energy stored in food materials is released and converted into the molecules that fuel life processes. It must also be remembered, however, that many living organisms (usually bacteria and protozoa) cannot tolerate oxygen; they form ATP from inorganic phosphate and ADP by substrate-level phosphorylations (the addition of a phosphate group) that do not involve the establishment and collapse of proton gradients across membranes. It must also be borne in mind that the fuels of life and the cellular "furnace" in which they are "burned" are made of the same types of material: if the fires burn too brightly, not only the fuel but also the furnace is consumed. It is therefore essential to release energy at small, discrete, readily utilizable intervals. The relative complexity of the catabolic pathways (by which food materials are broken down) and the complexity of the anabolic pathways (by which cell components are synthesized) reflect this need and offer the possibility for simple feedback systems to control the rate at which materials travel along these sequences of enzymic reactions.

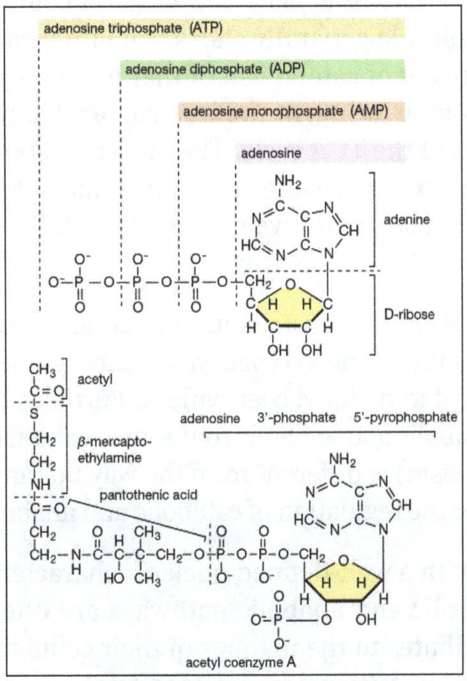

Biological energy carriers.

Catabolism

The release of chemical energy from food materials essentially occurs in three phases. In the first phase (phase I), the large molecules that make up the bulk of food materials are broken down into small constituent units: proteins are converted to the 20 or so different amino acids of which they are composed; carbohydrates (polysaccharides such as starch in plants and glycogen in animals) are degraded to sugars such as glucose; and fats (lipids) are broken down into fatty acids and glycerol. The amounts of energy liberated in phase I are relatively small: only about 0.6 percent of the free, or useful, energy of proteins and carbohydrates, and about 0.1 percent of that of fats, is released during this phase. Because this energy is liberated largely as heat, it cannot be used by the cell. The purpose of the reactions of phase I, which can be grouped under the term digestion and which, in animals, occur mainly in the intestinal tract and in tissues in which reserve materials are prepared, or mobilized, for energy production, is to prepare the foodstuffs for the energy-releasing processes.

Anabolism

Catabolic pathways effect the transformation of food materials into interconvertible intermediates. Anabolic pathways, on the other hand, are sequences of enzyme-catalyzed reactions in which the component building blocks of large molecules, or macromolecules (e.g., proteins, carbohydrates, and fats), are constructed from the same intermediates. Thus, catabolic routes have clearly defined beginnings but no unambiguously identifiable end products; anabolic routes, on the other hand, lead to clearly distinguishable end products from diffuse beginnings. The two types of pathway are linked through reactions of phosphate transfer, involving ADP, AMP, and ATP, and also through electron transfers, which enable reducing equivalents (i.e., hydrogen atoms or electrons), which have been released during catabolic reactions, to be utilized for biosynthesis. But, although catabolic and anabolic pathways are closely linked, and although the overall effect of one type of route is obviously the opposite of the other, they have few steps in common. The anabolic pathway for the synthesis of a particular molecule generally starts from intermediate compounds quite different from those produced as a result of catabolism of that molecule; for example, microorganisms catabolize aromatic (i.e., containing a ring, or cyclic, structure) amino acids to acetyl coenzyme A and an intermediate compound of the TCA cycle. The biosynthesis of these amino acids, however, starts with a compound derived from pyruvate and an intermediate compound of the metabolism of pentose (a general name for sugars with five carbon atoms). Similarly, histidine is synthesized from a pentose sugar but is catabolized to α-oxoglutarate.

Even in cases in which a product of catabolism is used in an anabolic pathway, differences emerge; thus, fatty acids, which are catabolized to acetyl coenzyme A, are synthesized not from acetyl coenzyme A directly but from a derivative of it, malonyl coenzyme A. Furthermore, even enzymes that catalyze apparently identical steps in catabolic and anabolic routes may exhibit different properties. In general, therefore, the way down (catabolism) is different from the way up (anabolism). These differences are important because they allow for the regulation of catabolic and anabolic processes in the cell.

In eukaryotic cells (i.e., those with a well-defined nucleus, characteristic of organisms higher than bacteria) the enzymes of catabolic and anabolic pathways are often located in different cellular compartments. This also contributes to the manner of their cellular control; for example, the formation of acetyl coenzyme A from fatty acids, referred to above, occurs in animal cells in small sausage-shaped components, or organelles, called mitochondria, which also contain the enzymes for terminal respiration and for oxidative phosphorylation. The biosynthesis of fatty acids from acetyl coenzyme A, on the other hand, occurs in the cytoplasm.

Metabolic Pathway

The processes of making and breaking down carbohydrate molecules illustrate two types of metabolic pathways. A metabolic pathway is a step-by-step series of interconnected biochemical reactions that convert a substrate molecule or molecules through a series of metabolic intermediates, eventually yielding a final product or products. For example, one metabolic pathway for carbohydrates breaks large molecules down into glucose. Another metabolic pathway might build glucose into large carbohydrate molecules for storage. The first of these processes requires energy and is referred to as anabolic. The second process produces energy and is referred to as catabolic. Consequently, metabolism is composed of these two opposite pathways:

1. Anabolism (building molecules)
2. Catabolism (breaking down molecules)

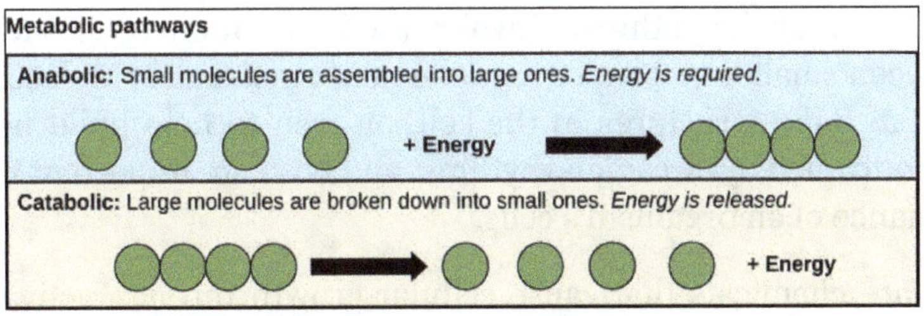

Anabolic and catabolic pathways: Anabolic pathways are those that require energy to synthesize larger molecules. Catabolic pathways are those that generate energy by breaking down larger molecules. Both types of pathways are required for maintaining the cell's energy balance.

Anabolic Pathways

Anabolic pathways require an input of energy to synthesize complex molecules from simpler ones. One example of an anabolic pathway is the synthesis of sugar from CO2. Other examples include the synthesis of large proteins from amino acid building blocks and the synthesis of new DNA strands from nucleic acid building blocks. These processes are critical to the life of the cell, take place constantly, and demand energy provided by ATP and other high-energy molecules like NADH (nicotinamide adenine dinucleotide) and NADPH.

Catabolic Pathways

Catabolic pathways involve the degradation of complex molecules into simpler ones, releasing the chemical energy stored in the bonds of those molecules. Some catabolic pathways can capture that energy to produce ATP, the molecule used to power all cellular processes. Other energy-storing molecules, such as lipids, are also broken down through similar catabolic reactions to release energy and make ATP.

Importance of Enzymes

Chemical reactions in metabolic pathways rarely take place spontaneously. Each reaction step is facilitated, or catalyzed, by a protein called an enzyme. Enzymes are important for catalyzing all types of biological reactions: those that require energy as well as those that release energy.

Anabolism

Anabolism collectively refers to all the processes of chemical reactions that build larger molecules out of smaller molecules or atoms; these processes are also known as anabolic processes or anabolic pathways. Anabolism and catabolism are the two types of metabolic pathways. Metabolic pathways are series of chemical reactions that take place in the cell. Anabolic pathways use energy, while catabolic pathways release energy.

Function of Anabolism

Anabolic pathways in a cell lead to the formation of larger, more complex molecules from smaller ones. First, catabolic pathways are needed to break down nutrient molecules from food into small building blocks. These smaller molecules are then joined together to make different, bigger molecules called macromolecules. Anabolic pathways involve the input of energy, which is needed to form chemical bonds between smaller molecules to make macromolecules. Molecules built from anabolism are then used to build structures in the cell, or even to help build new cells. Anabolism and catabolism are controlled by circadian rhythms, and both are important for the development, growth, and maintenance of an organism's cells.

Anabolic hormones are chemicals that cause cellular growth through activating anabolic pathways. Two examples of anabolic hormones are testosterone and insulin. Hormones can also be produced artificially in the lab; this is how anabolic steroids are created.

Examples of Anabolic Processes

Protein Synthesis

Proteins are macromolecules that carry out cellular activities encoded by an organism's genes. They have many different functions in the body, including DNA replication, aiding chemical reactions (as enzymes), transporting materials in the cell, cell growth and signaling, and providing physical structure. Each cell in the human body contains about 1 to 3 billion proteins.

Proteins are synthesized from smaller molecules called amino acids in a cell's ribosomes. Each protein is a chain of a specific sequence of amino acids. Since proteins are larger molecules put together from smaller ones, the process of protein synthesis is anabolic.

DNA Synthesis

Deoxyribonucleic acid, or DNA, is an organism's genetic material. It is macromolecule made up of smaller molecules called nucleic acids, which are themselves made up of a nucleotide base attached to a deoxyribose sugar and a phosphate molecule. DNA synthesis is an anabolic process that takes place in a cell's nucleus just before the cell divides. It involves unzipping the double strands of DNA and attaching new matching nucleotides to each half of the strand that has been unzipped, forming two new strands that each contain half of the old DNA strand.

Growth of Bones and Muscles

On a larger scale, the growth of body parts such as bones and muscles is anabolic. Bone growth, or ossification, occurs when bone is formed from cells called osteoclasts. It is then mineralized through cells called osteoblasts. This process is also anabolic; during mineralization, osteoblasts produce calcium phosphate crystals that are incorporated into the bone's structure, making the bones hard and sturdy.

Muscle growth, also called muscle hypertrophy, occurs when the cells of skeletal muscles, called myocytes, increase in size. It occurs through strength training exercises such as lifting weights. Factors like sex, age, and diet all affect hypertrophy. During hypertrophy, there is increased protein synthesis of actin and myosin, and the volume of sarcoplasmic fluid in the myocyte increases.

Anabolic Steroids

Anabolic steroids are androgenic hormones that are natural (e.g. testosterone) or produced synthetically and mimic the effects of testosterone. Androgens are anabolic; they increase protein within muscle cells, which causes an increase in muscle mass. Anabolic steroids are used to treat certain medical conditions like delayed puberty in boys, and they can also be used for muscle growth, to stimulate the appetite, to produce secondary sex characteristics in transgender males, and to help treat certain conditions that may cause a loss in muscle mass, such as cancer and AIDS.

However, sometimes anabolic steroids are also abused by athletes who want to quickly gain more muscle. Most major sports organizations ban doping, the use of anabolic steroids or other performance-enhancing drugs, in order to discourage steroid abuse. Although using steroids can increase muscle mass, it also has many adverse side effects. These range from acne to high blood pressure to aggression ("'roid rage"), and even psychosis. Steroids can affect the structure of the heart, which could lead to congestive heart failure or a heart attack. They also have adverse effects on cholesterol levels, may stunt height, and can lead to liver damage in high doses.

In females, steroids can cause masculinization, which is the development of male secondary sex characteristics such as a deeper voice, increases in body hair, an enlarged clitoris (the clitoris in females is homologous to the penis in males), and temporary disruption of the menstrual cycle. In males, however, steroids can cause feminization, including the development of breast tissue and decrease in testicle size. This occurs because testosterone can be converted to estradiol, a female sex hormone. The effects of masculinization and feminization are generally reduced/reversed when a person stops taking steroids and their hormone levels return to normal.

These bottles obtained from a drug raid contain anabolic steroid capsules.

Photosynthetic Carbohydrate Synthesis

The synthesis of carbohydrates in animal cells always employs precursors having at least three carbons, all at an oxidation state lower than that of carbon in CO_2. Photosynthetic organisms, by contrast, can make carbohydrates from CO_2 and water. They synthesize glucose, sucrose, and other carbohydrates by reducing CO_2 at the expense of energy furnished by the ATP and NADPH generated in photosynthetic electron transfer. This process represents a fundamental difference between autotrophic (phototrophic or chemotrophic) and heterotrophic organisms. Autotrophs can use CO_2 as the sole source of all the carbon atoms required for biosynthesis not only of cellulose and starch, but also of the lipids and proteins and all of the many other organic components of plant cells. By contrast, heterotrophic organisms in general are unable to bring about the net reduction of CO_2 to form "new" glucose in any significant amounts. Carbon dioxide can be taken up by animal tissues, as in the pyruvate carboxylase reaction during gluconeogenesis, but the CO_2 molecule incorporated into oxaloacetate is lost in a subsequent reaction step. Similarly, the CO_2 taken up by acetyl-CoA carboxylase during fatty acid synthesis in animal tissues or by carbamoyl phosphate synthetase I during urea formation is lost in later steps.

Melvin Calvin

Green plants contain in their chloroplasts unique enzymatic machinery to catalyze the conversion of CO_2 into simple (reduced) organic compounds, a process called CO_2 fixation, or carbon fixation. Plants convert these simple products of photosynthesis into more complex biomolecules, including sugars, polysaccharides, and metabolites derived from them, using metabolic pathways similar to those in animals. The reactions that result in CO_2 fixation make up a cyclic pathway in which key intermediates are constantly regenerated. The pathway was elucidated in the early 1950s by Melvin Calvin and coworkers, and is often called the Calvin cycle.

Carbon Dioxide Fixation Occurs in Three Stages

The first stage in the fixation of CO_2 into organic linkage is its condensation with a five-carbon acceptor, ribulose-1,5-bisphosphate, to form two molecules of 3-phosphoglycerate. Note that figure shows the number of molecules reacting to give net formation of one molecule of triose-this takes three molecules of CO_2. In the second stage the 3-phosphoglycerate is reduced to glyceraldehyde-3-phosphate: three molecules of CO_2 are fixed to three molecules of ribulose-1,5-bisphosphate to form six molecules of glyceraldehyde-3phosphate (18 carbons). One molecule of this triose phosphate (three carbons) can either be used for energy production via glycolysis and the citric acid cycle, or condensed to hexose phosphates to be used in the synthesis of starch or sucrose. In the third stage, five of the six molecules of glyceraldehyde-3-phosphate (15 carbons) are used to regenerate three molecules of ribulose-1,5-bisphosphate, the starting material.

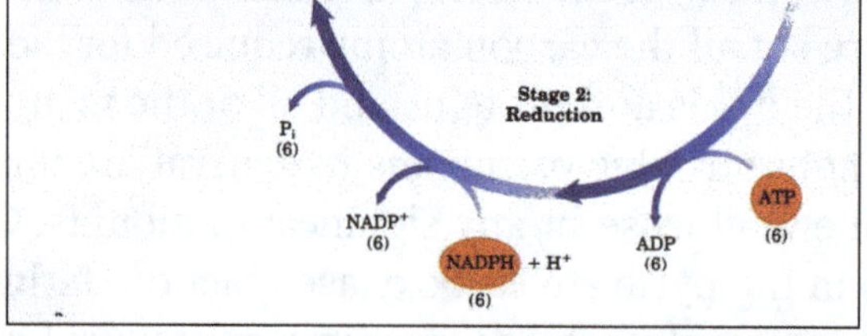

The three stages of CO_2 fixation in photosynthetic organisms. Stoichiometries of three key intermediates are shown tnumbers in parentheses) so that the fate of carbon atoms entering and leaving the cycle is apparent. As shown here, three CO_2 are fixed to permit the net synthesis of one molecule of glyceraldehyde-3-phosphate. This is called the photosynthetic carbon reduction cycle or the Calvin cycle.

Thus the overall process is cyclical and allows the continuous conversion of CO_2 into triose and hexose phosphates. Fructose-6-phosphate is a key intermediate in stage 3. The pathway from hexose phosphate to pentose bisphosphate involves the same reactions used in animal cells for the conversion of pentose phosphates to hexose phosphates during the operation of the pentose phosphate pathway, an alternative route for glucose oxidation. In photosynthetic fixation of CO_2, this pathway operates in the opposite direction, converting hexose phosphates into pentose phosphates.

Stage 1. Fixation of CO_2 into 3-Phosphoglycerate: An important clue to the nature of the CO_2 fixation mechanisms in photosynthetic organisms first came in the late 1940s when Calvin and his associates illuminated a suspension of green algae in the presence of radioactive carbon dioxide ($^{14}CO_2$) for only a few seconds. They quickly killed the cells, extracted their contents, and with the help of chromatographic methods searched for the metabolites in which the labeled carbon first appeared. The first compound that became labeled was 3-phosphoglycerate, with the ^{14}C predominantly located in the carboxyl carbon atom. This atom does not become labeled rapidly in animal tissues in the presence of radioactive CO_2. These experiments strongly suggested that 3-phosphoglycerate is an early intermediate in photosynthesis. The enzyme in green leaf extracts that catalyzes incorporation of CO_2 into organic form is ribulose-1,5-bisphosphate carboxylase or RuBP carboxylase/oxygenase (often called rubisco for short). As a carboxylase, rubisco catalyzes the covalent attachment of CO_2 to the five-carbon sugar ribulose-1,5-bisphosphate and the cleavage of the unstable six-carbon intermediate to form two molecules of 3-phosphoglycerate, one of which bears the new carbon introduced as CO_2 in its carboxyl group. The rubisco of plants (Mr 550,000) has a complex structure. There are eight large subunits (each of Mr 56,000), each containing an active site, and eight small subunits (each of Mr 14,000), whose function is not well understood. The subunit structure of rubisco of photosynthetic bacteria is quite different, with two subunits that resemble the large subunits of the plant enzyme in many respects. The plant enzyme is located in the chloroplast stroma, where it makes up about 50% of the total chloroplast protein. Rubisco, which does not occur in animals, is the most abundant enzyme in the biosphere; it is the key enzyme in the production of biomass from CO_2.

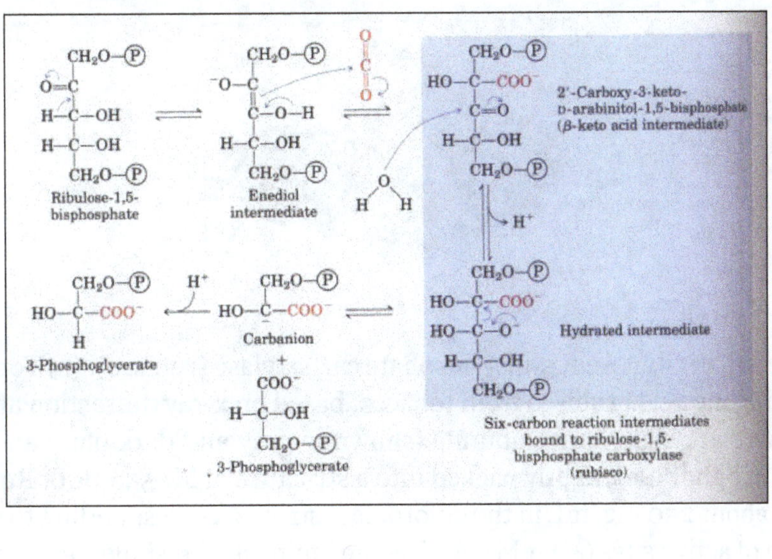

The first stage of CO_2 fixation is the reaction catalyzed by ribulose-1,5-bisphosphate carboxylase (rubisco). The carboxylated reaction intermediate is believed to be the enzyme-bound sixcarbon β-keto acid shown here, which is hydrolyzed and released from the enzyme surface as two identical three-carbon products, one of which contains the carbon atom from CO_2 (red).

Rubisco is subject to regulation by the covalent addition of a molecule of CO_2 to the e-amino group of a certain Lys residue to form a carbamate. This carbamate binds to a Mg^{2+} ion; the resulting complex is essential for activity, as we shall see later and may function directly in catalysis. Note that this is not the same molecule of CO_2 that is attached to ribulose-1,5-bisphosphate in the reaction catalyzed by this enzyme.

Stage 2. Conuersion of 3-Phosphoglycerate to Glyceraldehyde-3-Phosphate: 3-Phosphoglycerate is converted into glyceraldehyde-3-phosphate in two steps that are essentially the reversal of the corresponding steps in glycolysis, with one exception: the nucleotide cofactor for the reduction of 1,3-bisphosphoglycerate is NADPH, not NADH. The chloroplast stroma contains the full complement ofglycolytic enzymes. These stromal enzymes are isozymes of those found in the cytosol; both sets of enzymes catalyze the same reactions, but they are products of different genes.

In the first step of the sequence, 3-phosphoglycerate kinase in the stroma catalyzes the transfer of phosphate from ATP to 3-phosphoglycerate, yielding 1,3-bisphosphoglycerate. Then NADPH donates electrons in a reduction catalyzed by glyceraldehyde-3phosphate dehydrogenase, producing glyceraldehyde-3-phosphate. In addition to its role as an intermediate in CO_2 fixation, glyceraldehyde-3-phosphate has several possible fates in the plant cell. It may be oxidized via glycolysis for energy production or used for hexose synthesis.

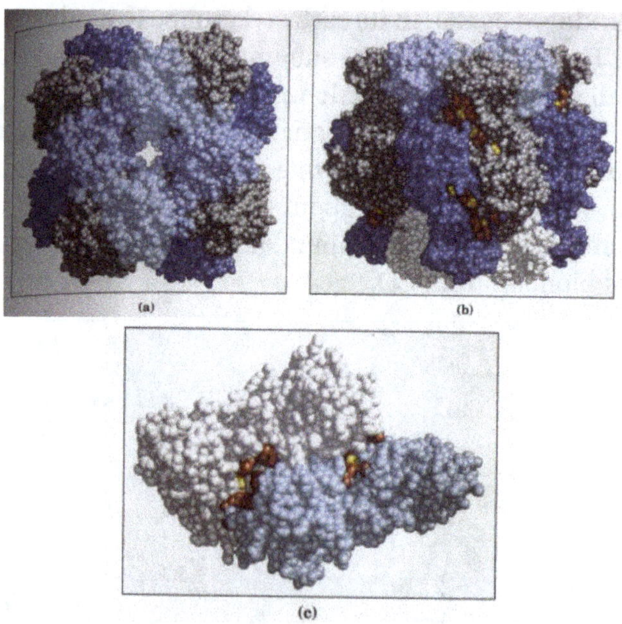

Structure and function of ribulose 1, 5-bisphosphate carboxylase (rubisco). (a) Top view and (b) side view of a space-filling model of rubisco from tobacco, based on x-ray diffraction analysis of the crystalline enzyme. There are eight large subunits (shown in gray and dark blue) and eight small ones (white and shades of light blue), tightly packed into a structure of Mr 550,000. Rubisco is present at a concentration of about 250 mg/mL in the chloroplast stroma, corresponding to an extraordinarily high concentration of active sites (~4 mM). Active-site amino acid residues are shown in red. Sulfate molecules bound to the active site in the crystal structure (these are not normal substrates) are in yellow. (c) Space-filling model of rubisco from the bacterium Rhodospirillum rubrum. The subunits are shown in white and light blue. Amino acid residues at the active site are in red. A Lys residue at the active site, which is carboxylated to form a carbamate in the active enzyme, is shown in yellow.

Stage 3: Regeneration of Ribulose-1,5-Bisphosphate from Triose Phosphates As we have seen, the first reaction in the fixation of CO_2 into triose phosphates consumes ribulose- 1,5- bisphosphate. For continuous flow of CO_2 into carbohydrate, ribulose-1,5 - bisphosphate must be constantly regenerated. Plant cells solve this problem with a series of reactions that, together with stages 1 and 2 form a cyclic pathway. By this pathway, the product of the first reaction (3-phosphoglycerate) passes through a series of transformations that eventually lead to the regeneration of the starting material, ribulose-1,5-bisphosphate.

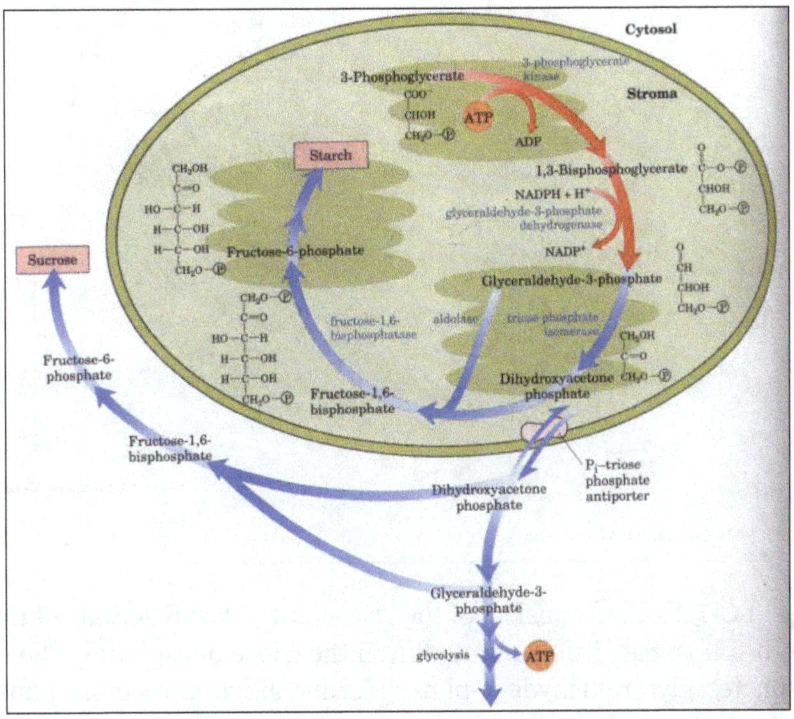

The second phase of CO_2 fixation is the conversion of 3-phosphoglycerate into glyceraldehyde-3-phosphate (red arrows). Alternative fates of the fixed carbon of glyceraldehyde-3-phosphate are also shown (blue arrows). Most is recycled to form ribulose-1,5-bisphosphate as shown in figure. The "extra" glyceraldehyde-3-phosphate may be used immediately as a source of energy, converted to sucrose for transport, or stored as starch for future use. If glyceraldehyde-3-phosphate is needed for starch synthesis, it condenses with dihydroxyacetone phosphate in the stroma and is converted to fructose-6-phosphate, a precursor of starch. In other situations it is converted to dihydroxyacetone phosphate, which leaves the chloroplast via a specific transporter. In the cytosol, dihydroxyacetone phosphate can be degraded via glycolysis to provide energy, or used to form fructose-6-phosphate and hence sucrose.

The regeneration of ribulose-1,5-bisphosphate involves rearrangements of the carbon skeletons of glyceraldehyde-3-phosphate and dihydroxyacetone phosphate produced in the first two stages of carbon fixation. The intermediates in the pathway include three-, four-, five-, six-, and seven-carbon sugars. In the following discussion, all step numbers refer to figure.

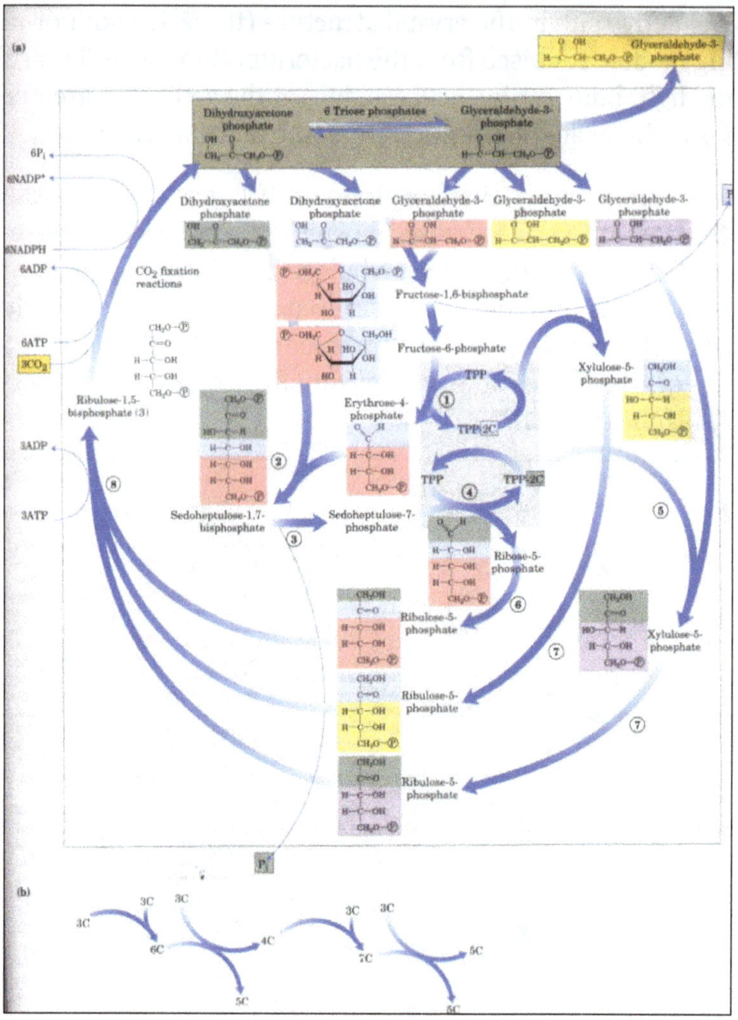

(a) The third stage of CO_2 fixation consists of the remaining set of reactions of the Calvin cycle, in which ribulose-1,5-bisphosphate is regenerated from the triose phosphates. The starting materials are the triose phosphates: glyceraldehyde-3-phosphate and dihydroxyacetone phosphate. Reactions catalyzed b aldolase (step 2) and transketolase (steps 1 , 4 and 5) produce pentose phosphates, all of which are eventually converted to ribulose-1 5-bisphosphate. The two-carbon group carried by TPP in steps l and 4(TPP-2C) is a ketol group: CH_2OH-CO-. The individual reactions are described in the text. (b) A simplified schematic diagram showing the interconversions of triose phosphates (three-carbon (3C) compounds) into pentose phosphates (five-carbon (5C) compounds).

Step l is catalyzed by the enzyme transketolase, which contains thiamine pyrophosphate (TPP) as its prosthetic group and requires Mg^{2+}. Transketolase catalyzes the reversible transfer of a ketol (CH_2OH-CO-) group from a ketose phosphate donor, fructose6-phosphate, to an aldose phosphate acceptor, glyceraldehyde-3-phosphate. The products are xylulose-5-phosphate (a pentose) and the four-carbon sugar erythrose-4-phosphate.

Step 2 is promoted by an aldolase similar to that which acts in glycolysis; it catalyzes the reversible condensation of an aldehyde, erythrose-4-phosphate, with dihydroxyacetone phosphate, yielding the seven-carbon sedoheptulose-1,7-bisphosphate. After removal of the C-1 phosphate group by sedoheptulose-1,7-bisphosphatase (step 3), the product sedoheptulose-7-phosphate is split by transketolase (step 4 into a pentose phosphate (ribose-5-phosphate) and a two-carbon fragment, carried on TPP. This two-carbon fragment is condensed with the three carbons of glyceraldehyde-3-phosphate (step 5) to form another molecule of xylulose-5-phosphate in a reaction also catalyzed by transketolase.

The pentose phosphates-ribose-5-phosphate and xylulose-5phosphate-are converted into ribulose-5-phosphate (steps 6 and 7), which in the final step of the cycle (step 8 is phosphorylated to ribulose-1,5-bisphosphate by ribulose-5-phosphate kinase. This pathway is essentially the reversal of the oxidative pentose phosphate pathway and employs the same enzymes to interconvert hexose and pentose phosphates. To highlight the similarity between the two pathways, the pathway of figure is sometimes called the reductive pentose phosphate pathway, even though no chemical reduction steps occur among these reactions. In the oxidative sequence, hexose phosphates are regenerated from pentose phosphates; here, hexose phosphates are converted into pentose phosphates.

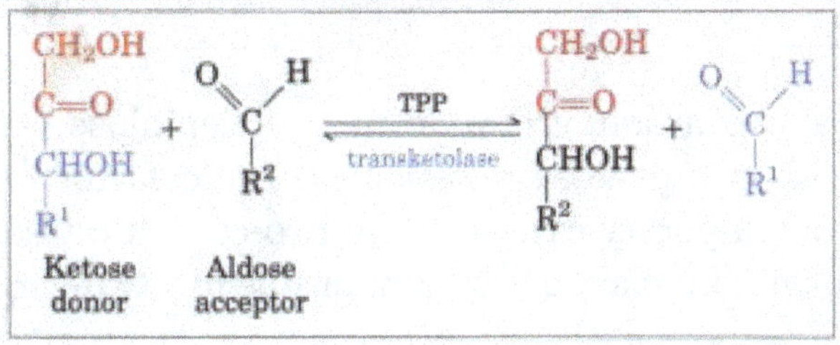

(a)

Transketolase catalyzes two very similar reactions in the Calvin cycle. (a) The general reaction catalyzed by transketolase is the transfer of a two-carbon group, carried temporarily on enzyme-bound TPP, from a ketose donor to an aldose acceptor. (b) Conversion of a hexose and a triose to a pentose and a four-carbon sugar by transketolase action. (c) Conversion of seven-carbon and three-carbon skeletons to two pentoses by transketolase.

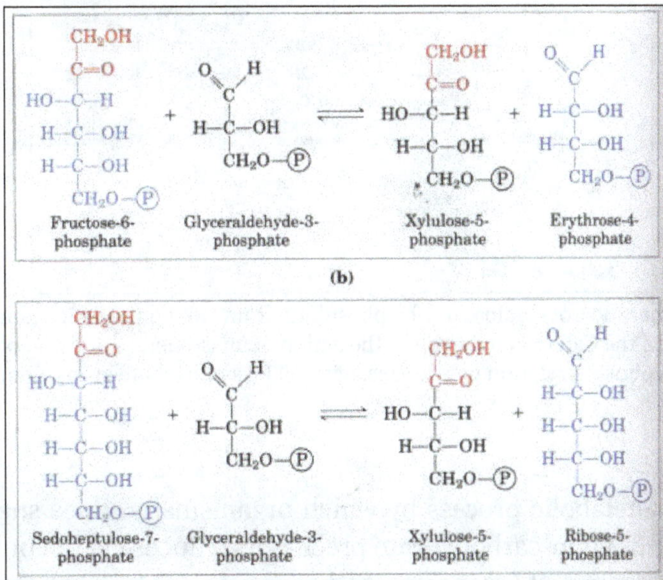

(c)

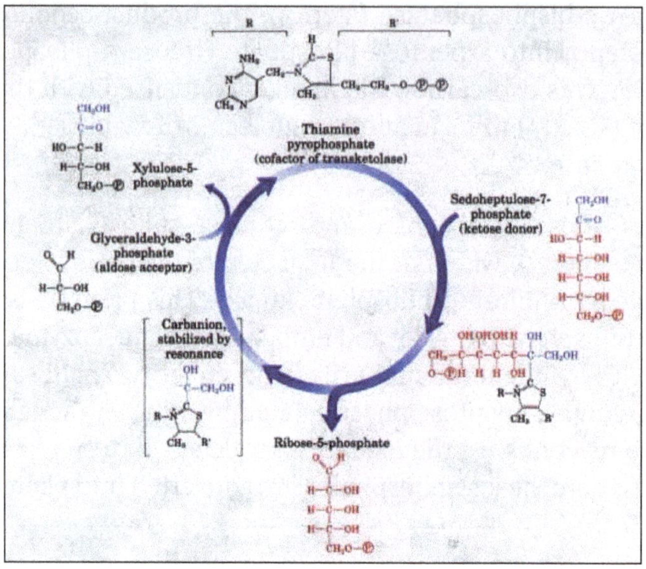

Transketolase transfers a twocarbon group from sedoheptulose-7-phosphate to glyceraldehyde-3-phosphate producing two pentose phosphates. The cofactor for transketolase, thiamine pyrophosphate, serves as a temporary carrier of the two-carbon unit and as an electron sink to facilitate the reactions shown. TPP plays a very similar role in a two-carbon group transfer in step 1 of figure.

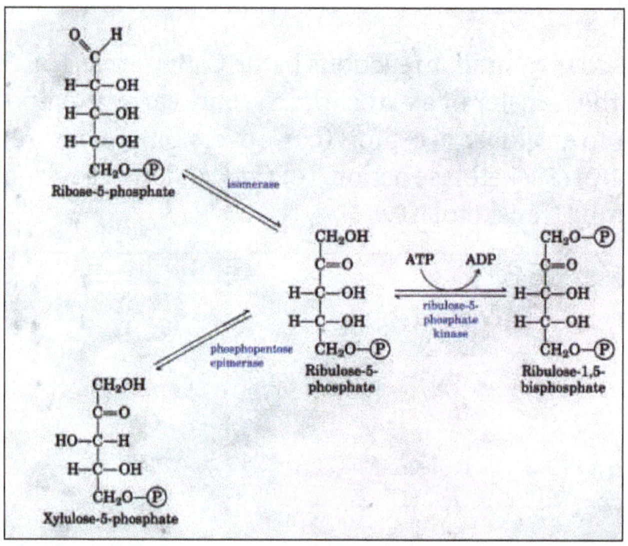

Regeneration of ribulose-1,5-bisphosphate from two pentose phosphates produced in the Calvin cycle involves the action of an isomerase and an epimerase, then phosphorylation by a kinase, with ATP as phosphate group donor.

Gluconeogenesis

Gluconeogenesis is the metabolic process by which organisms produce sugars (namely glucose) for catabolic reactions from non-carbohydrate precursors. Glucose is the only energy source used by the brain (with the exception of ketone bodies during times of fasting), testes, erythrocytes, and kidney medulla. In mammals this process occurs in the liver and kidneys.

The need for energy is important to sustain life. Organisms have evolved ways of producing substrates required for the catabolic reactions necessary to sustain life when desired substrates are unavailable. The main source of energy for eukaryotes is glucose. When glucose is unavailable, organisms are capable of metabolizing glucose from other non-carbohydrate precursors. The process that coverts pyruvate into glucose is called gluconeogenesis. Another way organisms derive glucose is from energy stores like glycogen and starch.

Gluconeogenesis is much like glycolysis only the process occurs in reverse. However, there are exceptions. In glycolysis there are three highly exergonic steps (steps 1,3,10). These are also regulatory steps which include the enzymes hexokinase, phosphofructokinase, and pyruvate kinase. Biological reactions can occur in both the forward and reverse direction. If the reaction occurs in the reverse direction the energy normally released in that reaction is now required. If gluconeogenesis were to simply occur in reverse the reaction would require too much energy to be profitable to that particular organism. In order to overcome this problem, nature has evolved three other enzymes to replace the glycolysis enzymes hexokinase, phosphofructokinase, and pyruvate kinase when going through the process of gluconeogenesis:

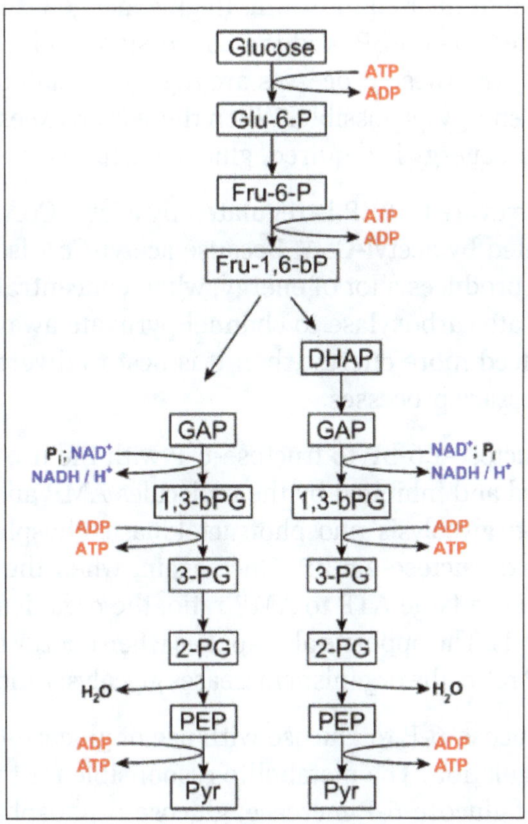

Glycolysis

1. The first step in gluconeogenesis is the conversion of pyruvate to phosphoenolpyruvic acid (PEP). In order to convert pyruvate to PEP there are several steps and several enzymes required. Pyruvate carboxylase, PEP carboxykinase and malate dehydrogenase are the three enzymes responsible for this conversion. Pyruvate carboxylase is found on the mitochondria and converts pyruvate into oxaloacetate. Because oxaloacetate cannot pass through

the mitochondria membranes it must be first converted into malate by malate dehydrogenase. Malate can then cross the mitochondria membrane into the cytoplasm where it is then converted back into oxaloacetate with another malate dehydrogenase. Lastly, oxaloacetate is converted into PEP via PEP carboxykinase. The next several steps are exactly the same as glycolysis only the process is in reverse.

2. The second step that differs from glycolysis is the conversion of fructose-1,6-bP to fructose-6-P with the use of the enzyme fructose-1,6-phosphatase. The conversion of fructose-6-P to glucose-6-P uses the same enzyme as glycolysis, phosphoglucoisomerase.
3. The last step that differs from glycolysis is the conversion of glucose-6-P to glucose with the enzyme glucose-6-phosphatase. This enzyme is located in the endoplasmic reticulum.

Regulation

Because it is important for organisms to conserve energy, they have derived ways to regulate those metabolic pathways that require and release the most energy. In glycolysis and gluconeogenesis seven of the ten steps occur at or near equilibrium. In gluconeogenesis the conversion of pyruvate to PEP, the conversion of fructose-1,6-bP, and the conversion of glucose-6-P to glucose all occur very spontaneously which is why these processes are highly regulated. It is important for the organism to conserve as much energy as possible. When there is an excess of energy available, gluconeogenesis is inhibited. When energy is required, gluconeogenesis is activated.

1. The conversion of pyruvate to PEP is regulated by acetyl-CoA. More specifically pyruvate carboxylase is activated by acetyl-CoA. Because acetyl-CoA is an important metabolite in the TCA cycle which produces a lot of energy, when concentrations of acetyl-CoA are high organisms use pyruvate carboxylase to channel pyruvate away from the TCA cycle. If the organism does not need more energy, then it is best to divert those metabolites towards storage or other necessary processes.
2. The conversion of fructose-1,6-bP to fructose-6-P with the use of fructose-1,6-phosphatase is negatively regulated and inhibited by the molecules AMP and fructose-2,6-bP. These are reciprocal regulators to glycolysis' phosphofructokinase. Phosphofructosekinase is positively regulated by AMP and fructose-2,6-bP. Once again, when the energy levels produced are higher than needed, i.e. a large ATP to AMP ratio, the organism increases gluconeogenesis and decreases glycolysis. The opposite also applies when energy levels are lower than needed, i.e. a low ATP to AMP ratio, the organism increases glycolysis and decreases gluconeogenesis.
3. The conversion of glucose-6-P to glucose with use of glucose-6-phosphatase is controlled by substrate level regulation. The metabolite responsible for this type of regulation is glucose-6-P. As levels of glucose-6-P increase, glucose-6-phosphatase increases activity and more glucose is produced. Thus glycolysis is unable to proceed.

Catabolism

Catabolism is the part of the metabolism responsible for breaking complex molecules down into smaller molecules. The other part of the metabolism, *anabolism*, builds simple molecules into more complex ones. During the catabolism energy is released from the bonds of the large molecules being broken down. Typically, that energy is then stored in the bonds of *adenosine triphosphate* (ATP). The catabolism increases the concentration of ATP in the cell as it breaks down nutrients and food. The ATP, in such high concentrations, becomes much more likely to give up its energy in the release of a phosphate. The anabolism then uses this energy to combine simple precursors into complex molecules that add to the cell and store energy for cell division.

Many pathways in the catabolism have similar versions in the anabolism. For example, large fat molecules in an organism's food must be broken down into the small fatty acids that it is comprised of. Then, for the organism to store energy for winter, large fat molecules must be created and stored. Catabolic reactions break the fats down, and anabolic pathways rebuild them. These metabolic pathways often use the same enzymes. To decrease the chance that the pathways will undo each other's progress, the pathways often inhibit each other and are separated into different organelles in eukaryotes.

Examples of Catabolism

Carbohydrate and Lipid Catabolism

Almost all organisms use the sugar glucose as a source of energy and carbon chains. Glucose is stored by organisms in larger molecules called polysaccharides. These polysaccharides can be starches, glycogen, or other simple sugars like sucrose. When an animal's cells need energy, it sends signals to the parts of the body that store glucose, or it consumes food. Glucose is released from the carbohydrates by special enzymes, in the first part of the catabolism. The glucose is then distributed into the body, for other cells to use as energy. The catabolic pathway glycolysis then breaks glucose down even further, releasing energy that is stored in ATP. From glucose, pyruvate molecules are made. Further catabolic pathways create acetate, which is a key metabolic intermediate molecule. Acetate can become a wide variety of molecules, from phospholipids, to pigment molecules, to hormones and vitamins.

Fats, which are large lipid molecules, are also degraded by the metabolism to produce energy and to create other molecules. Similar to carbohydrates, lipids are stored in large molecules, but can be broken down into individual fatty acids. These fatty acids are then converted through *beta-oxidation* into acetate. Again, acetate can be used by the anabolism, to produce larger molecules, or as part of the *citric acid cycle* which drives respiration and ATP production. Animals use fats to store large amount of energy for future use. Unlike starches and carbohydrates, lipids are hydrophobic, and exclude water. In this way, a lot of energy can be stored without the heavy weight of water slowing the organism down.

Most catabolic pathway are *convergent* in that they end in the same molecule. This enables organisms to consume and store energy in a variety of different forms, while still being able produce all the molecules it needs in the anabolic pathways. Other catabolic pathways, such as protein catabolism, create different intermediate molecules are precursors, known as *amino acids*, to build new proteins.

Protein Catabolism

All proteins in the known world are formed of the same 20 amino acids. That means that the proteins in plants, animals, and bacteria are all just different combinations of the 20 amino acids.

When an organism consumes a smaller organism, all of the protein in that organism must be digested in the catabolism. Enzymes known as *proteinases* break the bonds between the amino acids in each protein, until the acids are completely separated. Once separated, the amino acids can be distributed to the cells of the body. According to the organism's DNA, the amino acids will be recombined into new proteins.

If no source of glucose is present, or there are too many amino acids, the molecules will enter further catabolic pathways to be broken down into carbon skeletons. These small molecules can be combined in gluconeogenesis to create new glucose, which the cells can use as energy or store in large molecules. During starvation, cellular proteins can go through the catabolism to allow an organism to survive on its own tissues until more food is found. In this way, organisms can live with only small amounts of water for extremely long times. This makes them much more resilient to changing environmental conditions.

Cellular Respiration

Cellular respiration is a set of metabolic reactions and processes that take place in the cells of organisms to convert biochemical energy from nutrients into adenosine triphosphate (ATP), and then release waste products. The reactions involved in respiration are catabolic reactions, which break large molecules into smaller ones, releasing energy in the process, as weak so-called "high-energy" bonds are replaced by stronger bonds in the products. Respiration is one of the key ways a cell releases chemical energy to fuel cellular activity. Cellular respiration is considered an exothermic redox reaction which releases heat. The overall reaction occurs in a series of biochemical steps, most of which are redox reactions themselves. Although cellular respiration is technically a combustion reaction, it clearly does not resemble one when it occurs in a living cell because of the slow release of energy from the series of reactions.

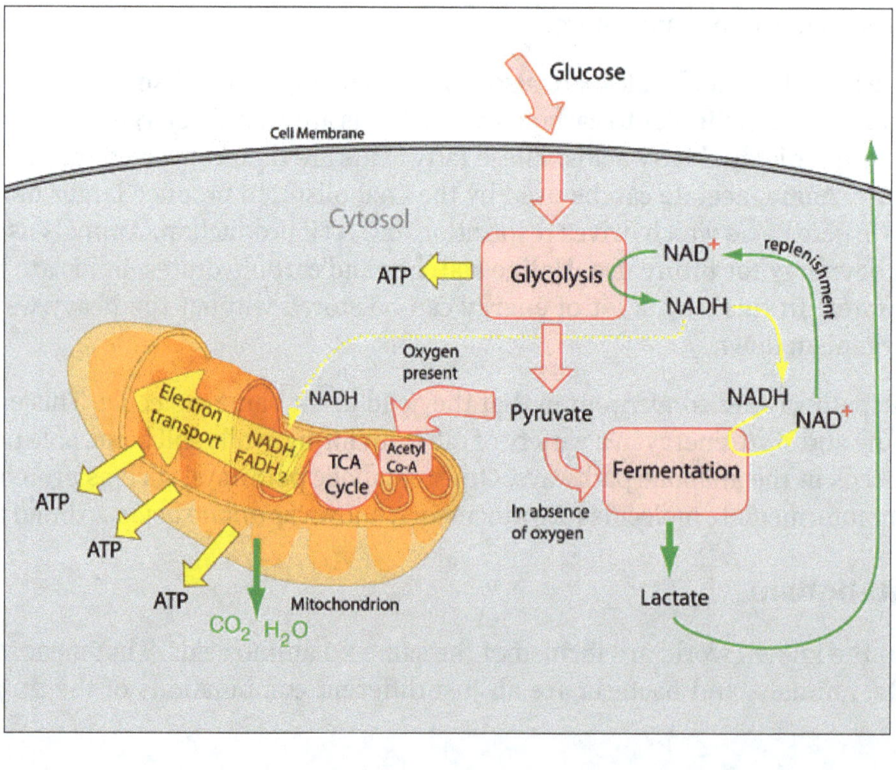

Typical Eukaryotic Cell

Nutrients that are commonly used by animal and plant cells in respiration include sugar, amino acids and fatty acids, and the most common oxidizing agent (electron acceptor) is molecular oxygen (O_2). The chemical energy stored in ATP (its third phosphate group is weakly bonded to the rest of the molecule and is cheaply broken allowing stronger bonds to form, thereby transferring energy for use by the cell) can then be used to drive processes requiring energy, including biosynthesis, locomotion or transportation of molecules across cell membranes.

Aerobic Respiration

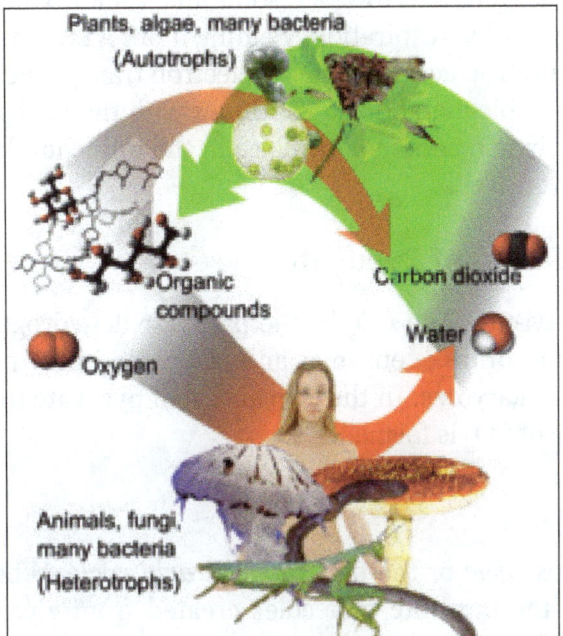

Aerobic respiration (red arrows) is the main means by which both fungi and animals utilize chemical energy in the form of organic compounds that were previously created through photosynthesis (green arrow).

Aerobic respiration requires oxygen (O_2) in order to create ATP. Although carbohydrates, fats, and proteins are consumed as reactants, it is the preferred method of pyruvate breakdown in glycolysis and requires that pyruvate enter the mitochondria in order to be fully oxidized by the Krebs cycle. The products of this process are carbon dioxide and water, but the energy transferred is used to break bonds in ADP as the third phosphate group is added to form ATP (adenosine triphosphate), by substrate-level phosphorylation, NADH and $FADH_2$

Simplified reaction :
$$C_6H_{12}O_6(s) + 6\ O_2(g) \rightarrow 6\ CO_2(g) + 6\ H_2O(l) + \text{heat}$$
$$\Delta G = -2880\ \text{kJ per mol of}\ C_6H_{12}O_6$$

The negative ΔG indicates that the reaction can occur spontaneously.

The potential of NADH and $FADH_2$ is converted to more ATP through an electron transport chain with oxygen as the "terminal electron acceptor". Most of the ATP produced by aerobic cellular respiration is made by oxidative phosphorylation. This works by the energy released in the consumption of pyruvate being used to create a chemiosmotic potential by pumping protons across a membrane. This potential is then used to drive ATP synthase and produce ATP from ADP and a phosphate group. Biology textbooks often state that 38 ATP molecules can be made per oxidized glucose molecule during cellular respiration (2 from glycolysis, 2 from the Krebs cycle, and about 34 from the electron transport system). However, this maximum yield is never quite reached because of losses due to leaky membranes as well as the cost of moving pyruvate and ADP into the mitochondrial matrix, and current estimates range around 29 to 30 ATP per glucose.

Aerobic metabolism is up to 15 times more efficient than anaerobic metabolism (which yields 2 molecules ATP per 1 molecule glucose). However, some anaerobic organisms, such as methanogens are able to continue with anaerobic respiration, yielding more ATP by using other inorganic molecules (not oxygen) as final electron acceptors in the electron transport chain. They share the initial pathway of glycolysis but aerobic metabolism continues with the Krebs cycle and oxidative phosphorylation. The post-glycolytic reactions take place in the mitochondria in eukaryotic cells, and in the cytoplasm in prokaryotic cells.

Oxidative Decarboxylation of Pyruvate

Pyruvate is oxidized to acetyl-CoA and CO_2 by the pyruvate dehydrogenase complex (PDC). The PDC contains multiple copies of three enzymes and is located in the mitochondria of eukaryotic cells and in the cytosol of prokaryotes. In the conversion of pyruvate to acetyl-CoA, one molecule of NADH and one molecule of CO_2 is formed.

Citric Acid Cycle

This is also called the *Krebs cycle* or the *tricarboxylic acid cycle*. When oxygen is present, acetyl-CoA is produced from the pyruvate molecules created from glycolysis. Once acetyl-CoA is formed, aerobic or anaerobic respiration can occur. When oxygen is present, the mitochondria will undergo aerobic respiration which leads to the Krebs cycle. However, if oxygen is not present, fermentation of the pyruvate molecule will occur. In the presence of oxygen, when acetyl-CoA is produced, the molecule then enters the citric acid cycle (Krebs cycle) inside the mitochondrial matrix, and is oxidized to CO_2 while at the same time reducing NAD to NADH. NADH can be used by the electron transport chain to create further ATP as part of oxidative phosphorylation. To fully oxidize the equivalent of one glucose molecule, two acetyl-CoA must be metabolized by the Krebs cycle. Two waste products, H_2O and CO_2, are created during this cycle.

The citric acid cycle is an 8-step process involving 18 different enzymes and co-enzymes. During the cycle, acetyl-CoA (2 carbons) + oxaloacetate (4 carbons) yields citrate (6 carbons), which is rearranged to a more reactive form called isocitrate (6 carbons). Isocitrate is modified to become α-ketoglutarate (5 carbons), succinyl-CoA, succinate, fumarate, malate, and, finally, oxaloacetate.

The net gain of high-energy compounds from one cycle is 3 NADH, 1 $FADH_2$, and 1 GTP; the GTP may subsequently be used to produce ATP. Thus, the total yield from 1 glucose molecule (2 pyruvate molecules) is 6 NADH, 2 $FADH_2$, and 2 ATP.

Efficiency of ATP Production

The table below describes the reactions involved when one glucose molecule is fully oxidized into carbon dioxide. It is assumed that all the reduced coenzymes are oxidized by the electron transport chain and used for oxidative phosphorylation.

Step	Coenzyme yield	ATP yield	Source of ATP
Glycolysis preparatory phase		−2	Phosphorylation of glucose and fructose 6-phosphate uses two ATP from the cytoplasm.
Glycolysis pay-off phase		4	Substrate-level phosphorylation
	2 NADH	3 or 5	Oxidative phosphorylation : Each NADH produces net 1.5 ATP (instead of usual 2.5) due to NADH transport over the mitochondrial membrane
Oxidative decarboxylation of pyruvate	2 NADH	5	Oxidative phosphorylation
Krebs cycle		2	Substrate-level phosphorylation
	6 NADH	15	Oxidative phosphorylation
	2 FADH2	3	Oxidative phosphorylation
Total yield		30 or 32 ATP	From the complete oxidation of one glucose molecule to carbon dioxide and oxidation of all the reduced coenzymes.

Although there is a theoretical yield of 38 ATP molecules per glucose during cellular respiration, such conditions are generally not realized because of losses such as the cost of moving pyruvate (from glycolysis), phosphate, and ADP (substrates for ATP synthesis) into the mitochondria. All are actively transported using carriers that utilize the stored energy in the proton electrochemical gradient.

- Pyruvate is taken up by a specific, low Km transporter to bring it into the mitochondrial matrix for oxidation by the pyruvate dehydrogenase complex.
- The phosphate carrier (PiC) mediates the electroneutral exchange (antiport) of phosphate ($H_2PO_4^-$; P_i) for OH^- or symport of phosphate and protons (H^+) across the inner membrane, and the driving force for moving phosphate ions into the mitochondria is the proton motive force.
- The ATP-ADP translocase (also called adenine nucleotide translocase, ANT) is an antiporter and exchanges ADP and ATP across the inner membrane. The driving force is due to the ATP (−4) having a more negative charge than the ADP (−3), and thus it dissipates some of the electrical component of the proton electrochemical gradient.

The outcome of these transport processes using the proton electrochemical gradient is that more than 3 H^+ are needed to make 1 ATP. Obviously this reduces the theoretical efficiency of the whole process and the likely maximum is closer to 28–30 ATP molecules. In practice the efficiency may be even lower because the inner membrane of the mitochondria is slightly leaky to protons. Other factors may also dissipate the proton gradient creating an apparently leaky mitochondria. An uncoupling protein known as thermogenin is expressed in some cell types and is a channel that can transport protons. When this protein is active in the inner membrane it short circuits the coupling between the electron transport chain and ATP synthesis. The potential energy from the proton gradient is not used to make ATP but generates heat. This is particularly important in brown fat thermogenesis of newborn and hibernating mammals.

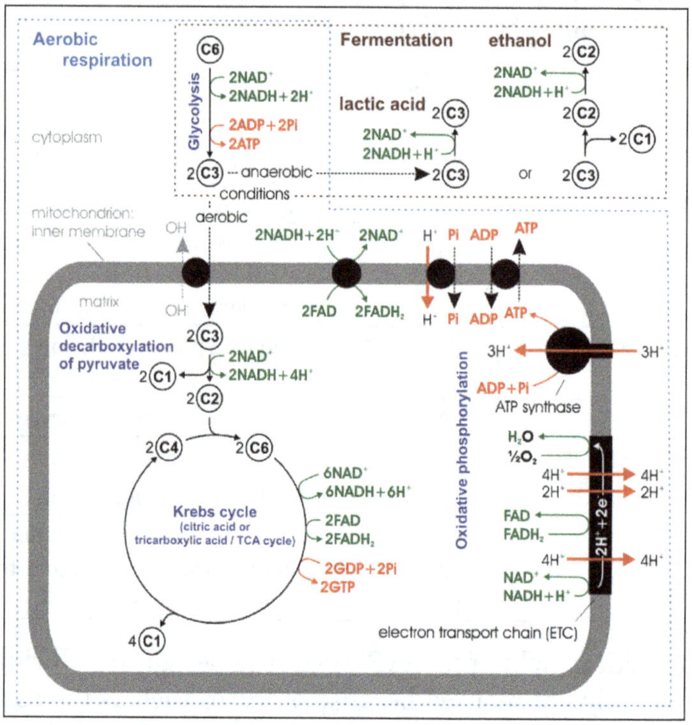

Stoichiometry of aerobic respiration and most known fermentation types in eucaryotic cell. Numbers in circles indicate counts of carbon atoms in molecules, C_6 is glucose $C_6H_{12}O_6$, C_1 carbon dioxide CO_2. Mitochondrial outer membrane is omitted.

According to some of newer sources the ATP yield during aerobic respiration is not 36–38, but only about 30–32 ATP molecules / 1 molecule of glucose, because:

- ATP: NADH+H^+ and ATP: $FADH_2$ ratios during the oxidative phosphorylation appear to be not 3 and 2, but 2.5 and 1.5 respectively. Unlike in the substrate-level phosphorylation, the stoichiometry here is difficult to establish.
    - ATP synthase produces 1 ATP / 3 H^+. However the exchange of matrix ATP for cytosolic ADP and Pi (antiport with OH^- or symport with H^+) mediated by ATP–ADP translocase and phosphate carrier consumes 1 H^+ / 1 ATP as a result of regeneration of the transmembrane potential changed during this transfer, so the net ratio is 1 ATP: 4 H^+.
    - The mitochondrial electron transport chain proton pump transfers across the inner membrane 10 H^+ / 1 NADH+H^+ (4 + 2 + 4) or 6 H^+ / 1 $FADH_2$ (2 + 4).

So the final stoichiometry is:

1 NADH+H^+: 10 H^+: 10/4 ATP = 1 NADH+H^+: 2.5 ATP

1 $FADH_2$: 6 H^+: 6/4 ATP = 1 $FADH_2$: 1.5 ATP

- ATP: NADH+H^+ coming from glycolysis ratio during the oxidative phosphorylation is:
    - 1.5, as for $FADH_2$, if hydrogen atoms ($2H^+$+$2e^-$) are transferred from cytosolic NADH+H^+ to mitochondrial FAD by the glycerol phosphate shuttle located in the inner mitochondrial membrane.
    - 2.5 in case of malate-aspartate shuttle transferring hydrogen atoms from cytosolic NADH+H^+ to mitochondrial NAD^+.

So finally we have, per molecule of glucose:

- Substrate-level phosphorylation: 2 ATP from glycolysis + 2 ATP (directly GTP) from Krebs cycle.
- Oxidative phosphorylation:
    - 2 NADH+H^+ from glycolysis: 2 × 1.5 ATP (if glycerol phosphate shuttle transfers hydrogen atoms) or 2 × 2.5 ATP (malate-aspartate shuttle).
    - 2 NADH+H^+ from the oxidative decarboxylation of pyruvate and 6 from Krebs cycle: 8 × 2.5 ATP.
    - 2 $FADH_2$ from the Krebs cycle: 2 × 1.5 ATP.

Altogether this gives 4 + 3 (or 5) + 20 + 3 = 30 (or 32) ATP per molecule of glucose.

These figures may still require further tweaking as new structural details become available. The above value of 3 H+/ATP for the synthase assumes that the synthase translocates 9 protons, and produces 3 ATP, per rotation. The number of protons depends on the number of c subunits in the Fo c-ring, and it is now known that this is 10 in yeast Fo and 8 for vertebrates. Including one H+ for the transport reactions, this means that synthesis of one ATP requires 1+10/3=4.33 protons in yeast and 1+8/3 = 3.67 in vertebrates. This would imply that in human mitochondria the 10 protons from oxidizing NADH would produce 2.72 ATP (instead of 2.5) and the 6 protons from oxidizing succinate or ubiquinol would produce 1.64 ATP (instead of 1.5).

The total ATP yield in ethanol or lactic acid fermentation is only 2 molecules coming from glycolysis, because pyruvate is not transferred to the mitochondrion and finally oxidized to the carbon dioxide (CO_2), but reduced to ethanol or lactic acid in the cytoplasm.

Anaerobic Respiration

Cellular respiration is the process by which biological fuels are oxidised in the presence of an inorganic electron acceptor (such as oxygen) to produce large amounts of energy, to drive the bulk production of ATP.

Anaerobic respiration is used by some microorganisms in which neither oxygen (aerobic respiration) nor pyruvate derivatives (fermentation) is the final electron acceptor. Rather, an inorganic acceptor such as sulfate or nitrate is used. Such organisms are typically found in unusual places such as underwater caves or near hydrothermal vents at the bottom of the ocean.

In July 2019, a scientific study of Kidd Mine in Canada discovered sulfur-breathing organisms which live 7900 feet below the surface, and which breathe sulfur in order to survive. These organisms are also remarkable due to eating rocks such as pyrite as their regular food source.

Glycolysis

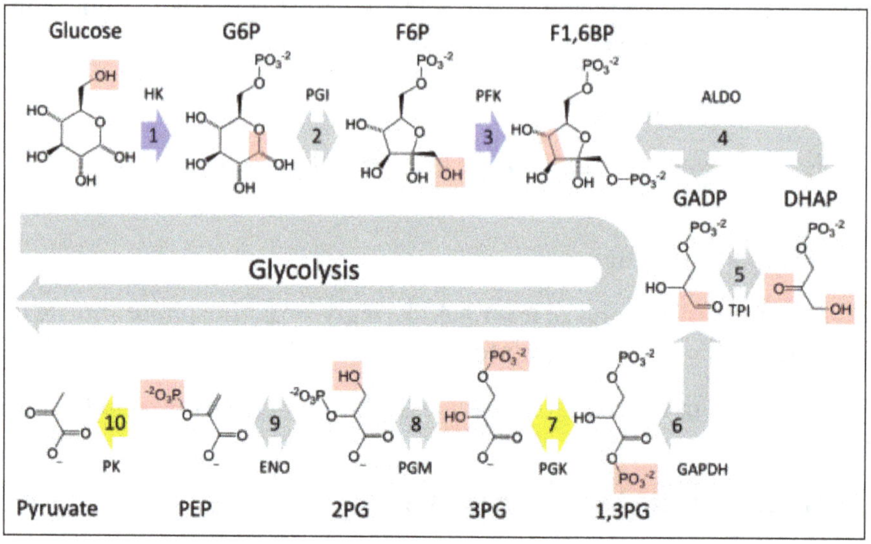

The metabolic pathway of glycolysis converts glucose to pyruvate by via a series of intermediate metabolites. Each chemical modification (red box) is performed by a different enzyme. Steps 1 and 3 consume ATP (blue) and steps 7 and 10 produce ATP (yellow). Since steps 6-10 occur twice per glucose molecule, this leads to a net production of ATP.

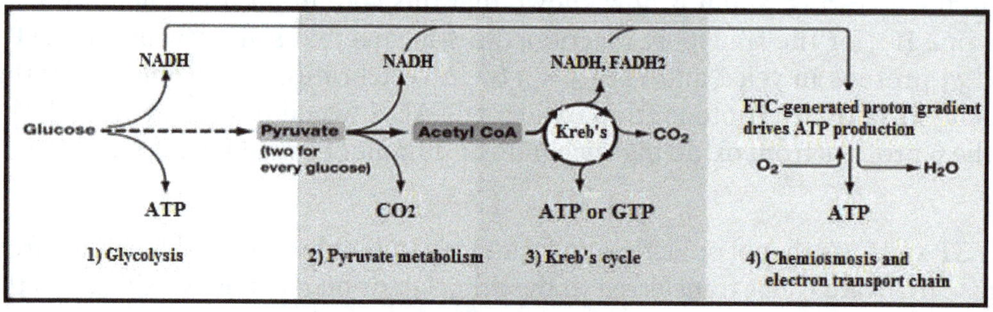

Aerobic respiration

Glycolysis (from *glycose*, an older term for glucose + *-lysis* degradation) is the metabolic pathway that converts glucose $C_6H_{12}O_6$, into pyruvate, $CH_3COCOO^- + H^+$. The free energy released in this process is used to form the high-energy molecules ATP (adenosine triphosphate) and NADH (reduced nicotinamide adenine dinucleotide). Glycolysis is a sequence of ten enzyme-catalyzed reactions. Most monosaccharides, such as fructose and galactose, can be converted to one of these intermediates. The intermediates may also be directly useful rather than just utilized as steps in the overall reaction. For example, the intermediate dihydroxyacetone phosphate (DHAP) is a source of the glycerol that combines with fatty acids to form fat.

Glycolysis is an oxygen-independent metabolic pathway. The wide occurrence of glycolysis indicates that it is an ancient metabolic pathway. Indeed, the reactions that constitute glycolysis and its parallel pathway, the pentose phosphate pathway, occur metal-catalyzed under the oxygen-free conditions of the Archean oceans, also in the absence of enzymes.

In most organisms, glycolysis occurs in the cytosol. The most common type of glycolysis is the *Embden–Meyerhof–Parnas (EMP pathway)*, which was discovered by Gustav Embden, Otto Meyerhof, and Jakub Karol Parnas. Glycolysis also refers to other pathways, such as the *Entner–Doudoroff pathway* and various heterofermentative and homofermentative pathways. However, the discussion here will be limited to the Embden–Meyerhof–Parnas pathway.

The glycolysis pathway can be separated into two phases:

- The Preparatory/Investment Phase – wherein ATP is consumed.
- The Pay Off Phase – wherein ATP is produced.

The overall reaction of glycolysis is:

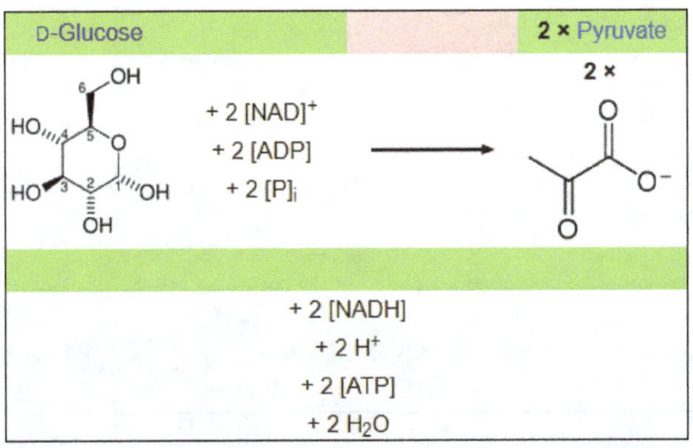

The use of symbols in this equation makes it appear unbalanced with respect to oxygen atoms, hydrogen atoms, and charges. Atom balance is maintained by the two phosphate (P_i) groups:

- Each exists in the form of a hydrogen phosphate anion (HPO_4^{2-}), dissociating to contribute 2 H^+ overall.
- Each liberates an oxygen atom when it binds to an ADP (adenosine diphosphate) molecule, contributing 2 O overall.

Charges are balanced by the difference between ADP and ATP. In the cellular environment, all three hydroxyl groups of ADP dissociate into $-O^-$ and H^+, giving ADP^{3-}, and this ion tends to exist in an ionic bond with Mg^{2+}, giving $ADPMg^-$. ATP behaves identically except that it has four hydroxyl groups, giving $ATPMg^{2-}$. When these differences along with the true charges on the two phosphate groups are considered together, the net charges of −4 on each side are balanced.

For simple fermentations, the metabolism of one molecule of glucose to two molecules of pyruvate has a net yield of two molecules of ATP. Most cells will then carry out further reactions to 'repay' the used NAD^+ and produce a final product of ethanol or lactic acid. Many bacteria use inorganic compounds as hydrogen acceptors to regenerate the NAD^+.

Cells performing aerobic respiration synthesize much more ATP, but not as part of glycolysis. These further aerobic reactions use pyruvate and NADH + H^+ from glycolysis. Eukaryotic aerobic respiration produces approximately 34 additional molecules of ATP for each glucose molecule, however most of these are produced by a vastly different mechanism to the substrate-level phosphorylation in glycolysis.

The lower-energy production, per glucose, of anaerobic respiration relative to aerobic respiration, results in greater flux through the pathway under hypoxic (low-oxygen) conditions, unless alternative sources of anaerobically oxidizable substrates, such as fatty acids, are found.

Sequence of Reactions

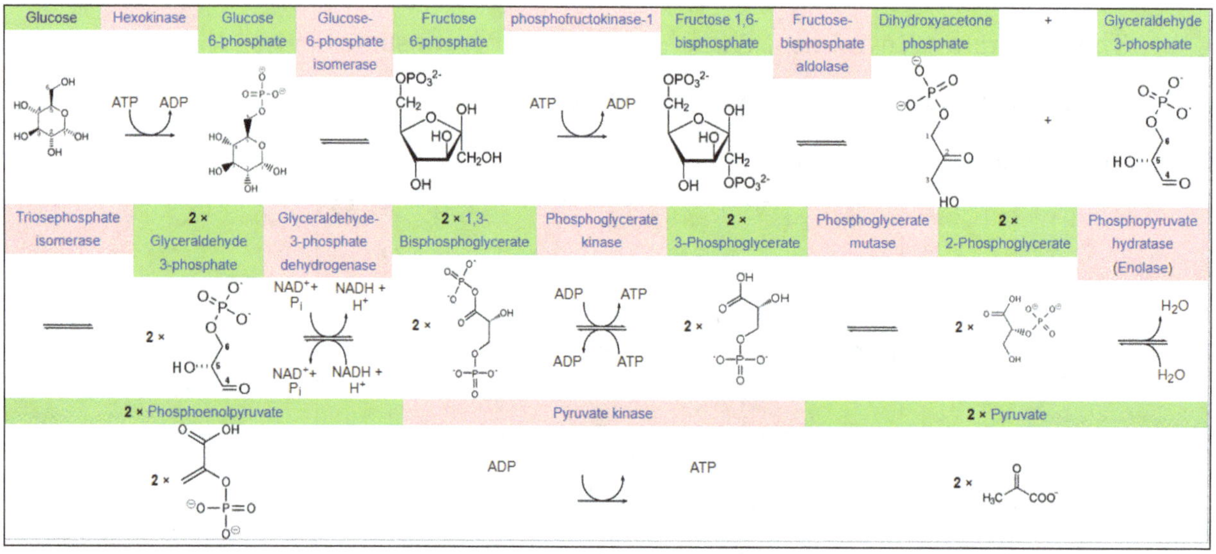

Preparatory Phase

The first five steps are regarded as the preparatory (or investment) phase, since they consume energy to convert the glucose into two three-carbon sugar phosphates (G3P).

D-Glucose (Glc)	Hexokinase (HK)a transferase	α-D-Glucose-6-phosphate (G6P)

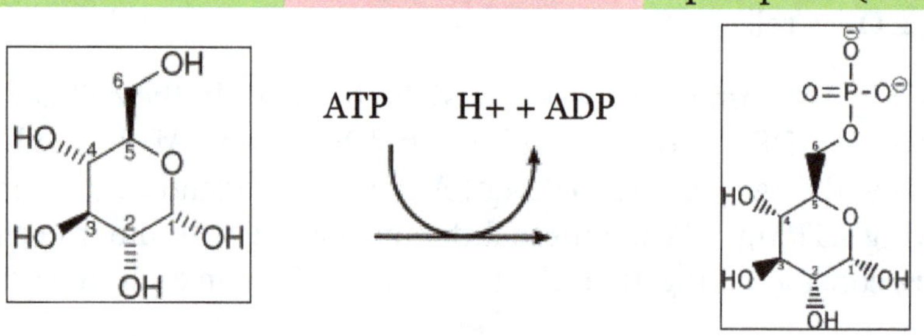

The first step in glycolysis is phosphorylation of glucose by a family of enzymes called hexokinases to form glucose 6-phosphate (G6P). This reaction consumes ATP, but it acts to keep the glucose concentration low, promoting continuous transport of glucose into the cell through the plasma membrane transporters. In addition, it blocks the glucose from leaking out – the cell lacks transporters for G6P, and free diffusion out of the cell is prevented due to the charged nature of G6P. Glucose may alternatively be formed from the phosphorolysis or hydrolysis of intracellular starch or glycogen.

In animals, an isozyme of hexokinase called glucokinase is also used in the liver, which has a much lower affinity for glucose (K_m in the vicinity of normal glycemia), and differs in regulatory properties. The different substrate affinity and alternate regulation of this enzyme are a reflection of the role of the liver in maintaining blood sugar levels.

Cofactors: Mg^{2+}

α-D-Glucose 6-phosphate (**G6P**)	Phosphoglucose isomerase (**PGI**) *an isomerase*	β-D-Fructose 6-phosphate (**F6P**)

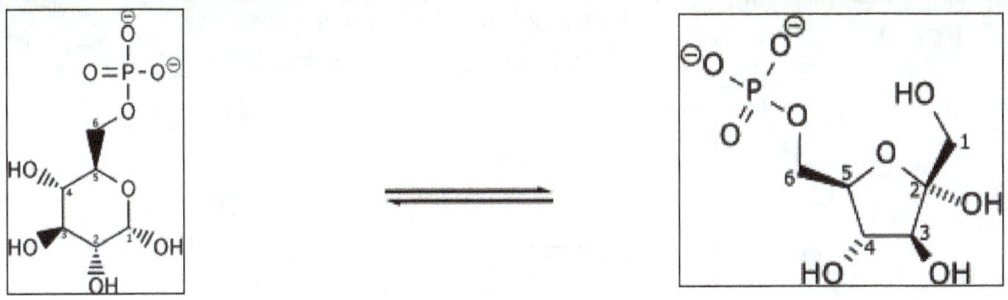

G6P is then rearranged into fructose 6-phosphate (F6P) by glucose phosphate isomerase. Fructose can also enter the glycolytic pathway by phosphorylation at this point.

The change in structure is an isomerization, in which the G6P has been converted to F6P. The reaction requires an enzyme, phosphoglucose isomerase, to proceed. This reaction is freely reversible under normal cell conditions. However, it is often driven forward because of a low concentration of F6P, which is constantly consumed during the next step of glycolysis. Under conditions of high F6P concentration, this reaction readily runs in reverse. This phenomenon can be explained through Le Chatelier's Principle. Isomerization to a keto sugar is necessary for carbanion stabilization in the fourth reaction step (below).

β-D-Fructose 6-phosphate (**F6P**)	phosphofructokinase (**PFK-1**)*a transferase*	β-D-Fructose 1,6-bisphosphate (**F1,6BP**)

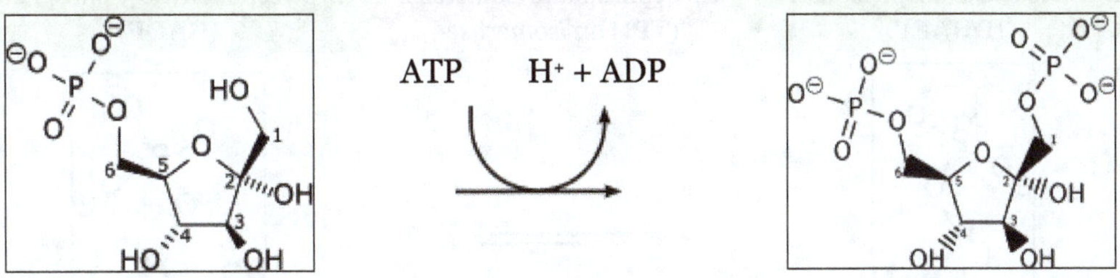

The energy expenditure of another ATP in this step is justified in 2 ways: The glycolytic process (up to this step) becomes irreversible, and the energy supplied destabilizes the molecule. Because the reaction catalyzed by Phosphofructokinase 1 (PFK-1) is coupled to the hydrolysis of ATP (an energetically favorable step) it is, in essence, irreversible, and a different pathway must be used to do the reverse conversion during gluconeogenesis. This makes the reaction a key regulatory point. This is also the rate-limiting step.

Furthermore, the second phosphorylation event is necessary to allow the formation of two charged groups (rather than only one) in the subsequent step of glycolysis, ensuring the prevention of free diffusion of substrates out of the cell.

The same reaction can also be catalyzed by pyrophosphate-dependent phosphofructokinase (PFP or PPi-PFK), which is found in most plants, some bacteria, archea, and protists, but not in animals. This enzyme uses pyrophosphate (PPi) as a phosphate donor instead of ATP. It is a reversible reaction, increasing the flexibility of glycolytic metabolism. A rarer ADP-dependent PFK enzyme variant has been identified in archaean species.

Cofactors: Mg^{2+}

β-D-Fructose 1,6-bisphosphate **(F1,6BP)**	fructose-bisphosphate aldolase **(ALDO)***a lyase*	D-glyceraldehyde 3-phosphate **(GADP)**	Dihydroxyacetone phosphate **(DHAP)**

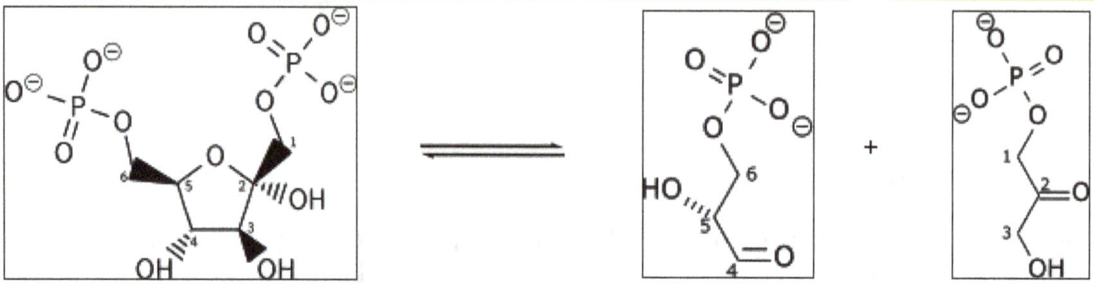

Destabilizing the molecule in the previous reaction allows the hexose ring to be split by aldolase into two triose sugars: dihydroxyacetone phosphate (a ketose), and glyceraldehyde 3-phosphate (an aldose). There are two classes of aldolases: class I aldolases, present in animals and plants, and class II aldolases, present in fungi and bacteria; the two classes use different mechanisms in cleaving the ketose ring.

Electrons delocalized in the carbon-carbon bond cleavage associate with the alcohol group. The resulting carbanion is stabilized by the structure of the carbanion itself via resonance charge distribution and by the presence of a charged ion prosthetic group.

Dihydroxyacetone phosphate **(DHAP)**	triosephosphate isomerase **(TPI)** *an isomerase*	D-glyceraldehyde 3-phosphate **(GADP)**

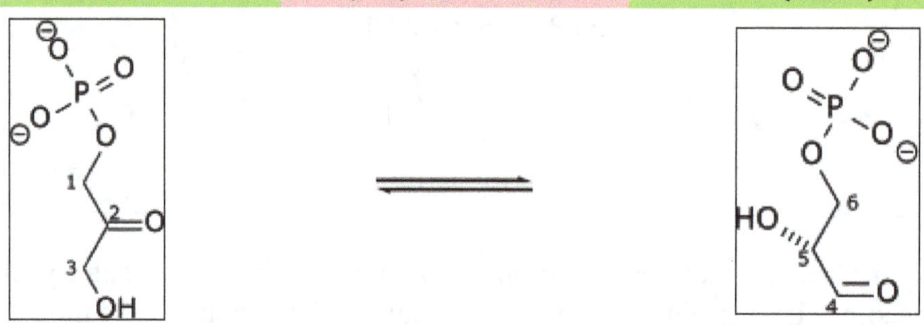

Triosephosphate isomerase rapidly interconverts dihydroxyacetone phosphate with glyceraldehyde 3-phosphate (GADP) that proceeds further into glycolysis. This is advantageous, as it directs dihydroxyacetone phosphate down the same pathway as glyceraldehyde 3-phosphate, simplifying regulation.

Pay-off Phase

The second half of glycolysis is known as the pay-off phase, characterised by a net gain of the energy-rich molecules ATP and NADH. Since glucose leads to two triose sugars in the preparatory phase, each reaction in the pay-off phase occurs twice per glucose molecule. This yields 2 NADH molecules and 4 ATP molecules, leading to a net gain of 2 NADH molecules and 2 ATP molecules from the glycolytic pathway per glucose.

glyceraldehyde 3-phosphate (**GADP**)	glyceraldehyde phosphate dehydrogenase (**GAPDH**) *an oxidoreductase*	D-1,3-bisphosphoglycerate (**1,3BPG**)

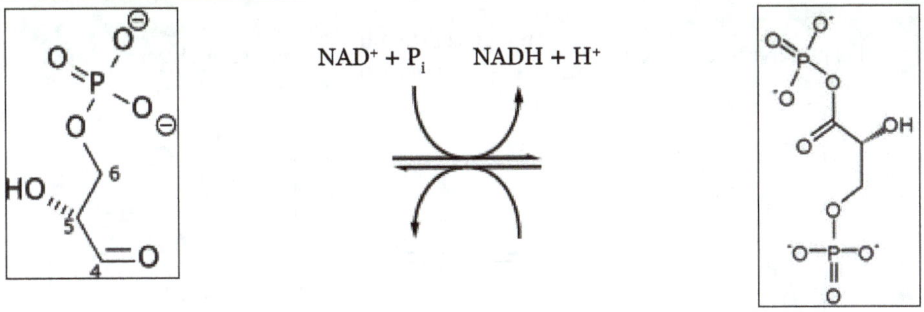

The aldehyde groups of the triose sugars are oxidised, and inorganic phosphate is added to them, forming 1,3-bisphosphoglycerate.

The hydrogen is used to reduce two molecules of NAD^+, a hydrogen carrier, to give NADH + H^+ for each triose.

Hydrogen atom balance and charge balance are both maintained because the phosphate (P_i) group actually exists in the form of a hydrogen phosphate anion (HPO_4^{2-}), which dissociates to contribute the extra H^+ ion and gives a net charge of -3 on both sides.

Here, Arsenate (AsO_4^{3-}), an anion akin to inorganic phosphate may replace phosphate as a substrate to form 1-arseno-3-phosphoglycerate. This, however, is unstable and readily hydrolyzes to form 3-phosphoglycerate, the intermediate in the next step of the pathway. As a consequence of bypassing this step, the molecule of ATP generated from 1-3 bisphosphoglycerate in the next reaction will not be made, even though the reaction proceeds. As a result, arsenate is an uncoupler of glycolysis.

1,3-bisphosphoglycerate (**1,3BPG**)	phosphoglycerate kinase (**PGK**) *a transferase*	3-phosphoglycerate (**3PG**)

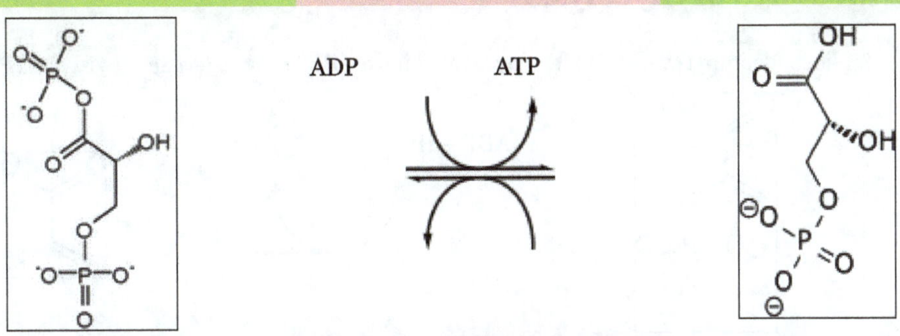

phosphoglycerate kinase (**PGK**)

This step is the enzymatic transfer of a phosphate group from 1,3-bisphosphoglycerate to ADP by phosphoglycerate kinase, forming ATP and 3-phosphoglycerate. At this step, glycolysis has reached the break-even point: 2 molecules of ATP were consumed, and 2 new molecules have now been synthesized. This step, one of the two substrate-level phosphorylation steps, requires ADP; thus, when the cell has plenty of ATP (and little ADP), this reaction does not occur. Because ATP decays relatively quickly when it is not metabolized, this is an important regulatory point in the glycolytic pathway.

ADP actually exists as ADPMg⁻, and ATP as ATPMg²⁻, balancing the charges at -5 both sides.

Cofactors: Mg^{2+}

3-phosphoglycerate (**3PG**)	phosphoglycerate mutase (**PGM**) *a mutase*	2-phosphoglycerate (**2PG**)

Phosphoglycerate mutase isomerises 3-phosphoglycerate into 2-phosphoglycerate.

2-phosphoglycerate (**2PG**)	enolase (**ENO**) *a lyase*	phosphoenolpyruvate (**PEP**)

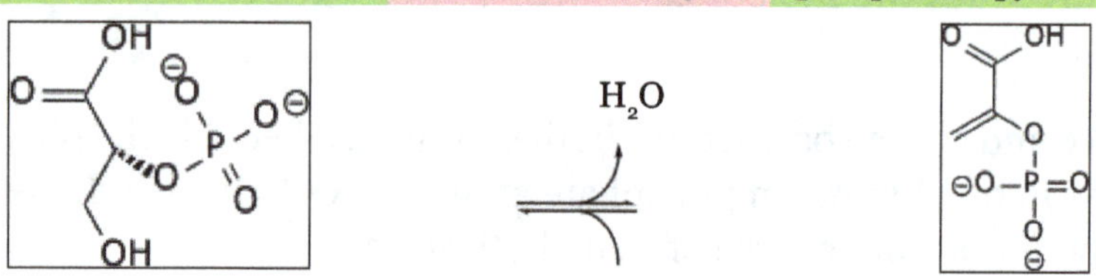

enolase (**ENO**)

Enolase next converts 2-phosphoglycerate to phosphoenolpyruvate. This reaction is an elimination reaction involving an E1cB mechanism.

Cofactors: 2 Mg^{2+}: one "conformational" ion to coordinate with the carboxylate group of the substrate, and one "catalytic" ion that participates in the dehydration.

The Intermediates of Glycolysis

phosphoenolpyruvate (**PEP**)	pyruvate kinase (**PK**) *a transferase*	pyruvate (**Pyr**)

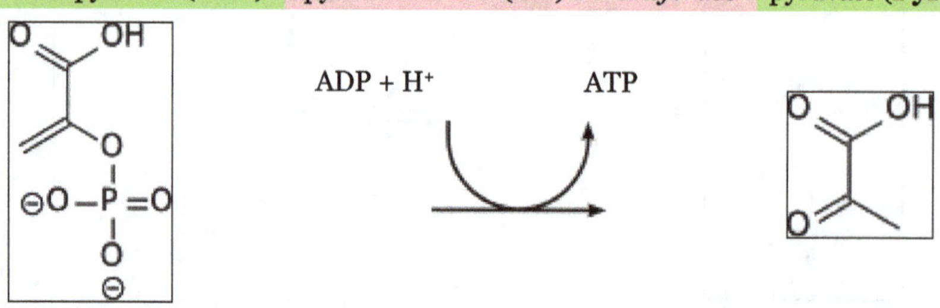

A final substrate-level phosphorylation now forms a molecule of pyruvate and a molecule of ATP by means of the enzyme pyruvate kinase. This serves as an additional regulatory step, similar to the phosphoglycerate kinase step.

Cofactors: Mg^{2+}

Biochemical Logic

The existence of more than one point of regulation indicates that intermediates between those points enter and leave the glycolysis pathway by other processes. For example, in the first regulated step, hexokinase converts glucose into glucose-6-phosphate. Instead of continuing through the glycolysis pathway, this intermediate can be converted into glucose storage molecules, such as glycogen or starch. The reverse reaction, breaking down, e.g., glycogen, produces mainly glucose-6-phosphate; very little free glucose is formed in the reaction. The glucose-6-phosphate so produced can enter glycolysis *after* the first control point.

In the second regulated step (the third step of glycolysis), phosphofructokinase converts fructose-6-phosphate into fructose-1,6-bisphosphate, which then is converted into glyceraldehyde-3-phosphate and dihydroxyacetone phosphate. The dihydroxyacetone phosphate can be removed from glycolysis by conversion into glycerol-3-phosphate, which can be used to form triglycerides. Conversely, triglycerides can be broken down into fatty acids and glycerol; the latter, in turn, can be converted into dihydroxyacetone phosphate, which can enter glycolysis *after* the second control point.

Free Energy Changes

Concentrations of metabolites in erythrocytes	
Compound	Concentration/mM
Glucose	5.0
Glucose-6-phosphate	0.083
Fructose-6-phosphate	0.014
Fructose-1,6-bisphosphate	0.031
Dihydroxyacetone phosphate	0.14
Glyceraldehyde-3-phosphate	0.019
1,3-Bisphosphoglycerate	0.001
2,3-Bisphosphoglycerate	4.0
3-Phosphoglycerate	0.12
2-Phosphoglycerate	0.03
Phosphoenolpyruvate	0.023
Pyruvate	0.051
ATP	1.85
ADP	0.14
P_i	1.0

The change in free energy, ΔG, for each step in the glycolysis pathway can be calculated using $\Delta G = \Delta G^{o\prime} + RT\ln Q$, where Q is the reaction quotient. This requires knowing the concentrations of the metabolites. All of these values are available for erythrocytes, with the exception of the concentrations of NAD^+ and NADH. The ratio of NAD^+ to NADH in the cytoplasm is approximately 1000, which makes the oxidation of glyceraldehyde-3-phosphate (step 6) more favourable.

Using the measured concentrations of each step, and the standard free energy changes, the actual free energy change can be calculated.

Change in free energy for each step of glycolysis		
Reaction	$\Delta G°'$ / (kJ/mol)	ΔG / (kJ/mol)
Glucose + ATP^{4-} → Glucose-6-phosphate^{2-} + ADP^{3-} + H^+	-16.7	-34
Glucose-6-phosphate^{2-} → Fructose-6-phosphate^{2-}	1.67	-2.9
Fructose-6-phosphate^{2-} + ATP^{4-} → Fructose-1,6-bisphosphate^{4-} + ADP^{3-} + H^+	-14.2	-19
Fructose-1,6-bisphosphate^{4-} → Dihydroxyacetone phosphate^{2-} + Glyceraldehyde-3-phosphate^{2-}	23.9	-0.23
Dihydroxyacetone phosphate^{2-} → Glyceraldehyde-3-phosphate^{2-}	7.56	2.4
Glyceraldehyde-3-phosphate^{2-} + P_i^{2-} + NAD^+ → 1,3-Bisphosphoglycerate^{4-} + NADH + H^+	6.30	-1.29
1,3-Bisphosphoglycerate^{4-} + ADP^{3-} → 3-Phosphoglycerate^{3-} + ATP^{4-}	-18.9	0.09
3-Phosphoglycerate^{3-} → 2-Phosphoglycerate^{3-}	4.4	0.83
2-Phosphoglycerate^{3-} → Phosphoenolpyruvate^{3-} + H_2O	1.8	1.1
Phosphoenolpyruvate^{3-} + ADP^{3-} + H^+ → Pyruvate$^-$ + ATP^{4-}	-31.7	-23.0

From measuring the physiological concentrations of metabolites in an erythrocyte it seems that about seven of the steps in glycolysis are in equilibrium for that cell type. Three of the steps — the ones with large negative free energy changes — are not in equilibrium and are referred to as *irreversible*; such steps are often subject to regulation.

Step 5 in the figure is shown behind the other steps, because that step is a side-reaction that can decrease or increase the concentration of the intermediate glyceraldehyde-3-phosphate. That compound is converted to dihydroxyacetone phosphate by the enzyme triose phosphate isomerase, which is a catalytically perfect enzyme; its rate is so fast that the reaction can be assumed to be in equilibrium. The fact that ΔG is not zero indicates that the actual concentrations in the erythrocyte are not accurately known.

Regulation of the Rate Limiting Enzymes

The four regulatory enzymes are hexokinase (or glucokinase in the liver), phosphofructokinase, and pyruvate kinase. The flux through the glycolytic pathway is adjusted in response to conditions both inside and outside the cell. The internal factors that regulate glycolysis do so primarily to provide ATP in adequate quantities for the cell's needs. The external factors act primarily on the liver, fat tissue, and muscles, which can remove large quantities of glucose from the blood after meals (thus preventing hyperglycemia by storing the excess glucose as fat or glycogen, depending on the tissue type). The liver is also capable of releasing glucose into the blood between meals, during fasting, and exercise thus preventing hypoglycemia by means of glycogenolysis and gluconeogenesis. These latter reactions coincide with the halting of glycolysis in the liver.

In animals, regulation of blood glucose levels by the pancreas in conjunction with the liver is a vital part of homeostasis. The beta cells in the pancreatic islets are sensitive to the blood glucose concentration. A rise in the blood glucose concentration causes them to release insulin into the blood, which has an effect particularly on the liver, but also on fat and muscle cells, causing these tissues to remove glucose from the blood. When the blood sugar falls the pancreatic beta cells cease insulin production, but, instead, stimulate the neighboring pancreatic alpha cells to release glucagon into the blood. This, in turn, causes the liver to release glucose into the blood by breaking down stored glycogen, and by means of gluconeogenesis. If the fall in the blood glucose level is particularly rapid or severe, other glucose sensors cause the release of epinephrine from the adrenal glands into the blood. This has the same action as glucagon on glucose metabolism, but its effect is more pronounced. In the liver glucagon and epinephrine cause the phosphorylation of the key, rate limiting enzymes of glycolysis, fatty acid synthesis, cholesterol synthesis, gluconeogenesis, and glycogenolysis. Insulin has the opposite effect on these enzymes. The phosphorylation and dephosphorylation of these enzymes (ultimately in response to the glucose level in the blood) is the dominant manner by which these pathways are controlled in the liver, fat, and muscle cells. Thus the phosphorylation of phosphofructokinase inhibits glycolysis, whereas its dephosphorylation through the action of insulin stimulates glycolysis.

In addition hexokinase and glucokinase act independently of the hormonal effects as controls at the entry points of glucose into the cells of different tissues. Hexokinase responds to the glucose-6-phosphate (G6P) level in the cell, or, in the case of glucokinase, to the blood sugar level in the blood to impart entirely intracellular controls of the glycolytic pathway in different tissues.

When glucose has been converted into G6P by hexokinase or glucokinase, it can either be converted to glucose-1-phosphate (G1P) for conversion to glycogen, or it is alternatively converted by glycolysis to pyruvate, which enters the mitochondrion where it is converted into acetyl-CoA and then into citrate. Excess citrate is exported from the mitochondrion back into the cytosol, where ATP citrate lyase regenerates acetyl-CoA and oxaloacetate (OAA). The acetyl-CoA is then used for fatty acid synthesis and cholesterol synthesis, two important ways of utilizing excess glucose when its concentration is high in blood. The rate limiting enzymes catalyzing these reactions perform these functions when they have been dephosphorylated through the action of insulin on the liver cells. Between meals, during fasting, exercise or hypoglycemia, glucagon and epinephrine are released into the blood. This causes liver glycogen to be converted back to G6P, and then converted to glucose by the liver-specific enzyme glucose 6-phosphatase and released into the blood. Glucagon and epinephrine also stimulate gluconeogenesis, which coverts non-carbohydrate substrates into G6P, which joins the G6P derived from glycogen, or substitutes for it when the liver glycogen store have been depleted. This is critical for brain function, since the brain utilizes glucose as an energy source under most conditions. The simultaneously phosphorylation of, particularly, phosphofructokinase, but also, to a certain extent pyruvate kinase, prevents glycolysis occurring at the same time as gluconeogenesis and glycogenolysis.

Hexokinase and Glucokinase

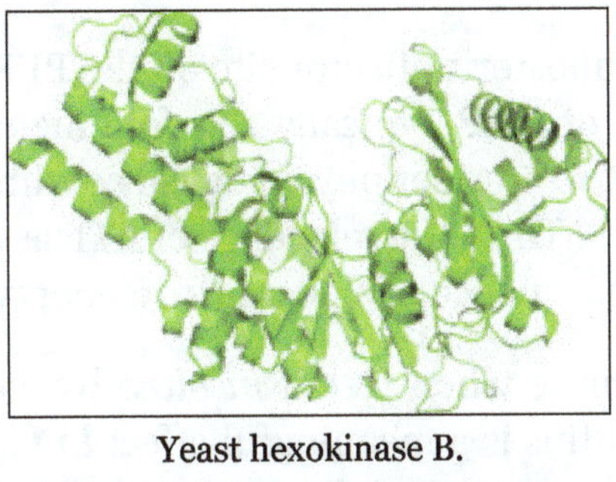

Yeast hexokinase B.

All cells contain the enzyme hexokinase, which catalyzes the conversion of glucose that has entered the cell into glucose-6-phosphate (G6P). Since the cell membrane is impervious to G6P, hexokinase essentially acts to transport glucose into the cells from which it can then no longer escape. Hexokinase is inhibited by high levels of G6P in the cell. Thus the rate of entry of glucose into cells partially depends on how fast G6P can be disposed of by glycolysis, and by glycogen synthesis (in the cells which store glycogen, namely liver and muscles).

Glucokinase, unlike hexokinase, is not inhibited by G6P. It occurs in liver cells, and will only phosphorylate the glucose entering the cell to form glucose-6-phosphate (G6P), when the sugar in the blood is abundant. This being the first step in the glycolytic pathway in the liver, it therefore imparts an additional layer of control of the glycolytic pathway in this organ.

Phosphofructokinase

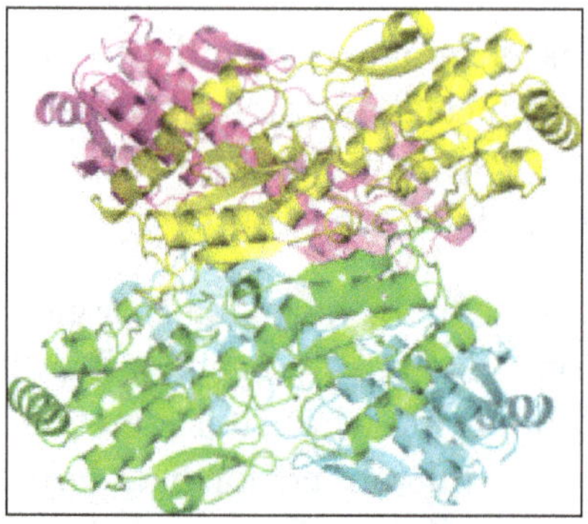

Bacillus stearothermophilus phosphofructokinase.

Phosphofructokinase is an important control point in the glycolytic pathway, since it is one of the irreversible steps and has key allosteric effectors, AMP and fructose 2,6-bisphosphate (F2,6BP).

Fructose 2,6-bisphosphate (F2, 6BP) is a very potent activator of phosphofructokinase (PFK-1) that is synthesized when F6P is phosphorylated by a second phosphofructokinase (PFK2). In the liver, when blood sugar is low and glucagon elevates cAMP, PFK2 is phosphorylated by protein kinase A. The phosphorylation inactivates PFK2, and another domain on this protein becomes active as fructose bisphosphatase-2, which converts F2,6BP back to F6P. Both glucagon and epinephrine cause high levels of cAMP in the liver. The result of lower levels of liver fructose-2,6-bisphosphate is a decrease in activity of phosphofructokinase and an increase in activity of fructose 1,6-bisphosphatase, so that gluconeogenesis (in essence, "glycolysis in reverse") is favored. This is consistent with the role of the liver in such situations, since the response of the liver to these hormones is to release glucose to the blood.

ATP competes with AMP for the allosteric effector site on the PFK enzyme. ATP concentrations in cells are much higher than those of AMP, typically 100-fold higher, but the concentration of ATP does not change more than about 10% under physiological conditions, whereas a 10% drop in ATP results in a 6-fold increase in AMP. Thus, the relevance of ATP as an allosteric effector is questionable. An increase in AMP is a consequence of a decrease in energy charge in the cell.

Citrate inhibits phosphofructokinase when tested *in vitro* by enhancing the inhibitory effect of ATP. However, it is doubtful that this is a meaningful effect *in vivo*, because citrate in the cytosol is utilized mainly for conversion to acetyl-CoA for fatty acid and cholesterol synthesis.

TIGAR, a p53 induced enzyme, is responsible for the regulation of Phosphofructokinase and acts to protect against oxidative stress. TIGAR is a single enzyme with dual function that regulates F2,6BP. It can behave as a phosphatase (Fructuose-2,6-Bisphosphatase) which cleaves the phosphate at carbon-2 producing F6P. It can also behave as a kinase (PFK2) adding a phosphate onto carbon-2 of F6P which produces F2,6BP. In humans, the TIGAR protein is encoded by *C12orf5* gene. The TIGAR enzyme will hinder the forward progression of glycolysis, by creating a build up of fructose-6-phosphate (F6P) which is isomerized into glucose-6-phosphate (G6P). The accumulation of G6P will shunt carbons into the pentose phosphate pathway.

Pyruvate Kinase

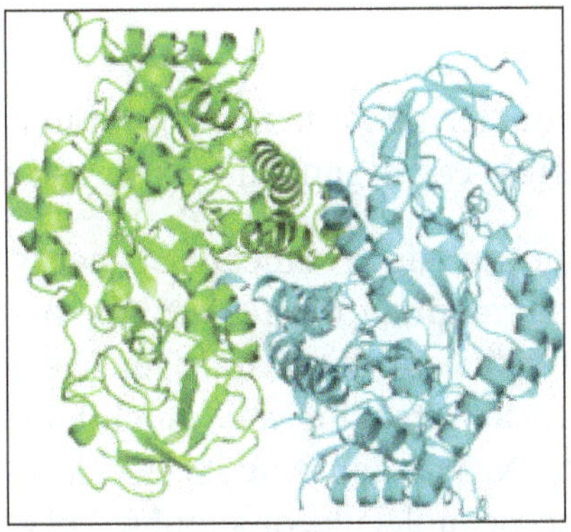

Yeast pyruvate kinase (PDB: 1A3W).

Pyruvate kinase enzyme catalyzes the last step of glycolysis, in which pyruvate and ATP are formed. Pyruvate kinase catalyzes the transfer of a phosphate group from phosphoenolpyruvate (PEP) to ADP, yielding one molecule of pyruvate and one molecule of ATP.

Liver pyruvate kinase is indirectly regulated by epinephrine and glucagon, through protein kinase A. This protein kinase phosphorylates liver pyruvate kinase to deactivate it. Muscle pyruvate kinase is not inhibited by epinephrine activation of protein kinase A. Glucagon signals fasting (no glucose available). Thus, glycolysis is inhibited in the liver but unaffected in muscle when fasting. An increase in blood sugar leads to secretion of insulin, which activates phosphoprotein phosphatase I, leading to dephosphorylation and activation of pyruvate kinase. These controls prevent pyruvate kinase from being active at the same time as the enzymes that catalyze the reverse reaction (pyruvate carboxylase and phosphoenolpyruvate carboxykinase), preventing a futile cycle.

Post-glycolysis Processes

The overall process of glycolysis is:

$$\text{Glucose} + 2\ \text{NAD}^+ + 2\ \text{ADP} + 2\ \text{P}_i \rightarrow 2\ \text{Pyruvate} + 2\ \text{NADH} + 2\ \text{H}^+ + 2\ \text{ATP}$$

If glycolysis were to continue indefinitely, all of the NAD^+ would be used up, and glycolysis would stop. To allow glycolysis to continue, organisms must be able to oxidize NADH back to NAD^+. How this is performed depends on which external electron acceptor is available.

Anoxic Regeneration of NAD^+

One method of doing this is to simply have the pyruvate do the oxidation; in this process, pyruvate is converted to lactate (the conjugate base of lactic acid) in a process called lactic acid fermentation:

$$\text{Pyruvate} + \text{NADH} + \text{H}^+ \rightarrow \text{Lactate} + \text{NAD}^+$$

This process occurs in the bacteria involved in making yogurt (the lactic acid causes the milk to curdle). This process also occurs in animals under hypoxic (or partially anaerobic) conditions, found, for example, in overworked muscles that are starved of oxygen. In many tissues, this is a cellular last resort for energy; most animal tissue cannot tolerate anaerobic conditions for an extended period of time.

Some organisms, such as yeast, convert NADH back to NAD^+ in a process called ethanol fermentation. In this process, the pyruvate is converted first to acetaldehyde and carbon dioxide, and then to ethanol.

Lactic acid fermentation and ethanol fermentation can occur in the absence of oxygen. This anaerobic fermentation allows many single-cell organisms to use glycolysis as their only energy source.

Anoxic regeneration of NAD^+ is only an effective means of energy production during short, intense exercise in vertebrates, for a period ranging from 10 seconds to 2 minutes during a maximal effort in humans. (At lower exercise intensities it can sustain muscle activity in diving animals, such as seals, whales and other aquatic vertebrates, for very much longer periods of time.) Under these conditions NAD^+ is replenished by NADH donating its electrons to pyruvate to form lactate. This produces 2 ATP molecules per glucose molecule, or about 5% of glucose's energy potential (38 ATP molecules in bacteria). But the speed at which ATP is produced in this manner is about 100 times that of oxidative phosphorylation. The pH in the cytoplasm quickly drops when hydrogen ions accumulate in the muscle, eventually inhibiting the enzymes involved in glycolysis.

The burning sensation in muscles during hard exercise can be attributed to the release of hydrogen ions during the shift to glucose fermentation from glucose oxidation to carbon dioxide and water, when aerobic metabolism can no longer keep pace with the energy demands of the muscles. These hydrogen ions form a part of lactic acid. The body falls back on this less efficient but faster method of producing ATP under low oxygen conditions. This is thought to have been the primary means of energy production in earlier organisms before oxygen reached high concentrations in the atmosphere between 2000 and 2500 million years ago, and thus would represent a more ancient form of energy production than the aerobic replenishment of NAD^+ in cells.

The liver in mammals gets rid of this excess lactate by transforming it back into pyruvate under aerobic conditions.

Fermentation of pyruvate to lactate is sometimes also called "anaerobic glycolysis", however, glycolysis ends with the production of pyruvate regardless of the presence or absence of oxygen.

In the above two examples of fermentation, NADH is oxidized by transferring two electrons to pyruvate. However, anaerobic bacteria use a wide variety of compounds as the terminal electron acceptors in cellular respiration: nitrogenous compounds, such as nitrates and nitrites; sulfur compounds, such as sulfates, sulfites, sulfur dioxide, and elemental sulfur; carbon dioxide; iron compounds; manganese compounds; cobalt compounds; and uranium compounds.

Aerobic Regeneration of NAD^+ and Disposal of Pyruvate

In aerobic organisms, a complex mechanism has been developed to use the oxygen in air as the final electron acceptor.

- Firstly, the NADH + H^+ generated by glycolysis has to be transferred to the mitochondrion to be oxidized, and thus to regenerate the NAD^+ necessary for glycolysis to continue. However the inner mitochondrial membrane is impermeable to NADH and NAD^+. Use is therefore made of two "shuttles" to transport the electrons from NADH across the mitochondrial membrane. They are the malate-aspartate shuttle and the glycerol phosphate shuttle. In the former the electrons from NADH are transferred to cytosolic oxaloacetate to form malate. The malate then traverses the inner mitochondrial membrane into the mitochondrial matrix, where it is reoxidized by NAD^+ forming intra-mitochondrial oxaloacetate and NADH. The oxaloacetate is then re-cycled to the cytosol via its conversion to aspartate which is readily transported out of the mitochondrion. In the glycerol phosphate shuttle electrons from cytosolic NADH are transferred to dihydroxyacetone to form glycerol-3-phosphate which readily traverses the outer mitochondrial membrane. Glycerol-3-phosphate is then reoxidized to dihydroxyacetone, donating its electrons to FAD instead of NAD^+. This reaction takes place on the inner mitochondrial membrane, allowing $FADH_2$ to donate its electrons directly to coenzyme Q (ubiquinone) which is part of the electron transport chain which ultimately transfers electrons to molecular oxygen (O_2), with the formation of water, and the release of energy eventually captured in the form of ATP.
- The glycolytic end-product, pyruvate (plus NAD^+) is converted to acetyl-CoA, CO_2 and NADH + H^+ within the mitochondria in a process called pyruvate decarboxylation.
- The resulting acetyl-CoA enters the citric acid cycle (or Krebs Cycle), where the acetyl group of the acetyl-CoA is converted into carbon dioxide by two decarboxylation reactions with the formation of yet more intra-mitochondrial NADH + H^+.
- The intra-mitochondrial NADH + H^+ is oxidized to NAD^+ by the electron transport chain, using oxygen as the final electron acceptor to form water. The energy released during this process is used to create a hydrogen ion (or proton) gradient across the inner membrane of the mitochondrion.
- Finally, the proton gradient is used to produce about 2.5 ATP for every NADH + H^+ oxidized in a process called oxidative phosphorylation.

Conversion of Carbohydrates into Fatty Acids and Cholesterol

The pyruvate produced by glycolysis is an important intermediary in the conversion of carbohydrates into fatty acids and cholesterol. This occurs via the conversion of pyruvate into acetyl-CoA in the mitochondrion. However, this acetyl CoA needs to be transported into cytosol where the synthesis of fatty acids and cholesterol occurs. This cannot occur directly. To obtain cytosolic acetyl-CoA, citrate (produced by the condensation of acetyl CoA with oxaloacetate) is removed from the citric acid cycle and carried across the inner mitochondrial membrane into the cytosol. There it is cleaved by ATP citrate lyase into acetyl-CoA and oxaloacetate. The oxaloacetate is returned to mitochondrion as malate (and then back into oxaloacetate to transfer more acetyl-CoA out of the mitochondrion). The cytosolic acetyl-CoA can be carboxylated by acetyl-CoA carboxylase into malonyl CoA, the first committed step in the synthesis of fatty acids, or it can be combined with acetoacetyl-CoA to form 3-hydroxy-3-methylglutaryl-CoA (HMG-CoA) which is the rate limiting step controlling the synthesis of cholesterol. Cholesterol can be used as is, as a structural component of cellular membranes, or it can be used to synthesize the steroid hormones, bile salts, and vitamin D.

Conversion of Pyruvate into Oxaloacetate for the Citric Acid Cycle

Pyruvate molecules produced by glycolysis are actively transported across the inner mitochondrial membrane, and into the matrix where they can either be oxidized and combined with coenzyme A to form CO_2, acetyl-CoA, and NADH, or they can be carboxylated (by pyruvate carboxylase) to form oxaloacetate. This latter reaction "fills up" the amount of oxaloacetate in the citric acid cycle, and is therefore an anaplerotic reaction, increasing the cycle's capacity to metabolize acetyl-CoA when the tissue's energy needs (e.g. in heart and skeletal muscle) are suddenly increased by activity. In the citric acid cycle all the intermediates (e.g. citrate, iso-citrate, alpha-ketoglutarate, succinate, fumarate, malate and oxaloacetate) are regenerated during each turn of the cycle. Adding more of any of these intermediates to the mitochondrion therefore means that that additional amount is retained within the cycle, increasing all the other intermediates as one is converted into the other. Hence the addition of oxaloacetate greatly increases the amounts of all the citric acid intermediates, thereby increasing the cycle's capacity to metabolize acetyl CoA, converting its acetate component into CO_2 and water, with the release of enough energy to form 11 ATP and 1 GTP molecule for each additional molecule of acetyl CoA that combines with oxaloacetate in the cycle.

To cataplerotically remove oxaloacetate from the citric cycle, malate can be transported from the mitochondrion into the cytoplasm, decreasing the amount of oxaloacetate that can be regenerated. Furthermore, citric acid intermediates are constantly used to form a variety of substances such as the purines, pyrimidines and porphyrins.

Intermediates for other Pathways

This article concentrates on the catabolic role of glycolysis with regard to converting potential chemical energy to usable chemical energy during the oxidation of glucose to pyruvate. Many of the metabolites in the glycolytic pathway are also used by anabolic pathways, and, as a consequence, flux through the pathway is critical to maintain a supply of carbon skeletons for biosynthesis.

The following metabolic pathways are all strongly reliant on glycolysis as a source of metabolites: and many more.

- Pentose phosphate pathway, which begins with the dehydrogenation of glucose-6-phosphate, the first intermediate to be produced by glycolysis, produces various pentose sugars, and NADPH for the synthesis of fatty acids and cholesterol.
- Glycogen synthesis also starts with glucose-6-phosphate at the beginning of the glycolytic pathway.
- Glycerol, for the formation of triglycerides and phospholipids, is produced from the glycolytic intermediate glyceraldehyde-3-phosphate.
- Various post-glycolytic pathways:
    - Fatty acid synthesis,
    - Cholesterol synthesis.
- The citric acid cycle which in turn leads to:
    - Amino acid synthesis,

- Nucleotide synthesis,
- Tetrapyrrole synthesis.

Although gluconeogenesis and glycolysis share many intermediates the one is not functionally a branch or tributary of the other. There are two regulatory steps in both pathways which, when active in the one pathway, are automatically inactive in the other. The two processes can therefore not be simultaneously active. Indeed, if both sets of reactions were highly active at the same time the net result would be the hydrolysis of four high energy phosphate bonds (two ATP and two GTP) per reaction cycle.

NAD^+ is the oxidizing agent in glycolysis, as it is in most other energy yielding metabolic reactions (e.g. beta-oxidation of fatty acids, and during the citric acid cycle). The NADH thus produced is primarily used to ultimately transfer electrons to O_2 to produce water, or, when O_2 is not available, to produced compounds such as lactate or ethanol. NADH is rarely used for synthetic processes, the notable exception being gluconeogenesis. During fatty acid and cholesterol synthesis the reducing agent is NADPH. This difference exemplifies a general principle that NADPH is consumed during biosynthetic reactions, whereas NADH is generated in energy-yielding reactions. The source of the NADPH is two-fold. When malate is oxidatively decarboxylated by "$NADP^+$-linked malic enzyme" pyruvate, CO_2 and NADPH are formed. NADPH is also formed by the pentose phosphate pathway which converts glucose into ribose, which can be used in synthesis of nucleotides and nucleic acids, or it can be catabolized to pyruvate.

Glycolysis in Disease

Diabetes

Cellular uptake of glucose occurs in response to insulin signals, and glucose is subsequently broken down through glycolysis, lowering blood sugar levels. However, the low insulin levels seen in diabetes result in hyperglycemia, where glucose levels in the blood rise and glucose is not properly taken up by cells. Hepatocytes further contribute to this hyperglycemia through gluconeogenesis. Glycolysis in hepatocytes controls hepatic glucose production, and when glucose is overproduced by the liver without having a means of being broken down by the body, hyperglycemia results.

Genetic Diseases

Glycolytic mutations are generally rare due to importance of the metabolic pathway, this means that the majority of occurring mutations result in an inability for the cell to respire, and therefore cause the death of the cell at an early stage. However, some mutations are seen with one notable example being Pyruvate kinase deficiency, leading to chronic hemolytic anemia.

Cancer

Malignant tumor cells perform glycolysis at a rate that is ten times faster than their noncancerous tissue counterparts. During their genesis, limited capillary support often results in hypoxia (decreased O_2 supply) within the tumor cells. Thus, these cells rely on anaerobic metabolic processes such as glycolysis for ATP (adenosine triphosphate). Some tumor cells overexpress specific glycolytic enzymes which result in higher rates of glycolysis. Often these enzymes are Isoenzymes, of traditional glycolysis enzymes, that vary in their susceptibility to traditional feedback inhibition. The increase in glycolytic activity ultimately counteracts the effects of hypoxia by generating sufficient ATP from this anaerobic pathway. This phenomenon was first described in 1930 by Otto Warburg and is referred to as the Warburg effect. The Warburg hypothesis claims that cancer is primarily caused by dysfunctionality in mitochondrial metabolism, rather than because of the uncontrolled growth of cells. A number of theories have been advanced to explain the Warburg effect. One such theory suggests that the increased glycolysis is a normal protective process of the body and that malignant change could be primarily caused by energy metabolism.

This high glycolysis rate has important medical applications, as high aerobic glycolysis by malignant tumors is utilized clinically to diagnose and monitor treatment responses of cancers by imaging uptake of 2-^{18}F-2-deoxyglucose (FDG) (a radioactive modified hexokinase substrate) with positron emission tomography (PET).

There is ongoing research to affect mitochondrial metabolism and treat cancer by reducing glycolysis and thus starving cancerous cells in various new ways, including a ketogenic diet.

Pathway Map

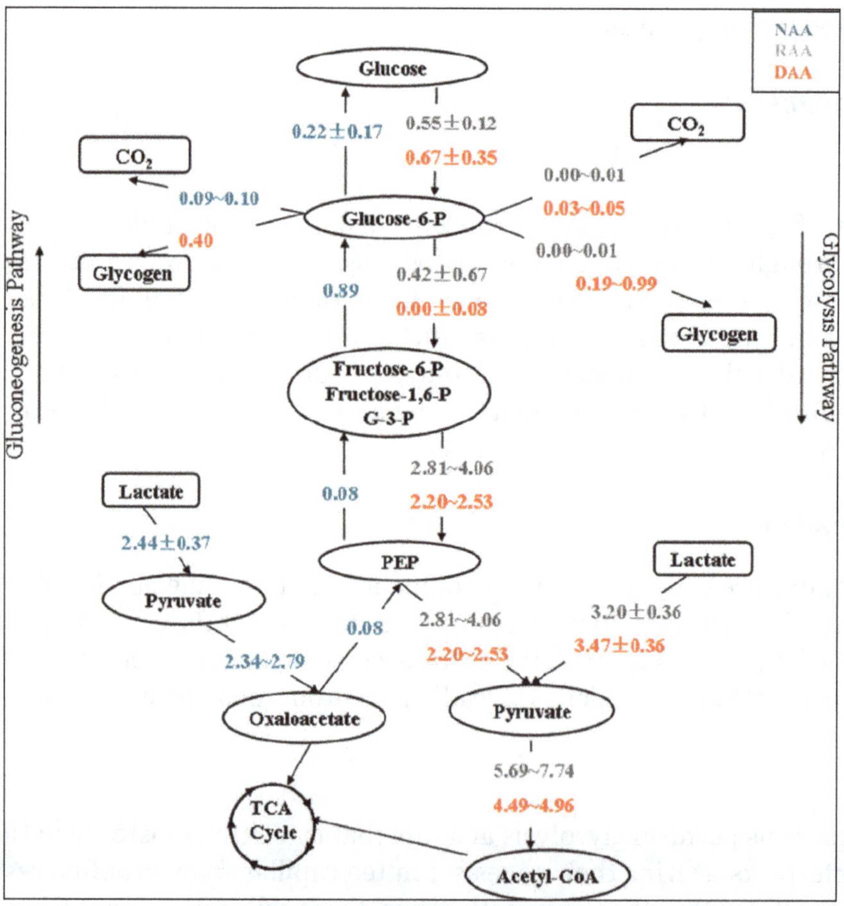

Glycolysis and Gluconeogenesis.

Alternative Nomenclature

Some of the metabolites in glycolysis have alternative names and nomenclature. In part, this is because some of them are common to other pathways, such as the Calvin cycle.

This article		Alternative names	Alternative nomenclature
Glucose	Glc	Dextrose	
Glucose-6-phosphate	G6P		
Fructose-6-phosphate	F6P		
Fructose-1,6-bisphosphate	F1,6BP	Fructose 1,6-diphosphate	FBP, FDP, F1,6DP
Dihydroxyacetone phosphate	DHAP	Glycerone phosphate	
Glyceraldehyde-3-phosphate	GADP	3-Phosphoglyceraldehyde	PGAL, G3P, GALP,GAP,TP
1,3-Bisphosphoglycerate	1,3BPG	Glycerate-1,3-bisphosphate, glycerate-1,3-diphosphate, 1,3-diphosphoglycerate	PGAP, BPG, DPG
3-Phosphoglycerate	3PG	Glycerate-3-phosphate	PGA, GP
2-Phosphoglycerate	2PG	Glycerate-2-phosphate	
Phosphoenolpyruvate	PEP		
Pyruvate	Pyr	Pyruvic acid	

Oxidative Phosphorylation

Oxidative Phosphorylation, also known as OXPHOS, refers to the redox reactions involving the flow of electrons along a series of membrane-bound proteins, coupled with the generation of Adenosine triphosphate (ATP).

Oxidative phosphorylation is the fourth and final step in cellular respiration. While respiration can be represented as the simple 'combustion' of carbohydrates to form carbon dioxide and water, the cell cannot afford to release all the chemical energy stored in carbohydrates in a single step, since it would irreversibly damage the intricate balance and homeostasis of the organism.

$$C_m(H_2O)_n + xO_2 \rightarrow mCO_2 + yH_2O + \text{energy}$$

Equation Represents the Combustion of Carbohydrates

Instead, nutrients are first digested and assimilated. They undergo metabolism in the cytoplasm and their end products are transported into the mitochondria, to participate in the Kreb's cycle, also known as the citric acid cycle. Here, the chemical energy in the organic molecules is released gradually, through step-wise oxidation to carbon dioxide. The process also generates a number of high-energy electrons, which are harnessed by special molecules called electron carriers. The most common electron carriers associated with oxidative phosphorylation are nicotinamide adenine dinucleotide (NAD^+) and flavin adenine dinucleotide (FAD). When NAD^+ and FAD accept electrons, they become reducing agents (NADH and $FADH_2$) that are capable of transferring these electrons to molecules that have a high affinity for them.

Oxidative phosphorylation begins with the oxidation of NADH and $FADH_2$. The energy of the electrons released by these two molecules is harnessed in a step-wise manner and used to create a proton gradient across the inner mitochondrial membrane in eukaryotes. This proton gradient then powers the formation of ATP from ADP, catalyzed by the enzyme ATP synthase. The ultimate acceptor of these high-energy electrons is oxygen and therefore oxidative phosphorylation generates both ATP and water.

ATP as Energy Currency

ATP has three phosphate groups in close physical proximity to one another. The addition of every phosphate group needs to be coupled with some other energy-releasing or exergonic reaction, since the natural repulsion between the negatively charged phosphate groups needs to be overcome. Once ATP is formed, it becomes a ready store of energy, because the terminal phosphate bond can be quickly hydrolyzed to power some other process within the cell. This makes ATP behave like the energy currency of the cell.

The presence of ATP allows the cell to conduct respiration, store the energy temporarily in the high-energy bonds of the molecule, and use it at different locations and points in time. Without ATP, every single endergonic (energy-requiring) reaction would have to be coupled in time and space with the oxidation of nutrients, severely limiting the complexity possible within a living organism.

The importance of this molecule is underscored by the fact that ATP is found in all living cells and ATP synthase is largely conserved across the living world. Additionally, any process that fundamentally interrupts the generation of this molecule leads to a very quick death.

Energy Conversion

The original energy source for nearly the entire biosphere is the nuclear reactions within the sun. From plants and other autotrophs, to single celled protozoa and the largest mammals, the energy to sustain life is derived from the sun through a series of energy conversions.

In autotrophs, solar radiation is first used to generate high-energy electrons, which are then used to pump protons against their concentration gradient, creating a proton-motive force across a membrane. The potential energy in such an electrochemical gradient is channeled to generate ATP, which, in turn, facilitates the formation of complex macromolecules. Nutrients created by autotrophs are consumed by heterotrophs, digested and then metabolized within their cells. The chemical bond energy in these molecules behaves like a storage system for the energy initially harnessed from the sun. When nutrients are oxidized, the bond energy is released – both as ATP and as high-energy electrons. In a process that parallels the initial reactions within chloroplasts, these electrons are used to gradually create an electrochemical gradient that, once again, powers the formation of ATP. ATP is repeatedly generated and utilized to sustain the living processes of the organism.

Energy from the sun, therefore, is transmuted from one form to another, as the energy in electrons, the potential energy in proton gradients and the bond energy of macromolecules.

Structure of Mitochondria

The mitochondria is said to have evolved from ancient bacteria that became endosymbionts within eukaryotes, creating the first nucleated cells that could undergo aerobic respiration. The structure of mitochondrial membranes reflects this origin; these organelles have a distinct genome, independent protein translation machinery (tRNA, ribosomes, and associated proteins) and respiratory chain complexes that mirror aerobic respiration in bacteria.

Mitochondria have two membranes – the inner and outer membrane – and the outer membrane is structurally similar to eukaryotic plasma membranes. However, they also contain a number of transmembrane channels called porins. The largest molecules to diffuse freely through porins are about 6000 Daltons in atomic mass, making the membrane permeable to most small molecules and proteins. The inner membrane is impermeable to nearly every molecule with the exception of oxygen, carbon dioxide and water. This property is important, because it allows the organelle to regulate the flow of ions and molecules across the membrane and use the differential concentration of these substances to power the formation of ATP.

The inner mitochondrial membrane has a number of invaginations called cristae that increase its surface area. The protein complexes involved in the redox reactions of oxidative phosphorylation are embedded in cristae. Similarly, ATP synthase is also present as a transmembrane protein on cristae. The inner mitochondrial membrane encloses a protein-rich matrix that receives pyruvate molecules from the cytoplasm and contains enzymes that generate acetyl coenzyme A. The matrix is also the site where the reactions of the citric acid cycle occur. The matrix is so dense that it resembles a protein crystal, with a protein concentration of 500 mg/ml.

The impermeability of the inner mitochondrial membrane to ions allows it to maintain an electrochemical potential of 180 mV generated primarily by the pumping of protons from the matrix into the inter-membrane space. This makes the matrix of the mitochondrion slightly alkaline, with a pH of about 8.

Electron Transport Chain

The electron transport chain of oxidative phosphorylation involves four major protein complexes, (numbered I, II, III and IV) each with increasing reduction potentials. In essence, as the electron moves from one protein complex to the next, it travels towards molecules that have a greater affinity for electrons. The electron loses energy in this process, and this energy is used to pump protons into the inter-membrane space.

Between the two electron carriers, NADH has a lower reduction potential, and releases electrons to complex I. Complex I is also known as NADH:quinone oxidoreductase and is instrumental in the transfer the electrons from NADH to a protein called ubiquinone. This enormous membrane-bound complex consists of 46 polypeptide chains and can combine the acceptance of two electrons from NADH with the movement of 4 H^+ ions from the matrix to the inter-membrane space. Each of these four protons is pumped through a separate, dedicated channel.

On the other hand, $FADH_2$ donates its electrons through complex II, which is also known as succinate dehydrogenase. This relatively smaller complex consists of four subunits and does not facilitate the tandem movement of protons across the membrane. Of the four subunits, two act as hydrophobic anchors to the inner mitochondrial membrane. The third subunit has a covalently attached FAD molecule. The four subunits together form a ubiquinone-binding site. Complex II participates in both the citric acid cycle and in oxidative phosphorylation. Therefore, it becomes a parallel route for electrons to reach the quinone pool, and FAD receives high-energy electrons from the products of fatty acid metabolism.

Thereafter, the electrons travel through the same pathway, being fed into complex III through ubiquinone, which acts as a mobile electron carrier within the membrane. Complex III is also involved in pumping H^+ ions into the inter-membrane space. From complex III, another mobile carrier called cytochrome C carries electrons to complex IV. Complex IV acts as the site for the final step in these reactions involving the splitting of an oxygen molecule and the formation of water. Water is the final resting spot for the electrons that entered the system through NADH and $FADH_2$ and is either used in the cell's metabolic pathways or is excreted from the body.

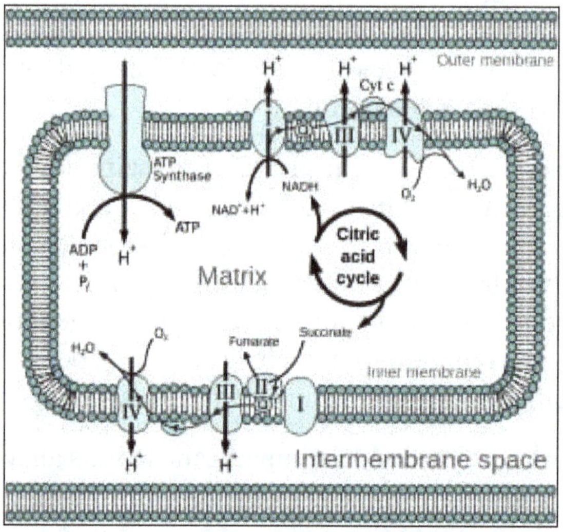

This is a schematic representation of the complexes involved in oxidative phosphorylation and ATP generation.

At the end of the electron transport chain, NAD^+ and FAD are regenerated and the electrochemical gradient is created.

ATP Synthase and ATP Generation

The inner mitochondrial membrane is impermeable to ions and has one dedicated channel for the flow of protons back into the matrix. This is the membrane-bound ATP synthase enzyme. It consists of two parts – the F_0 and F_1 regions. F_0 forms the proton pore and is embedded within the membrane. When protons flow back into the matrix from the inter-membrane space, the catalytic activity of the enzyme uses the potential energy released due to chemiosmosis to synthesis ATP from ADP and P_i.

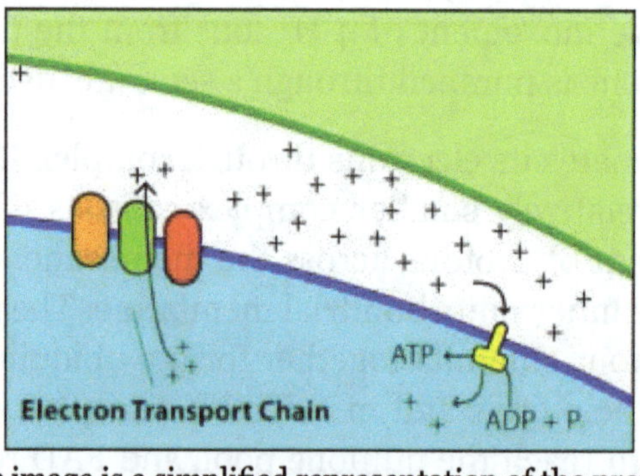

The image is a simplified representation of the proton motive force being used to generate ATP.

Carbohydrate Metabolism

Carbohydrate metabolism refers to various biochemical processes responsible for the metabolic formation, breakdown, and interconversion of carbohydrates in living organisms.

Carbohydrates are central to many essential metabolic pathways. Plants synthesize carbohydrates from carbon dioxide and water through photosynthesis, allowing them to store energy absorbed from sunlight internally. When animals and fungi consume plants, they use cellular respiration to break down these stored carbohydrates to make energy available to cells. Both animals and plants temporarily store the released energy in the form of high energy molecules, such as ATP, for use in various cellular processes.

Although humans consume a variety of carbohydrates, digestion breaks down complex carbohydrates into a few simple monomers (monosaccharides) for metabolism: glucose, fructose, and galactose. Glucose constitutes about 80% of the products, and is the primary structure that is distributed to cells in the tissues, where it is broken down or stored as glycogen. In aerobic respiration, the main form of cellular respiration used by humans, glucose and oxygen are metabolized to release energy, with carbon dioxide and water as byproducts. Most of the fructose and galactose travel to the liver, where they can be converted to glucose.

Some simple carbohydrates have their own enzymatic oxidation pathways, as do only a few of the more complex carbohydrates. The disaccharide lactose, for instance, requires the enzyme lactase to be broken into its monosaccharide components, glucose and galactose.

Metabolic Pathways

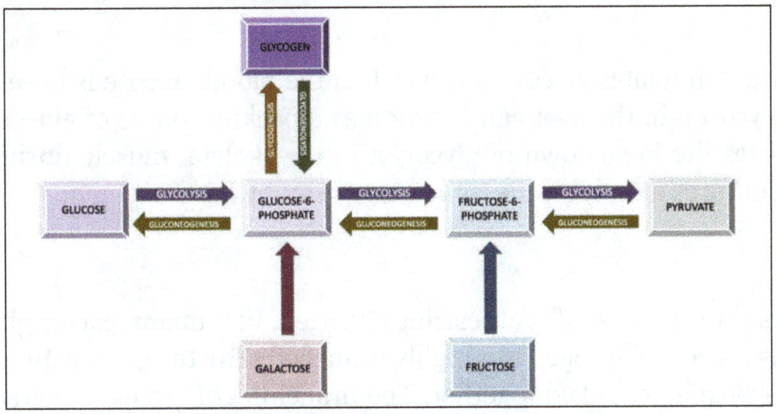

Connections between metabolic processes.

Glycolysis

Glycolysis is the process of breaking down a glucose molecule into two pyruvate molecules, while storing energy released during this process as ATP and NADH. Nearly all organisms that break down glucose utilize glycolysis. Glucose regulation and product use are the primary categories in which these pathways differ between organisms. In some tissues and organisms, glycolysis is the sole method of energy production. This pathway is common to both anaerobic and aerobic respiration.

Glycolysis consists of ten steps, split into two phases. During the first phase, it requires the breakdown of two ATP molecules. During the second phase, chemical energy from the intermediates is transferred into ATP and NADH. The breakdown of one molecule of glucose results in two molecules of pyruvate, which can be further oxidized to access more energy in later processes.

Gluconeogenesis

Gluconeogenesis is the reverse process of glycolysis. It involves the conversion of non-carbohydrate molecules into glucose. The non-carbohydrate molecules that are converted in this pathway include pyruvate, lactate, glycerol, alanine, and glutamine. This process occurs when there are lowered amounts of glucose. The production of glucose by this pathway is important to tissues that cannot use any other fuels, such as the brain. The liver is the primary location of gluconeogenesis, but some also occurs in the kidney.

This pathway is regulated by multiple different molecules. Glucagon, adrenocorticotropic hormone, and ATP encourage gluconeogenesis. Gluconeogenesis is inhibited by AMP, ADP, and insulin.

Glycogenolysis

Glycogenolysis refers to the breakdown of glycogen. In the liver, muscles, and the kidney, this process occurs to provide glucose when necessary. A single glucose molecule is cleaved from a branch of glycogen, and is transformed into glucose-1-phosphate during this process. This molecule can then be converted to glucose-6-phosphate, an intermediate in the glycolysis pathway.

Glucose-6-phosphate can then progress through glycolysis. Glycolysis only requires the input of one molecule of ATP when the glucose originates in glycogen. Alternatively, glucose-6-phosphate can be converted back into glucose in the liver and the kidneys, allowing it to raise blood glucose levels if necessary.

Glucagon in the liver stimulates glycogenolysis when the blood glucose is lowered, known as hypoglycemia. The glycogen in the liver can function as a backup source of glucose between meals. Adrenaline stimulates the breakdown of glycogen in the skeletal muscle during exercise. In the muscles, glycogen ensures a rapidly accessible energy source for movement.

Glycogenesis

Glycogenesis refers to the process of synthesizing glycogen. In humans, excess glucose is converted to glycogen via this process. Glycogen is a highly branched structure, consisting of glucose, in the form of glucose-6-phosphate, linked together. The branching of glycogen increases its solubility, and allows for a higher number of glucose molecules to be accessible for breakdown. Glycogenesis occurs primarily in the liver, skeletal muscles, and kidney.

Pentose Phosphate Pathway

The pentose phosphate pathway is an alternative method of oxidizing glucose. It occurs in the liver, adipose tissue, adrenal cortex, testis, milk glands, phagocyte cells, and red blood cells. It produces products that are used in other cell processes, while reducing NADP to NADPH. This pathway is regulated through changes in the activity of glucose-6-phosphate dehydrogenase.

Fructose Metabolism

Fructose must undergo certain extra steps in order to enter the glycolysis pathway. Enzymes located in certain tissues can add a phosphate group to fructose. This phosphorylation creates fructose-6-phosphate, an intermediate in the glycolysis pathway that can be broken down directly in those tissues. This pathway occurs in the muscles, adipose tissue, and kidney. In the liver, enzymes produce fructose-1-phosphate, which enters the glycolysis pathway and is later cleaved into glyceraldehyde and dihydroxyacetone phosphate.

Galactose Metabolism

Lactose, or milk sugar, consists of one molecule of glucose and one molecule of galactose. After separation from glucose, galactose travels to the liver for conversion to glucose. Galactokinase uses one molecule of ATP to phosphorylate galactose. The phosphorylated galactose is then converted to glucose-1-phosphate, and then eventually glucose-6-phosphate, which can be broken down in glycolysis.

Energy Production

Many steps of carbohydrate metabolism allow the cells to access energy, and store it more transiently in ATP. The cofactors NAD^+ and FAD are sometimes reduced during this process to form NADH and $FADH_2$, which drive the creation of ATP in other processes. A molecule of NADH can produce 1.5–2.5 molecules of ATP, whereas a molecule of $FADH_2$ will yield 1.5 molecules of ATP.

Energy produced during metabolism of one glucose molecule						
Pathway	ATP Input	ATP Output	Net ATP	NADH Output	$FADH_2$ Output	ATP Final Yield
Glycolysis (aerobic)	2	4	2	2	0	5-7
Citric Acid Cycle	0	2	2	8	2	17-25

Typically, the complete breakdown of one molecule of glucose via aerobic respiration (i.e. involving both glycolysis and the Citric Acid Cycle) is usually about 30–32 molecules of ATP. Oxidation of one gram of carbohydrate yields approximately 4 kcal of energy.

Hormonal Regulation

Glucoregulation is the maintenance of steady levels of glucose in the body.

Hormones released from the pancreas regulate the overall metabolism of glucose. Insulin and glucagon are the primary hormones involved in maintaining a steady level of glucose in the blood, and the release of each is controlled by the amount of nutrients currently available. The amount of insulin released in the blood and sensitivity of the cells to the insulin both determine the amount of glucose that cells break down. Increased levels of glucagon activates the enzymes that catalyze glycogenolysis, and inhibits the enzymes that catalyze glycogenesis. Conversely, glycogenesis is enhanced and glycogenolysis inhibited when there are high levels of insulin in the blood.

The level of circulatory glucose (known informally as "blood sugar") is the most important factor determining the amount of glucagon or insulin produced. The release of glucagon is precipitated by low levels of blood glucose, whereas high levels of blood glucose stimulates cells to produce insulin. Because the level of circulatory glucose is largely determined by the intake of dietary carbohydrates, diet controls major aspects of metabolism via insulin. In humans, insulin is made by beta cells in the pancreas, fat is stored in adipose tissue cells, and glycogen is both stored and released as needed by liver cells. Regardless of insulin levels, no glucose is released to the blood from internal glycogen stores from muscle cells.

Carbohydrates as Storage

Carbohydrates are typically stored as long polymers of glucose molecules with glycosidic bonds for structural support (e.g. chitin, cellulose) or for energy storage (e.g. glycogen, starch). However, the strong affinity of most carbohydrates for water makes storage of large quantities of carbohydrates inefficient due to the large molecular weight of the solvated water-carbohydrate complex. In most organisms, excess carbohydrates are regularly catabolised to form acetyl-CoA, which is a feed stock for the fatty acid synthesis pathway; fatty acids, triglycerides, and other lipids are commonly used for long-term energy storage. The hydrophobic character of lipids makes them a much more compact form of energy storage than hydrophilic carbohydrates. However, animals, including humans, lack the necessary enzymatic machinery and so do not synthesize glucose from lipids (with a few exceptions, e.g. glycerol).

In some animals (such as termites) and some microorganisms (such as protists and bacteria), cellulose can be disassembled during digestion and absorbed as glucose.

Lipid Metabolism

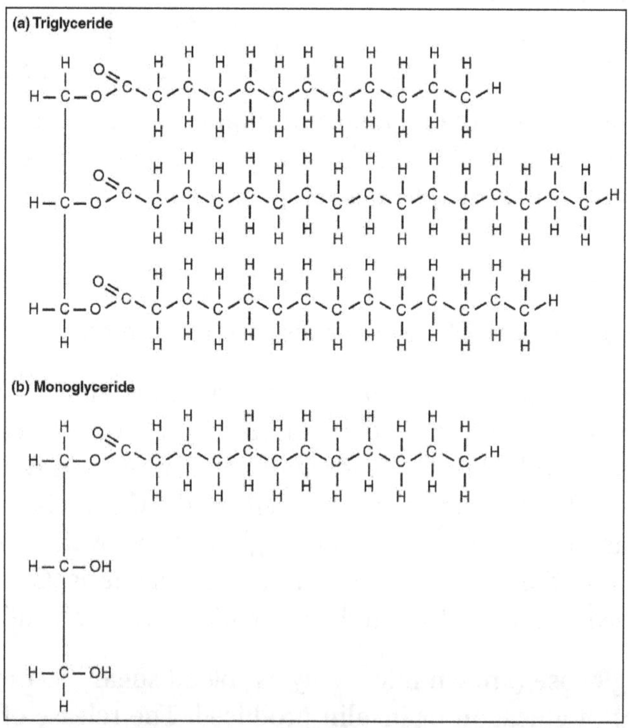

Triglyceride Broken Down into a Monoglyceride A triglyceride molecule (a) breaks down into a monoglyceride (b).

Fats (or triglycerides) within the body are ingested as food or synthesized by adipocytes or hepatocytes from carbohydrate precursors. Lipid metabolism entails the oxidation of fatty acids to either generate energy or synthesize new lipids from smaller constituent molecules. Lipid metabolism is associated with carbohydrate metabolism, as products of glucose (such as acetyl CoA) can be converted into lipids.

Lipid metabolism begins in the intestine where ingested triglycerides are broken down into smaller chain fatty acids and subsequently into monoglyceride molecules by pancreatic lipases, enzymes that break down fats after they are emulsified by bile salts. When food reaches the small intestine in the form of chyme, a digestive hormone called cholecystokinin (CCK) is released by intestinal cells in the intestinal mucosa. CCK stimulates the release of pancreatic lipase from the pancreas and stimulates the contraction of the gallbladder to release stored bile salts into the intestine. CCK also travels to the brain, where it can act as a hunger suppressant.

Together, the pancreatic lipases and bile salts break down triglycerides into free fatty acids. These fatty acids can be transported across the intestinal membrane. However, once they cross the membrane, they are recombined to again form triglyceride molecules. Within the intestinal cells, these triglycerides are packaged along with cholesterol molecules in phospholipid vesicles called chylomicrons. The chylomicrons enable fats and cholesterol to move within the aqueous environment of your lymphatic and circulatory systems. Chylomicrons leave the enterocytes by exocytosis and enter the lymphatic system via lacteals in the villi of the intestine. From the lymphatic system, the chylomicrons are transported to the circulatory system. Once in the circulation, they can either go to the liver or be stored in fat cells (adipocytes) that comprise adipose (fat) tissue found throughout the body.

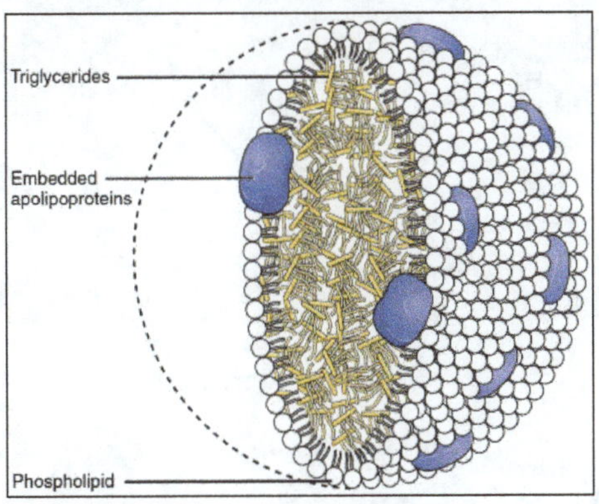

Chylomicrons. Chylomicrons contain triglycerides, cholesterol molecules, and other apolipoproteins (protein molecules). They function to carry these water-insoluble molecules from the intestine, through the lymphatic system, and into the bloodstream, which carries the lipids to adipose tissue for storage.

Lipolysis

To obtain energy from fat, triglycerides must first be broken down by hydrolysis into their two principal components, fatty acids and glycerol. This process, called lipolysis, takes place in the cytoplasm. The resulting fatty acids are oxidized by β-oxidation into acetyl CoA, which is used by the Krebs cycle. The glycerol that is released from triglycerides after lipolysis directly enters the glycolysis pathway as DHAP. Because one triglyceride molecule yields three fatty acid molecules with as much as 16 or more carbons in each one, fat molecules yield more energy than carbohydrates and are an important source of energy for the human body. Triglycerides yield more than twice the energy per unit mass when compared to carbohydrates and proteins. Therefore, when glucose levels are low, triglycerides can be converted into acetyl CoA molecules and used to generate ATP through aerobic respiration.

The breakdown of fatty acids, called fatty acid oxidation or beta (β)-oxidation, begins in the cytoplasm, where fatty acids are converted into fatty acyl CoA molecules. This fatty acyl CoA combines with carnitine to create a fatty acyl carnitine molecule, which helps to transport the fatty acid across the mitochondrial membrane. Once inside the mitochondrial matrix, the fatty acyl carnitine molecule is converted back into fatty acyl CoA and then into acetyl CoA. The newly formed acetyl CoA enters the Krebs cycle and is used to produce ATP in the same way as acetyl CoA derived from pyruvate.

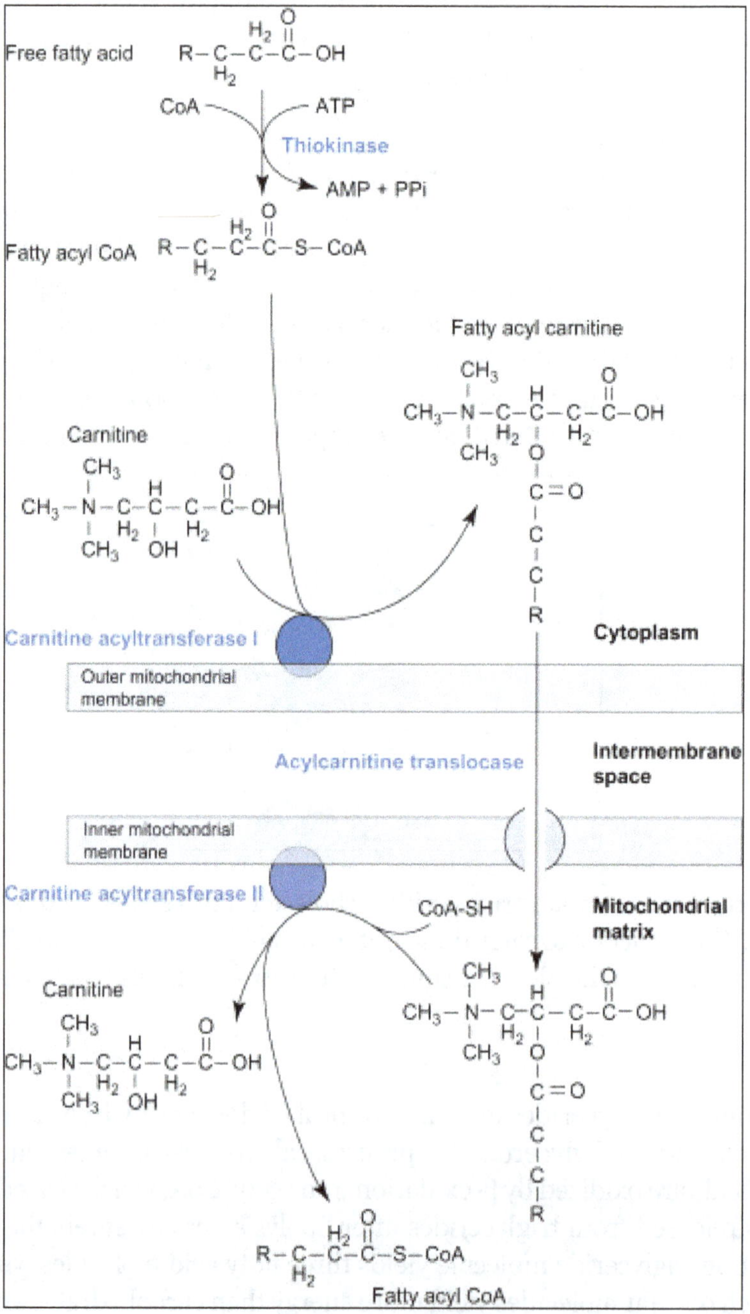

Ketogenesis

If excessive acetyl CoA is created from the oxidation of fatty acids and the Krebs cycle is overloaded and cannot handle it, the acetyl CoA is diverted to create ketone bodies. These ketone bodies can serve as a fuel source if glucose levels are too low in the body. Ketones serve as fuel in times of prolonged starvation or when patients suffer from uncontrolled diabetes and cannot utilize most of the circulating glucose. In both cases, fat stores are liberated to generate energy through the Krebs cycle and will generate ketone bodies when too much acetyl CoA accumulates.

In this ketone synthesis reaction, excess acetyl CoA is converted into hydroxymethylglutaryl CoA (HMG CoA). HMG CoA is a precursor of cholesterol and is an intermediate that is subsequently converted into β-hydroxybutyrate, the primary ketone body in the blood.

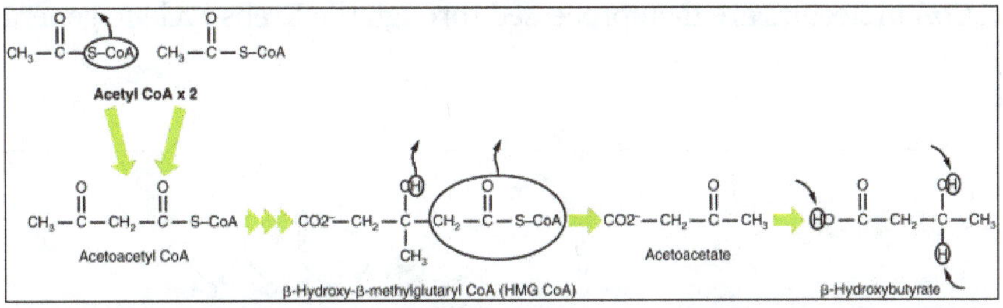

Ketogenesis. Excess acetyl CoA is diverted from the Krebs cycle to the ketogenesis pathway. This reaction occurs in the mitochondria of liver cells. The result is the production of β-hydroxybutyrate, the primary ketone body found in the blood.

Ketone Body Oxidation

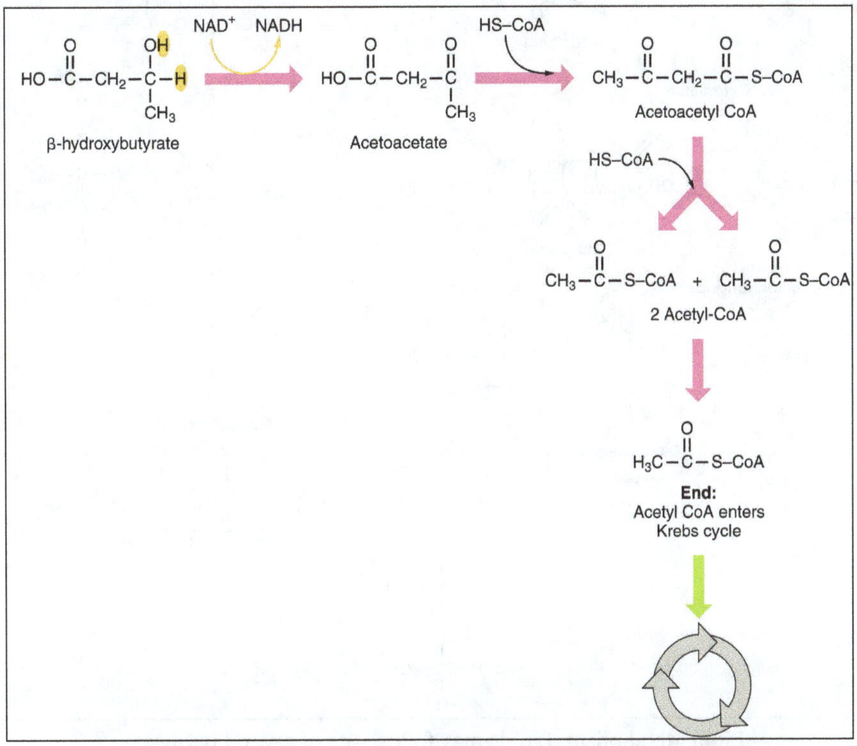

Ketone Oxidation. When glucose is limited, ketone bodies can be oxidized to produce acetyl CoA to be used in the Krebs cycle to generate energy.

Organs that have classically been thought to be dependent solely on glucose, such as the brain, can actually use ketones as an alternative energy source. This keeps the brain functioning when glucose is limited. When ketones are produced faster than they can be used, they can be broken down into CO_2 and acetone. The acetone is removed by exhalation. One symptom of ketogenesis is that the patient's breath smells sweet like alcohol. This effect provides one way of telling if a diabetic is properly controlling the disease. The carbon dioxide produced can acidify the blood, leading to diabetic ketoacidosis, a dangerous condition in diabetics.

Ketones oxidize to produce energy for the brain. beta (β)-hydroxybutyrate is oxidized to acetoacetate and NADH is released. An HS-CoA molecule is added to acetoacetate, forming acetoacetyl CoA. The carbon within the acetoacetyl CoA that is not bonded to the CoA then detaches, splitting the molecule in two. This carbon then attaches to another free HS-CoA, resulting in two acetyl CoA molecules. These two acetyl CoA molecules are then processed through the Krebs cycle to generate energy.

Lipogenesis

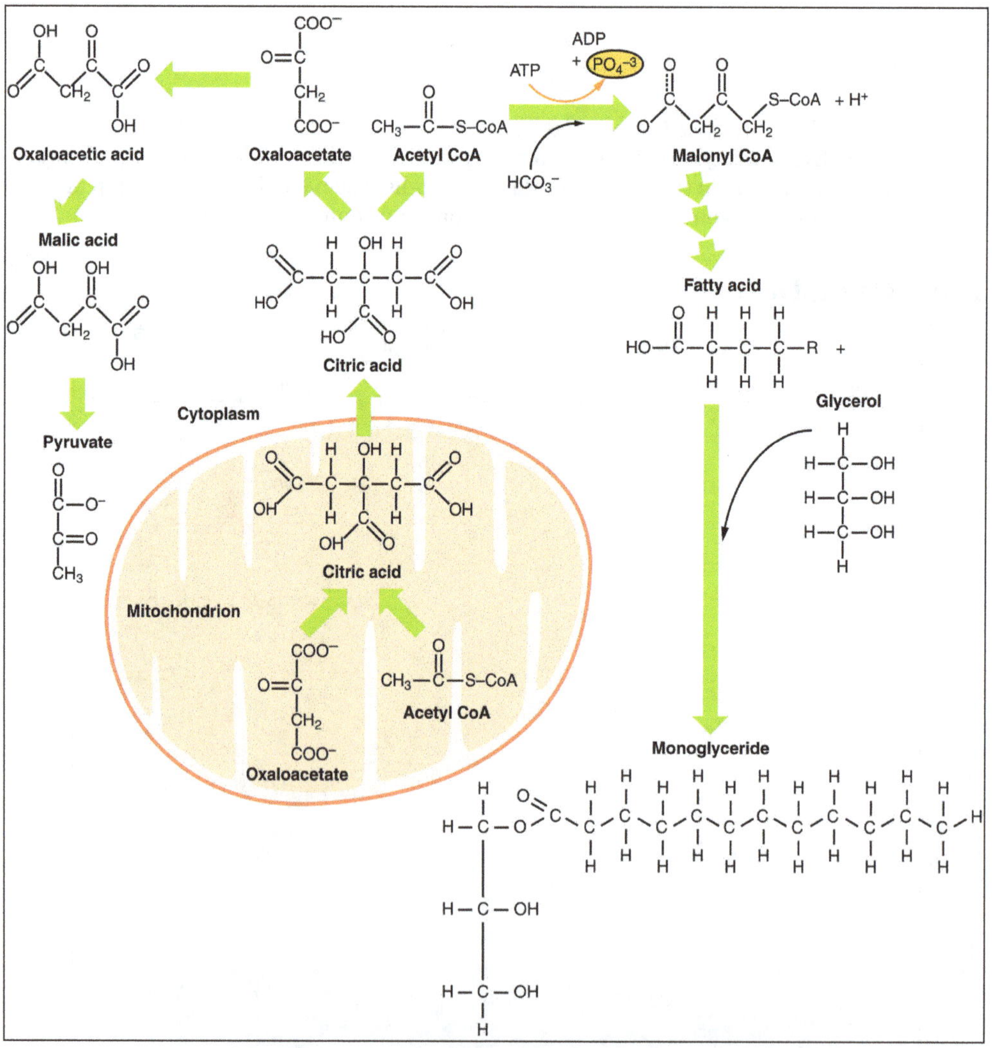

Lipid Metabolism. Lipids may follow one of several pathways during metabolism. Glycerol and fatty acids follow different pathways.

When glucose levels are plentiful, the excess acetyl CoA generated by glycolysis can be converted into fatty acids, triglycerides, cholesterol, steroids, and bile salts. This process, called lipogenesis, creates lipids (fat) from the acetyl CoA and takes place in the cytoplasm of adipocytes (fat cells) and hepatocytes (liver cells). When you eat more glucose or carbohydrates than your body needs, your system uses acetyl CoA to turn the excess into fat. Although there are several metabolic sources of acetyl CoA, it is most commonly derived from glycolysis. Acetyl CoA availability is significant, because it initiates lipogenesis. Lipogenesis begins with acetyl CoA and advances by the subsequent addition of two carbon atoms from another acetyl CoA; this process is repeated until fatty acids are the appropriate length. Because this is a bond-creating anabolic process, ATP is consumed. However, the creation of triglycerides and lipids is an efficient way of storing the energy available in carbohydrates. Triglycerides and lipids, high-energy molecules, are stored in adipose tissue until they are needed.

Although lipogenesis occurs in the cytoplasm, the necessary acetyl CoA is created in the mitochondria and cannot be transported across the mitochondrial membrane. To solve this problem, pyruvate is converted into both oxaloacetate and acetyl CoA. Two different enzymes are required for these conversions. Oxaloacetate forms via the action of pyruvate carboxylase, whereas the action of pyruvate dehydrogenase creates acetyl CoA. Oxaloacetate and acetyl CoA combine to form citrate, which can cross the mitochondrial membrane and enter the cytoplasm. In the cytoplasm, citrate is converted back into oxaloacetate and acetyl CoA. Oxaloacetate is converted into malate and then into pyruvate. Pyruvate crosses back across the mitochondrial membrane to wait for the next cycle of lipogenesis. The acetyl CoA is converted into malonyl CoA that is used to synthesize fatty acids. Figure summarizes the pathways of lipid metabolism.

Fatty Acid Metabolism

Fatty acid metabolism consists of catabolic processes that generate energy, and anabolic processes that create biologically important molecules (triglycerides, phospholipids, second messengers, local hormones and ketone bodies). Fatty acids are a family of molecules classified within the lipid macronutrient class. One role of fatty acids in animal metabolism is energy production, captured in the form of adenosine triphosphate (ATP). When compared to other macronutrient classes (carbohydrates and protein), fatty acids yield the most ATP on an energy per gram basis, when they are completely oxidized to CO_2 and water by beta oxidation and the citric acid cycle. Fatty acids (mainly in the form of triglycerides) are therefore the foremost storage form of fuel in most animals, and to a lesser extent in plants. In addition, fatty acids are important components of the phospholipids that form the phospholipid bilayers out of which all the membranes of the cell are constructed (the cell wall, and the membranes that enclose all the organelles within the cells, such as the nucleus, the mitochondria, endoplasmic reticulum, and the Golgi apparatus). Fatty acids can also be cleaved, or partially cleaved, from their chemical attachments in the cell membrane to form second messengers within the cell, and local hormones in the immediate vicinity of the cell. The prostaglandins made from arachidonic acid stored in the cell membrane, are probably the most well known group of these local hormones.

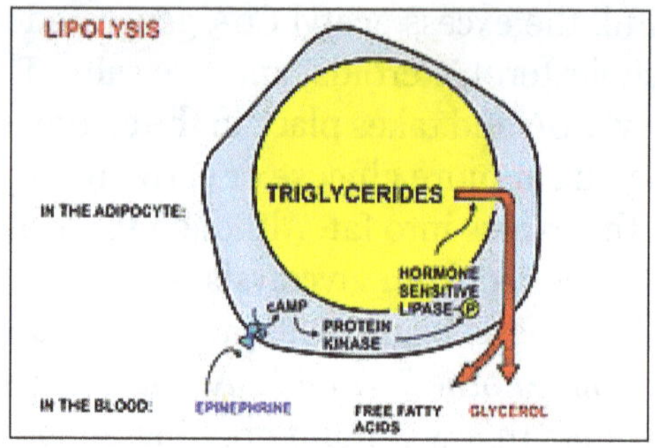

A diagrammatic illustration of the process of lipolysis (in a fat cell) induced by high epinephrine and low insulin levels in the blood. Epinephrine binds to a beta-adrenergic receptor in the cell membrane of the adipocyte, which causes cAMP to be generated inside the cell. The cAMP activates a protein kinase, which phosphorylates and thus, in turn, activates a hormone-sensitive lipase in the fat cell. This lipase cleaves free fatty acids from their attachment to glycerol in the fat stored in the fat droplet of the adipocyte. The free fatty acids and glycerol are then released into the blood. However more recent studies have shown that adipose triglyceride lipase has to first convert triacylglycerides to diacylglycerides, and that hormone-sensitive lipase converts the diacylglycerides to monoglycerides and free fatty acids. Monoglycerides are hydrolyzed by monoglyceride lipase. The activity of hormone sensitive lipase is regulated by the circulation hormones insulin, glucagon, norepinephrine, and epinephrine, as shown in the diagram.

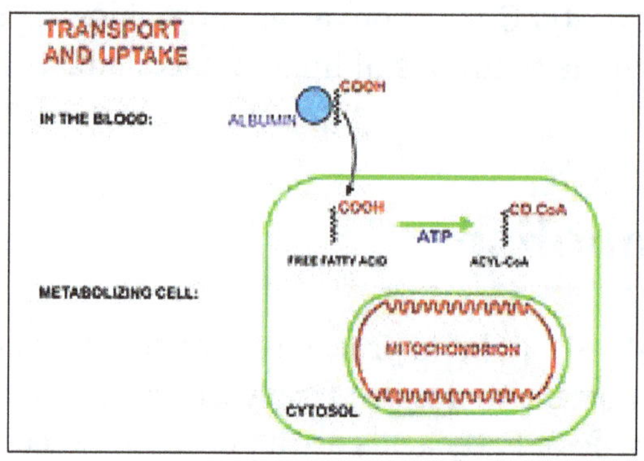

A diagrammatic illustration of the transport of free fatty acids in the blood attached to plasma albumin, its diffusion across the cell membrane using a protein transporter, and its activation, using ATP, to form acyl-CoA in the cytosol. The illustration is, for diagrammatic purposes, of a 12 carbon fatty acid. Most fatty acids in human plasma are 16 or 18 carbon atoms long.

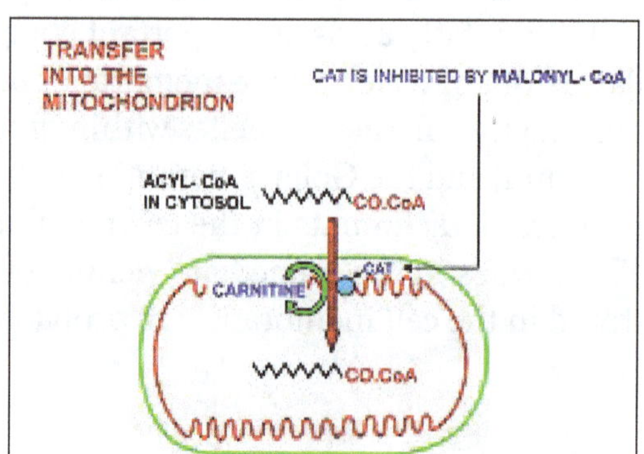

A diagrammatic illustration of the transfer of an acyl-CoA molecule across the inner membrane of the mitochondrion by carnitine-acyl-CoA transferase (CAT). The illustrated acyl chain is, for diagrammatic purposes, only 12 carbon atoms long. Most fatty acids in human plasma are 16 or 18 carbon atoms long. CAT is inhibited by high concentrations of malonyl-CoA (the first committed step in fatty acid synthesis) in the cytoplasm. This means that fatty acid synthesis and fatty acid catabolism cannot occur simultaneously in any given cell.

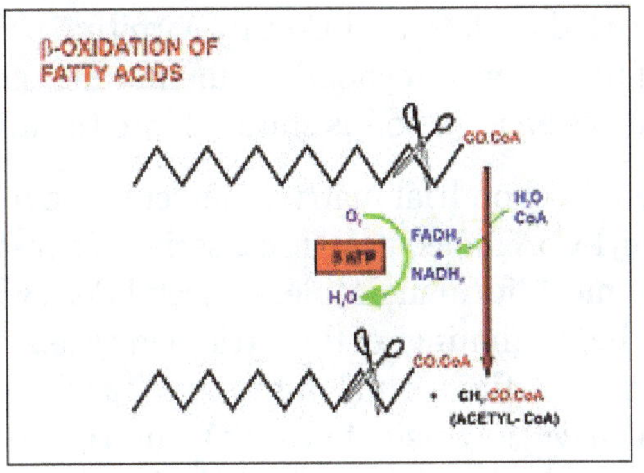

A diagrammatic illustration of the process of the beta-oxidation of an acyl-CoA molecule in the mitochodrial matrix. During this process an acyl-CoA molecule which is 2 carbons shorter than it was at the beginning of the process is formed. Acetyl-CoA, water and 5 ATP molecules are the other products of each beta-oxidative event, until the entire acyl-CoA molecule has been reduced to a set of acetyl-CoA molecules.

Fatty acids are released, between meals, from the fat depots in adipose tissue, where they are stored as triglycerides, as follows:

- Lipolysis, the removal of the fatty acid chains from the glycerol to which they are bound in their storage form as triglycerides (or fats), is carried out by lipases. These lipases are activated by high epinephrine and glucagon levels in the blood (or norepinephrine secreted by sympathetic nerves in adipose tissue), caused by declining blood glucose levels after meals, which simultaneously lowers the insulin level in the blood.
- Once freed from glycerol, the free fatty acids enter the blood, which transports them, attached to plasma albumin, throughout the body.
- Long chain free fatty acids enter the metabolizing cells (i.e. most living cells in the body except red blood cells and neurons in the central nervous system) through specific transport proteins, such as the SLC27 family fatty acid transport protein. Red blood cells do not contain mitochondria and are therefore incapable of metabolizing fatty acids; the tissues of the central nervous system cannot use fatty acids, despite containing mitochondria, because long chain fatty acids (as opposed to medium chain fatty acids) cannot cross the blood brain barrier into the interstitial fluids that bathe these cells.
- Once inside the cell long-chain-fatty-acid—CoA ligase catalyzes the reaction between a fatty acid molecule with ATP (which is broken down to AMP and inorganic pyrophosphate) to give a fatty acyl-adenylate, which then reacts with free coenzyme A to give a fatty acyl-CoA molecule.

In order for the acyl-CoA to enter the mitochondrion the carnitine shuttle is used:

- Acyl-CoA is transferred to the hydroxyl group of carnitine by carnitine palmitoyltransferase I, located on the cytosolic faces of the outer and inner mitochondrial membranes.
- Acyl-carnitine is shuttled inside by a carnitine-acylcarnitine translocase, as a carnitine is shuttled outside.
- Acyl-carnitine is converted back to acyl-CoA by carnitine palmitoyltransferase II, located on the interior face of the inner mitochondrial membrane. The liberated carnitine is shuttled back to the cytosol, as an acyl-CoA is shuttled into the mitochondrial matrix.
- Beta oxidation, in the mitochondrial matrix, then cuts the long carbon chains of the fatty acids (in the form of acyl-CoA molecules) into a series of two-carbon (acetate) units, which, combined with co-enzyme A, form molecules of acetyl CoA, which condense with oxaloacetate to form citrate at the "beginning" of the citric acid cycle. It is convenient to think of this reaction as marking the "starting point" of the cycle, as this is when fuel - acetyl-CoA - is added to the cycle, which will be dissipated as CO_2 and H_2O with the release of a substantial quantity of energy captured in the form of ATP, during the course of each turn of the cycle.

Briefly, the steps in beta oxidation (the initial breakdown of free fatty acids into acetyl-CoA) are as follows:

- Dehydrogenation by acyl-CoA dehydrogenase, yielding 1 $FADH_2$.
- Hydration by enoyl-CoA hydratase.
- Dehydrogenation by 3-hydroxyacyl-CoA dehydrogenase, yielding 1 NADH + H^+.
- Cleavage by thiolase, yielding 1 acetyl-CoA and a fatty acid that has now been shortened by 2 carbons (forming a new, shortened acyl-CoA).
- This beta oxidation reaction is repeated until the fatty acid has been completely reduced to acetyl-CoA or, in, the case of fatty acids with odd numbers of carbon atoms, acetyl-CoA and 1 molecule of propionyl-CoA per molecule of fatty acid. Each beta oxidative cut of the acyl-CoA molecule yields 5 ATP molecules.
- The acetyl-CoA produced by beta oxidation enters the citric acid cycle in the mitochondrion by combining with oxaloacetate to form citrate. This results in the complete combustion of the acetyl-CoA to CO_2 and water. The energy released in this process is captured in the form of 1 GTP and 11 ATP molecules per acetyl-CoA molecule oxidized. This is the fate of acetyl-CoA wherever beta oxidation of fatty acids occurs, except under certain circumstances in the liver.

In the liver oxaloacetate can be wholly or partially diverted into the gluconeogenic pathway during fasting, starvation, a low carbohydrate diet, prolonged strenuous exercise, and in uncontrolled type 1 diabetes mellitus. Under these circumstances oxaloacetate is hydrogenated to malate which is then removed from the mitochondria of the liver cells to be converted into glucose in the cytoplasm of the liver cells, from where it is released into the blood. In the liver, therefore, oxaloacetate is unavailable for condensation with acetyl-CoA when significant gluconeogenesis has been stimulated by low (or absent) insulin and high glucagon concentrations in the blood. Under these circumstances acetyl-CoA is diverted to the formation of acetoacetate and beta-hydroxybutyrate. Acetoacetate, beta-hydroxybutyrate, and their spontaneous breakdown product, acetone, are frequently, but confusingly, known as ketone bodies (as they are not "bodies" at all, but water-soluble chemical substances). The ketones are released by the liver into the blood. All cells with mitochondria can take ketones up from the blood and reconvert them into acetyl-CoA, which can then be used as fuel in their citric acid cycles, as no other tissue can divert its oxaloacetate into the gluconeogenic pathway in the way that this can occur in the liver. Unlike free fatty acids, ketones can cross the blood-brain barrier and are therefore available as fuel for the cells of the central nervous system, acting as a substitute for glucose, on which these cells normally survive. The occurrence of high levels of ketones in the blood during starvation, a low carbohydrate diet, prolonged heavy exercise and uncontrolled type 1 diabetes mellitus is known as ketosis, and, in its extreme form, in out-of-control type 1 diabetes mellitus, as ketoacidosis.

The glycerol released by lipase action is phosphorylated by glycerol kinase in the liver (the only tissue in which this reaction can occur), and the resulting glycerol 3-phosphate is oxidized to dihydroxyacetone phosphate. The glycolytic enzyme triose phosphate isomerase converts this compound to glyceraldehyde 3-phosphate, which is oxidized via glycolysis, or converted to glucose via gluconeogenesis.

Fatty Acids as an Energy Source

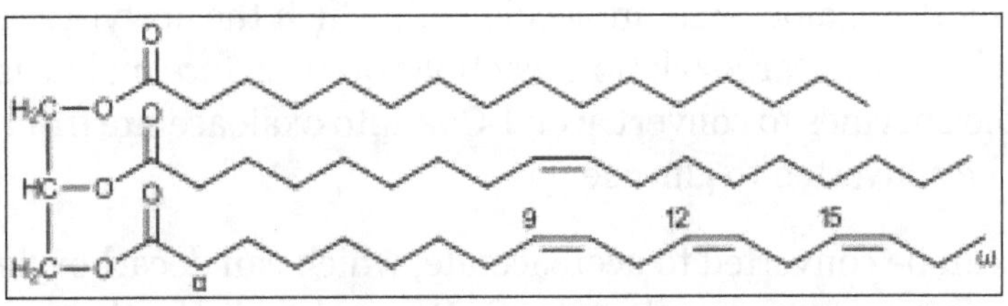

Example of an unsaturated fat triglyceride. Left part: glycerol, right part from top to bottom: palmitic acid, oleic acid, alpha-linolenic acid. Chemical formula: $C_{55}H_{98}O_6$

Fatty acids, stored as triglycerides in an organism, are an important source of energy because they are both reduced and anhydrous. The energy yield from a gram of fatty acids is approximately 9 kcal (37 kJ), compared to 4 kcal (17 kJ) for carbohydrates. Since the hydrocarbon portion of fatty acids is hydrophobic, these molecules can be stored in a relatively anhydrous (water-free) environment. Carbohydrates, on the other hand, are more highly hydrated. For example, 1 g of glycogen can bind approximately 2 g of water, which translates to 1.33 kcal/g (4 kcal/3 g). This means that fatty acids can hold more than six times the amount of energy per unit of storage mass. Put another way, if the human body relied on carbohydrates to store energy, then a person would need to carry 31 kg (67.5 lb) of hydrated glycogen to have the energy equivalent to 4.6 kg (10 lb) of fat.

Hibernating animals provide a good example for utilizing fat reserves as fuel. For example, bears hibernate for about 7 months, and, during this entire period, the energy is derived from degradation of fat stores. Migrating birds similarly build up large fat reserves before embarking on their intercontinental journeys.

Thus the young adult human's fat stores average between about 10–20 kg, but varies greatly depending on age, gender, and individual disposition. By contrast the human body stores only about

400 g of glycogen, of which 300 g is locked inside the skeletal muscles and is unavailable to the body as a whole. The 100 g or so of glycogen stored in the liver is depleted within one day of starvation. Thereafter the glucose that is released into the blood by the liver for general use by the body tissues, has to be synthesized from the glucogenic amino acids and a few other gluconeogenic substrates, which do not include fatty acids. Please note however that lipolysis releases glycerol which can enter the pathway of gluconeogenesis.

Animals and Plants Synthesize Carbohydrates from both Glycerol and Fatty Acids

Fatty acids are broken down to acetyl-CoA by means of beta oxidation inside the mitochondria, whereas fatty acids are synthesized from acetyl-CoA outside the mitochondria, in the cytosol. The two pathways are distinct, not only in where they occur, but also in the reactions that occur, and the substrates that are used. The two pathways are mutually inhibitory, preventing the acetyl-CoA produced by beta-oxidation from entering the synthetic pathway via the acetyl-CoA carboxylase reaction. It can also not be converted to pyruvate as the pyruvate dehydrogenase complex reaction is irreversible. Instead the acetyl-CoA produced by the beta-oxidation of fatty acids condenses with oxaloacetate, to enter the citric acid cycle. During each turn of the cycle, two carbon atoms leave the cycle as CO_2 in the decarboxylation reactions catalyzed by isocitrate dehydrogenase and alpha-ketoglutarate dehydrogenase. Thus each turn of the citric acid cycle oxidizes an acetyl-CoA unit while regenerating the oxaloacetate molecule with which the acetyl-CoA had originally combined to form citric acid. The decarboxylation reactions occur before malate is formed in the cycle. Only plants possess the enzymes to convert acetyl-CoA into oxaloacetate from which malate can be formed to ultimately be converted to glucose.

However acetyl-CoA can be converted to acetoacetate, which can decarboxylate to acetone (either spontaneously, or by acetoacetate decarboxylase). It can then be further metabolized to isopropanol which is excreted in breath/urine, or by CYP2E1 into hydroxyacetone (acetol). Acetol can be converted to propylene glycol. This converts to formate and acetate (the latter converting to glucose), or pyruvate (by two alternative enzymes), or propionaldehyde, or to L-lactaldehyde then L-lactate (the common lactate isomer). Another pathway turns acetol to methylglyoxal, then to pyruvate, or to D-lactaldehyde (via S-D-lactoyl-glutathione or otherwise) then D-lactate. D-lactate metabolism (to glucose) is slow or impaired in humans, so most of the D-lactate is excreted in the urine; thus D-lactate derived from acetone can contribute significantly to the metabolic acidosis associated with ketosis or isopropanol intoxication. L-Lactate can complete the net conversion of fatty acids into glucose. The first experiment to show conversion of acetone to glucose was carried out in 1951. This, and further experiments used carbon isotopic labelling. Up to 11% of the glucose can be derived from acetone during starvation in humans.

The glycerol released into the blood during the lipolysis of triglycerides in adipose tissue can only be taken up by the liver. Here it is converted into glycerol 3-phosphate by the action of glycerol kinase which hydrolyzes one molecule of ATP per glycerol molecule which is phosphorylated. Glycerol 3-phosphate is then oxidized to dihydroxyacetone phosphate, which is, in turn, converted into glyceraldehyde 3-phosphate by the enzyme triose phosphate isomerase. From here the three carbon atoms of the original glycerol can be oxidized via glycolysis, or converted to glucose via gluconeogenesis.

Other Functions and uses of Fatty Acids

Intracellular Signaling

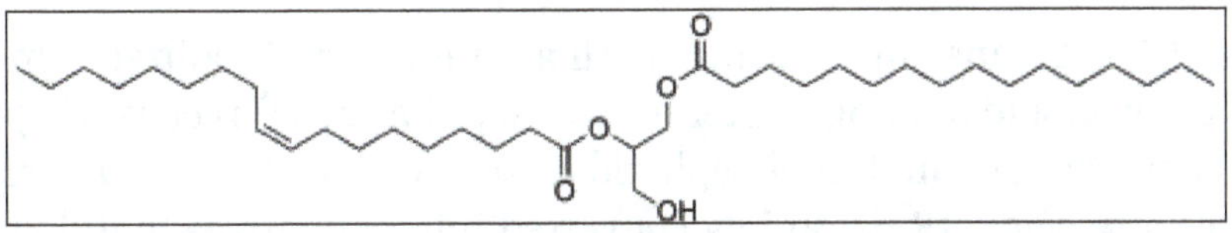

Chemical structure of the diglyceride 1-palmitoyl-2-oleoyl-glycerol.

Fatty acids are an integral part of the phospholipids that make up the bulk of the plasma membranes, or cell membranes, of cells. These phospholipids can be cleaved into diacylglycerol (DAG) and inositol trisphosphate (IP_3) through hydrolysis of the phospholipid, phosphatidylinositol 4,5-bisphosphate (PIP_2), by the cell membrane bound enzyme phospholipase C (PLC).

An example of a diacyl-glycerol shown on the right. This DAG is 1-palmitoyl-2-oleoyl-glycerol, which contains side-chains derived from palmitic acid and oleic acid. Diacylglycerols can also have many other combinations of fatty acids attached at either the C-1 and C-2 positions or the C-1 and C-3 positions of the glycerol molecule. 1,2 disubstituted glycerols are always chiral, 1,3 disubstituted glycerols are chiral if the substituents are different from each other.

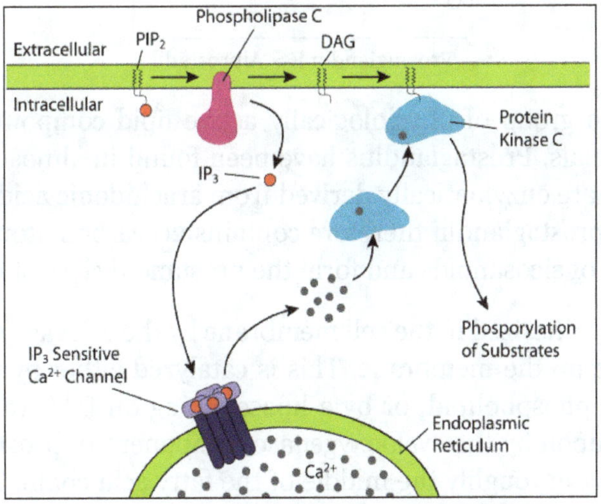

PIP_2 cleavage to IP_3 and DAG. IP_3 initiates intracellular calcium release, while DAG activates PKC (protein kinase C). PLC (phospholipase C) is not an intermediate, as possibly suggested by the diagram, but is the enzyme that catalyzes the IP3/DAG separation.

Inositol trisphosphate (IP_3) functions as an intracellular second messenger, which initiates the intracellular release of calcium ions (which activates intracellular enzymes, causes the release of hormones and neurotransmitters from the cells in which they are stored, and causes smooth muscle contraction when released by IP_3), and the activation of protein kinase C (PKC), which is then translocated from the cell cytoplasm to the cell membrane. Although inositol trisphosphate, (IP_3), diffuses into the cytosol, diacylglycerol (DAG) remains within the plasma membrane, due to its hydrophobic properties. IP_3 stimulates the release of calcium ions from the smooth endoplasmic reticulum, whereas DAG is a physiological activator of protein kinase C (PKC), promoting its translocation from the cytosol to the plasma membrane. PKC is a multifunctional protein kinase which phosphorylates serine and threonine residues in many target proteins. However PKC is only active in the presence of calcium ions, and it is DAG that increases the affinity of PKC for Ca^{2+} and thereby renders it active at the physiological intracellular levels of this ion.

Diacylglycerol and IP_3 act transiently because both are rapidly metabolized. This is important as their message function should not linger after the message has been" received" by their target molecules. DAG can be phosphorylated to phosphatidate or it can be it can be hydrolysed to glycerol and its constituent fatty acids. IP_3 is rapidly converted into derivatives that do not open calcium ion channels.

Eicosanoid Paracrine Hormones

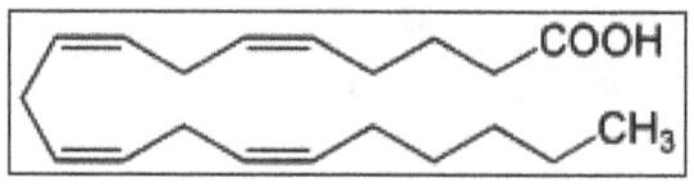

Arachidonic acid

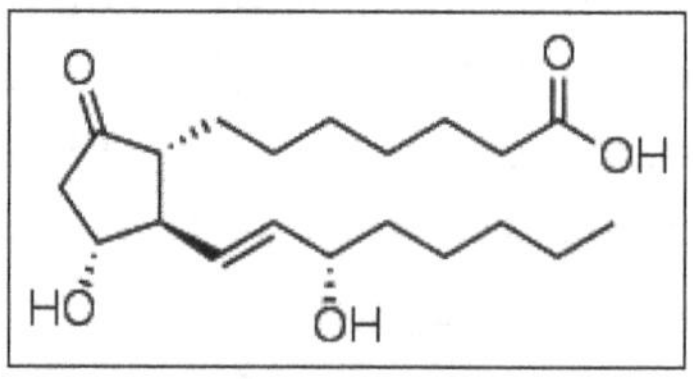

Prostaglandin E_1 - Alprostadil

The prostaglandins are a group of physiologically active lipid compounds having diverse hormone-like effects in animals. Prostaglandins have been found in almost every tissue in humans and other animals. They are enzymatically derived from arachidonic acid a 20-carbon polyunsaturated fatty acid. Every prostaglandin therefore contains 20 carbon atoms, including a 5-carbon ring. They are a subclass of eicosanoids and form the prostanoid class of fatty acid derivatives.

The prostaglandins are synthesized in the cell membrane by the cleavage of arachidonate from the phospholipids that make up the membrane. This is catalyzed either by phospholipase A_2 acting directly on a membrane phospholipid, or by a lipase acting on DAG (diacyl-glycerol). The arachidonate is then acted upon by the cyclooxygenase component of prostaglandin synthase. This forms a cyclopentane ring in roughly the middle of the fatty acid chain. The reaction also adds 4 oxygen atoms derived from two molecules of O_2. The resulting molecule is prostaglandin G_2 which is converted by the hydroperoxidase component of the enzyme complex into prostaglandin H_2. This highly unstable compound is rapidly transformed into other prostaglandins, prostacyclin and thromboxanes. These are then released into the interstitial fluids surrounding the cells that have manufactured the eicosanoid hormone.

If arachidonate is acted upon by a lipoxygenase instead of cyclooxygenase, Hydroxyeicosatetraenoic acids and leukotrienes are formed. They also act as local hormones.

Prostaglandins were originally believed to leave the cells via passive diffusion because of their high lipophilicity. The discovery of the prostaglandin transporter (PGT, SLCO2A1), which mediates the cellular uptake of prostaglandin, demonstrated that diffusion alone cannot explain the penetration of prostaglandin through the cellular membrane. The release of prostaglandin has now also been shown to be mediated by a specific transporter, namely the multidrug resistance protein 4 (MRP4,

ABCC4), a member of the ATP-binding cassette transporter superfamily. Whether MRP4 is the only transporter releasing prostaglandins from the cells is still unclear.

The structural differences between prostaglandins account for their different biological activities. A given prostaglandin may have different and even opposite effects in different tissues. The ability of the same prostaglandin to stimulate a reaction in one tissue and inhibit the same reaction in another tissue is determined by the type of receptor to which the prostaglandin binds. They act as autocrine or paracrine factors with their target cells present in the immediate vicinity of the site of their secretion. Prostaglandins differ from endocrine hormones in that they are not produced at a specific site but in many places throughout the human body.

Prostaglandins have two derivatives: prostacyclins and thromboxanes. Prostacyclins are powerful locally acting vasodilators and inhibit the aggregation of blood platelets. Through their role in vasodilation, prostacyclins are also involved in inflammation. They are synthesized in the walls of blood vessels and serve the physiological function of preventing needless clot formation, as well as regulating the contraction of smooth muscle tissue. Conversely, thromboxanes (produced by platelet cells) are vasoconstrictors and facilitate platelet aggregation. Their name comes from their role in clot formation (thrombosis).

Dietary Sources of Fatty Acids, Digestion, Absorption and Transport in the Blood and Storage

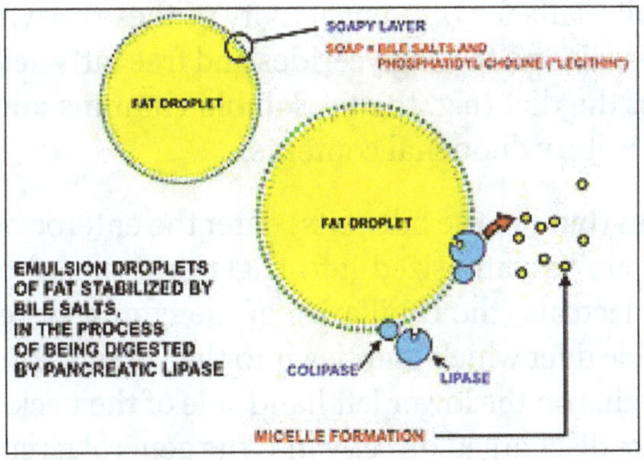

Dietary fats are emulsified in the duodenum by soaps in the form of bile salts and phospholipids, such as phosphatidylcholine. The fat droplets thus formed can be attacked by pancreatic lipase.

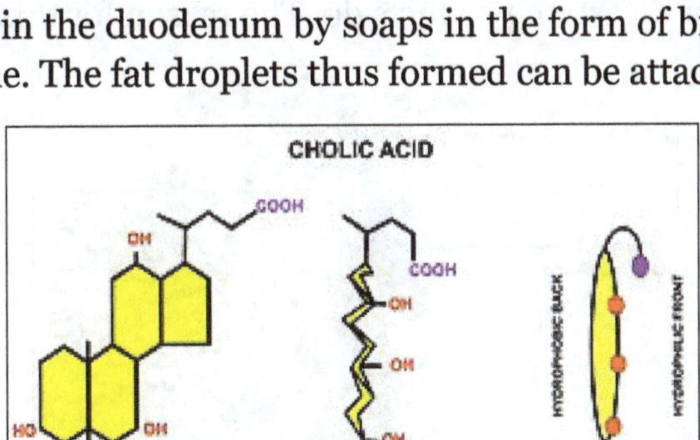

Structure of a bile acid (cholic acid), represented in the standard form, a semi-realistic 3D form, and a diagrammatic 3D form.

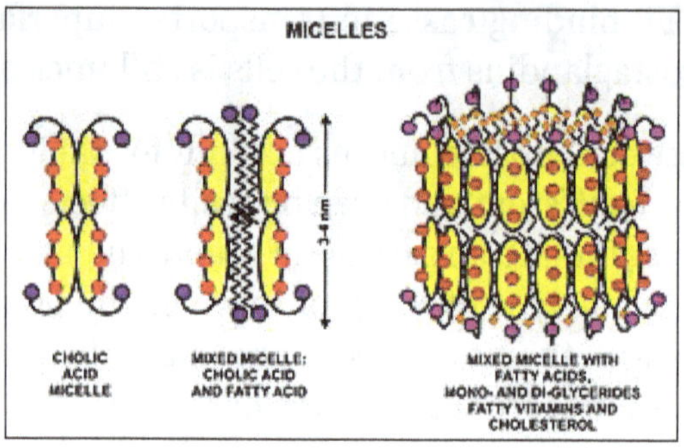

Diagrammatic illustration of mixed micelles formed in the duodenum in the presence of bile acids (e.g. cholic acid) and the digestion products of fats, the fat soluble vitamins and cholesterol.

A significant proportion of the fatty acids in the body are obtained from the diet, in the form of triglycerides of either animal or plant origin. The fatty acids in the fats obtained from land animals tend to be saturated, whereas the fatty acids in the triglycerides of fish and plants are often polyunsaturated and therefore present as oils.

These triglycerides, cannot be absorbed by the intestine. They are broken down into mono- and di-glycerides plus free fatty acids (but no free glycerol) by pancreatic lipase, which forms a 1:1 complex with a protein called colipase (also a constituent of pancreatic juice), which is necessary for its activity. The activated complex can work only at a water-fat interface. Therefore, it is essential that fats are first emulsified by bile salts for optimal activity of these enzymes. The digestion products consisting of a mixture of tri-, di- and monoglycerides and free fatty acids, which, together with the other fat soluble contents of the diet (e.g. the fat soluble vitamins and cholesterol) and bile salts form mixed micelles, in the watery duodenal contents.

The contents of these micelles (but not the bile salts) enter the enterocytes (epithelial cells lining the small intestine) where they are resynthesized into triglycerides, and packaged into chylomicrons which are released into the lacteals (the capillaries of the lymph system of the intestines). These lacteals drain into the thoracic duct which empties into the venous blood at the junction of the left jugular and left subclavian veins on the lower left hand side of the neck. This means that the fat soluble products of digestion are discharged directly into the general circulation, without first passing through the liver, as all other digestion products do. The reason for this peculiarity is unknown.

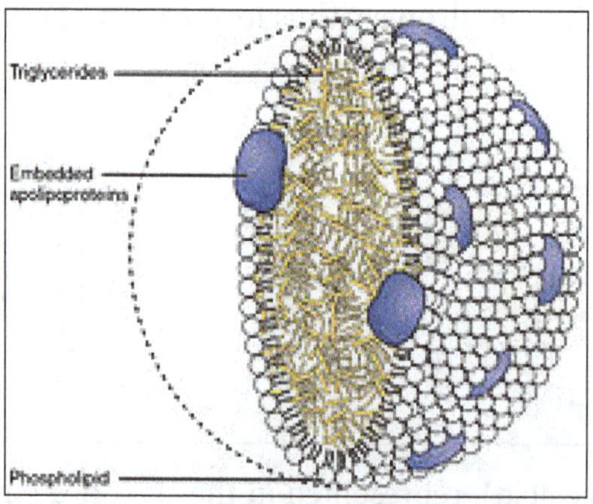

A schematic diagram of a chylomicron.

The chylomicrons circulate throughout the body, giving the blood plasma a milky, or creamy appearance after a fatty meal. Lipoprotein lipase on the endothelial surfaces of the capillaries, especially in adipose tissue, but to a lesser extent also in other tissues, partially digests the chylomicrons into free fatty acids, glycerol and chylomicron remnants. The fatty acids are absorbed by the adipocytes, but the glycerol and chylomicron remnants remain in the blood plasma, ultimately to be removed from the circulation by the liver. The free fatty acids released by the digestion of the chylomicrons are absorbed by the adipocytes, where they are resynthesized into triglycerides using glycerol derived from glucose in the glycolytic pathway. These triglycerides are stored, until needed for the fuel requirements of other tissues, in the fat droplet of the adipocyte.

The liver absorbs a proportion of the glucose from the blood in the portal vein coming from the intestines. After the liver has replenished its glycogen stores (which amount to only about 100 g of glycogen when full) much of the rest of the glucose is converted into fatty acids as described below. These fatty acids are combined with glycerol to form triglycerides which are packaged into droplets very similar to chylomicrons, but known as very low-density lipoproteins (VLDL). These VLDL droplets are handled in exactly the same manner as chylomicrons, except that the VLDL remnant is known as an intermediate-density lipoprotein (IDL), which is capable of scavenging cholesterol from the blood. This converts IDL into low-density lipoprotein (LDL), which is taken up by cells that require cholesterol for incorporation into their cell membranes or for synthetic purposes (e.g. the formation of the steroid hormones). The remainder of the LDLs is removed by the liver.

Adipose tissue and lactating mammary glands also take up glucose from the blood for conversion into triglycerides. This occurs in the same way as it does in the liver, except that these tissues do not release the triglycerides thus produced as VLDL into the blood. Adipose tissue cells store the triglycerides in their fat droplets, ultimately to release them again as free fatty acids and glycerol into the blood, when the plasma concentration of insulin is low, and that of glucagon and/or epinephrine is high. Mammary glands discharge the fat (as cream fat droplets) into the milk that they produce under the influence of the anterior pituitary hormone prolactin.

All cells in the body need to manufacture and maintain their membranes and the membranes of their organelles. Whether they rely for this entirely on free fatty acids absorbed from the blood, or are able to synthesize their own fatty acids from the blood glucose, is not known. The cells of the central nervous system will almost certainly have the capability of manufacturing their own fatty acids, as these molecules cannot reach them through the blood brain barrier, while, on the other hand, no cell in the body can manufacture the required essential fatty acids which have to be obtained from the diet and delivered to each cell via the blood.

Fatty Acid Synthesis

Much like beta-oxidation, straight-chain fatty acid synthesis occurs via the six recurring reactions shown below, until the 16-carbon palmitic acid is produced.

The diagrams presented show how fatty acids are synthesized in microorganisms and list the enzymes found in Escherichia coli. These reactions are performed by fatty acid synthase II (FASII), which in general contain multiple enzymes that act as one complex. FASII is present in prokaryotes, plants, fungi, and parasites, as well as in mitochondria.

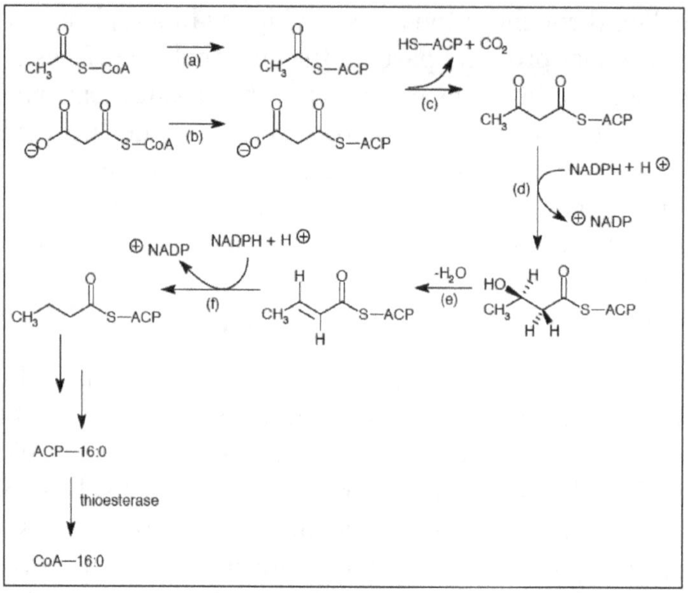

Synthesis of saturated fatty acids via Fatty Acid Synthase II in E. coli.

In animals, as well as some fungi such as yeast, these same reactions occur on fatty acid synthase I (FASI), a large dimeric protein that has all of the enzymatic activities required to create a fatty acid. FASI is less efficient than FASII; however, it allows for the formation of more molecules, including "medium-chain" fatty acids via early chain termination. Enzymes, acyltransferases and transacylases, incorporate fatty acids in phospholipids, triacylglycerols, etc. by transferring fatty acids between an acyl acceptor and donor. They also have the job of synthesizing bioactive lipids as well as their precursor molecules.

Once a 16:0 carbon fatty acid has been formed, it can undergo a number of modifications, resulting in desaturation and/or elongation. Elongation, starting with stearate (18:0), is performed mainly in the ER by several membrane-bound enzymes. The enzymatic steps involved in the elongation process are principally the same as those carried out by FAS, but the four principal successive steps of the elongation are performed by individual proteins, which may be physically associated.

Enzyme	Reaction	Description
Acetyl CoA:ACP transacylase	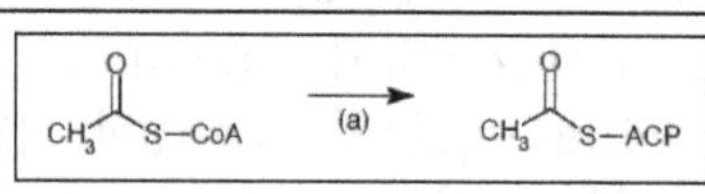	Activates acetyl CoA for reaction with malonyl-ACP.
Malonyl CoA:ACP transacylase	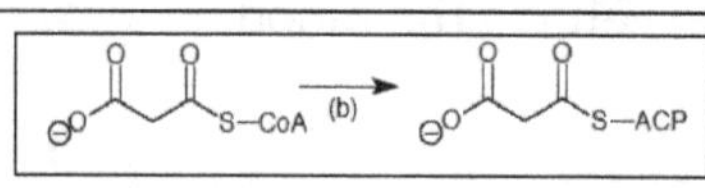	Activates malonyl CoA for reaction with acetyl-ACP.
3-ketoacyl-ACP synthase	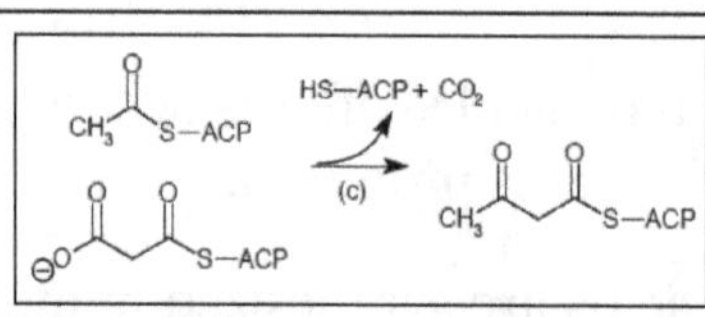	Reacts priming acetyl-ACP with chain-extending malonyl-ACP.
3-ketoacyl-ACP reductase	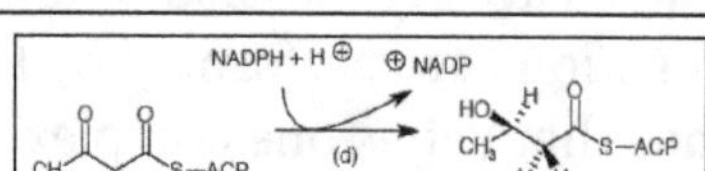	Reduces the carbon 3 ketone to a hydroxyl group.

3-Hydroxyacyl ACP dehydrase	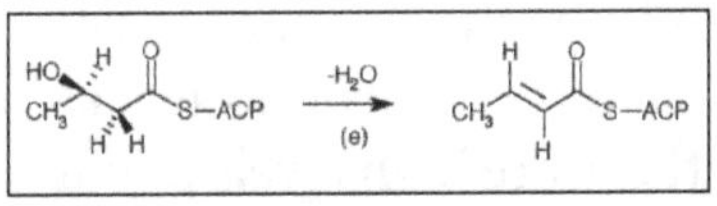	Removes water.
Enoyl-ACP reductase	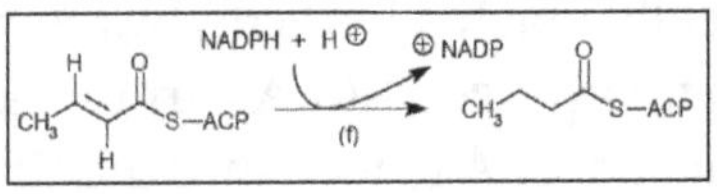	Reduces the C_2-C_3 double bond.

Abbreviations: ACP – Acyl carrier protein, CoA – Coenzyme A, NADP – Nicotinamide adenine dinucleotide phosphate.

Note that during fatty synthesis the reducing agent is NADPH, whereas NAD is the oxidizing agent in beta-oxidation (the breakdown of fatty acids to acetyl-CoA). This difference exemplifies a general principle that NADPH is consumed during biosynthetic reactions, whereas NADH is generated in energy-yielding reactions. (Thus NADPH is also required for the synthesis of cholesterol from acetyl-CoA; while NADH is generated during glycolysis.) The source of the NADPH is two-fold. When malate is oxidatively decarboxylated by "$NADP^+$-linked malic enzyme" pyruvate, CO_2 and NADPH are formed. NADPH is also formed by the pentose phosphate pathway which converts glucose into ribose, which can be used in synthesis of nucleotides and nucleic acids, or it can be catabolized to pyruvate.

Glycolytic End Products are used in the Conversion of Carbohydrates into Fatty Acids

In humans, fatty acids are formed from carbohydrates predominantly in the liver and adipose tissue, as well as in the mammary glands during lactation. The cells of the central nervous system probably also make most of the fatty acids needed for the phospholipids of their extensive membranes from glucose, as blood-born fatty acids cannot cross the blood brain barrier to reach these cells. However, how the essential fatty acids, which mammals cannot synthesize themselves, but are nevertheless important components of cell membranes (and other functions described above) reach them is unknown.

The pyruvate produced by glycolysis is an important intermediary in the conversion of carbohydrates into fatty acids and cholesterol. This occurs via the conversion of pyruvate into acetyl-CoA in the mitochondrion. However, this acetyl CoA needs to be transported into cytosol where the synthesis of fatty acids and cholesterol occurs. This cannot occur directly. To obtain cytosolic acetyl-CoA, citrate (produced by the condensation of acetyl CoA with oxaloacetate) is removed from the citric acid cycle and carried across the inner mitochondrial membrane into the cytosol. There it is cleaved by ATP citrate lyase into acetyl-CoA and oxaloacetate. The oxaloacetate is returned to mitochondrion as malate (and then converted back into oxaloacetate to transfer more acetyl-CoA out of the mitochondrion). The cytosolic acetyl-CoA is carboxylated by acetyl CoA carboxylase into malonyl CoA, the first committed step in the synthesis of fatty acids.

Regulation of Fatty Acid Synthesis

Acetyl-CoA is formed into malonyl-CoA by acetyl-CoA carboxylase, at which point malonyl-CoA is destined to feed into the fatty acid synthesis pathway. Acetyl-CoA carboxylase is the point of regulation in saturated straight-chain fatty acid synthesis, and is subject to both phosphorylation and allosteric regulation. Regulation by phosphorylation occurs mostly in mammals, while allosteric regulation occurs in most organisms. Allosteric control occurs as feedback inhibition by palmitoyl-CoA and activation by citrate. When there are high levels of palmitoyl-CoA, the final product of saturated fatty acid synthesis, it allosterically inactivates acetyl-CoA carboxylase to prevent a build-up of fatty acids in cells. Citrate acts to activate acetyl-CoA carboxylase under high levels, because high levels indicate that there is enough acetyl-CoA to feed into the Krebs cycle and produce energy.

High plasma levels of insulin in the blood plasma (e.g. after meals) cause the dephosphorylation and activation of acetyl-CoA carboxylase, thus promoting the formation of malonyl-CoA from acetyl-CoA, and consequently the conversion of carbohydrates into fatty acids, while epinephrine and glucagon (released into the blood during starvation and exercise) cause the phosphorylation of this enzyme, inhibiting lipogenesis in favor of fatty acid oxidation via beta-oxidation.

Disorders

Disorders of fatty acid metabolism can be described in terms of, for example, hypertriglyceridemia (too high level of triglycerides), or other types of hyperlipidemia. These may be familial or acquired.

Familial types of disorders of fatty acid metabolism are generally classified as inborn errors of lipid metabolism. These disorders may be described as fatty oxidation disorders or as a *lipid storage disorders*, and are any one of several inborn errors of metabolism that result from enzyme defects affecting the ability of the body to oxidize fatty acids in order to produce energy within muscles, liver, and other cell types.

Protein Metabolism

Protein metabolism denotes the various biochemical processes responsible for the synthesis of proteins and amino acids (anabolism), and the breakdown of proteins by catabolism.

The steps of protein synthesis include transcription, translation, and post translational modifications. During transcription, RNA polymerase transcribes a coding region of the DNA in a cell producing a sequence of RNA, specifically messenger RNA (mRNA). This mRNA sequence contains codons: 3 nucleotide long segments that code for a specific amino acid. Ribosomes translate the codons to their respective amino acids. In humans, non-essential amino acids are synthesized from intermediates in major metabolic pathways such as the Citric Acid Cycle. Essential amino acids must be consumed and are made in other organisms. The amino acids are joined by peptide bonds making a polypeptide chain. This polypeptide chain then goes through post translational modifications and is sometimes joined with other polypeptide chains to form a fully functional protein.

Dietary proteins are first broken down to individual amino acids by various enzymes and hydrochloric acid present in the gastrointestinal tract. These amino acids are further broken down to α-keto acids which can be recycled in the body for generation of energy, and production of glucose or fat or other amino acids. Proteins can be broken down by enzymes known as peptidases or can break down as a result of denaturation. Proteins can denature in environmental conditions the protein is not made for.

Protein Synthesis

Protein anabolism is the process by which protein are formed from amino acids. It relies on five processes: amino acid synthesis, transcription, translation, post translational modifications, and protein folding. Proteins are made from amino acids. In humans, some amino acids can be synthesized using already existing intermediates. These amino acids are known as non-essential amino acids. Essential amino acids require intermediates not present in the human body. These intermediates must be ingested, mostly from eating other organisms.

Amino Acid Synthesis

Pathways that form each amino acid		
Amino Acid	R-group‡	Pathway*
Glycine	H-	Serine + THF → Glycine (*hydroxymethyltransferase)*
Alanine	CH_3-	Pyruvate → Alanine (*aminotransferase)*
Valine	$(CH_3)_2$-CH-	Hydroxyethyl-TPP + Pyruvate → α-acetolactate → Valine
Leucine	$(CH_3)_2$-CH-CH_2-	Hydroxyethyl-TPP + Pyruvate → α-ketobutyrate → Leucine
Isoleucine	CH_3-CH_2-CH(CH_3)-	Hydroxyethyl-TPP + Pyruvate → α-acetolactate → Isoleucine
Methionine	CH_3-S-$(CH_2)_2$-	Homocysteine → Methionine (*methionine synthase)*
Proline	-$(CH_2)_3$-	Glutamic Acid → Glutamate-5-semialdehyde → Proline (*γ-glutamyl kinase)*
Phenylalanine	Ph-CH_2-	Phosphoenolpyruvate → 2-keto-3-deoxy arabino heptulosonate-7-phosphate → Chorismate → Phenylalanine
Tryptophan	Ph-NH-CH=C-CH_2-	Phosphoenolpyruvate → 2-keto-3-deoxy arabino heptulosonate-7-phosphate → Chorismate → Tryptophan
Tyrosine	HO-Ph-CH_2-	Phenylalanine → Tyrosine (*phenylalanine hydroxylase)*
Serine	HO-CH_2-	3-phosphoglycerate → 3-phosphohydroxypyruvate (*3-phosphoglycerate dehydrogenase)* → 3-phosphoserine (*aminotransferase)* → Serine (*phosphoserine phosphatase)*
Threonine	CH_3-CH(OH)-	Aspartate → β-aspartate-semialdehyde → Homoserine → Threonine
Cysteine	HS-CH_2-	Serine → Cystathionine → α-ketobutyrate → Cysteine
Asparagine	H_2N-CO-CH_2-	Aspartic Acid → Asparagine (*asparagine synthetase)*
Glutamine	H_2N-CO-$(CH_2)_2$-	Glutamic Acid → Glutamine (*glutamine synthetase)*
Arginine	$^+H_2$N=C(NH_2)-NH-$(CH_2)_3$-	Glutamate → Glutamate-5-semialdehyde (*γ-glutamyl kinase)* → Arginine
Histidine	NH-CH=N-CH=C-CH_2-	Glucose → Glucose-6-phosphate → Ribose-5-phosphate → Histidine
Lysine	$^+H_3$N-$(CH_2)_4$-	Aspartate → β-aspartate-semialdehyde → Homoserine + lysine
Aspartic Acid	$^-$OOC-CH_2-	Oxaloacetate → Aspartic Acid (*aminotransferase)*
Glutamic Acid	$^-$OOC-$(CH_2)_2$-	α-ketoglutarate → Glutamic Acid (*aminotransferase)*

Polypeptide Synthesis

Transcription

In transcription, RNA polymerase reads a DNA strand and produces an mRNA strand that can be further translated. In order to initiate transcription, the DNA segment that is to be transcribed must be accessible (i.e. it cannot be tightly packed). Once the DNA segment is accessible, the RNA polymerase can begin to transcribe the coding DNA strand by pairing RNA nucleotides to the template DNA strand. During the initial transcription phase, the RNA polymerase searches for a promoter region on the DNA template strand. Once the RNA polymerase binds to this region, it begins to "read" the template DNA strand in the 3' to 5' direction. RNA polymerase attaches RNA bases complementary to the template DNA strand (Uracil will be used instead of Thymine). The new nucleotide bases are bonded to each other covalently. The new bases eventually dissociate from the DNA bases but stay linked to each other, forming a new mRNA strand. This mRNA strand is synthesized in the 5' to 3' direction. Once the RNA reaches a terminator sequence, it dissociates from the DNA template strand and terminates the mRNA sequence as well.

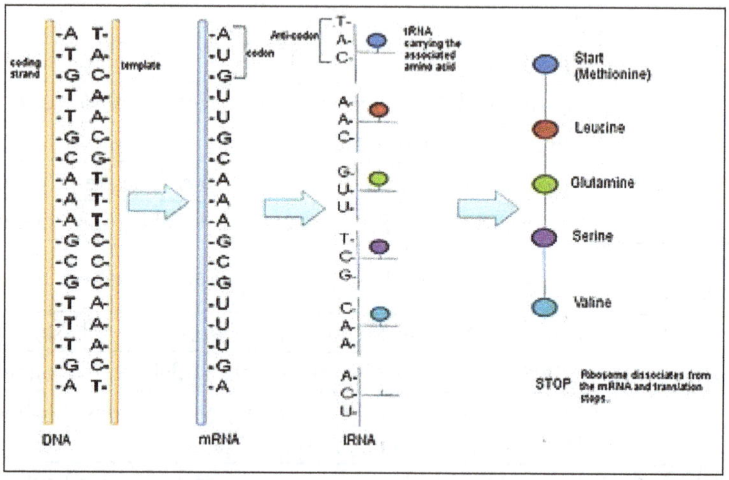

DNA is transcribed to mRNA which is translated into amino acids.

Transcription is regulated in the cell via transcription factors. Transcription factors are proteins that bind to regulatory sequences in the DNA strand such as promoter regions or operator regions. Proteins bound to these regions can either directly halt or allow RNA polymerase to read the DNA strand or can signal other proteins to halt or allow RNA polymerase reading.

Translation

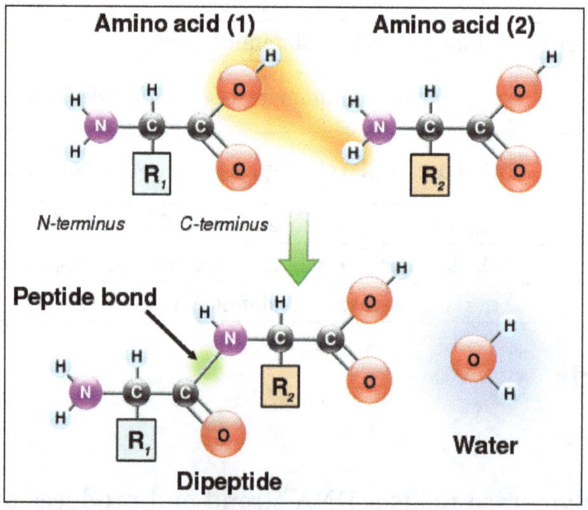

Formation of a dipeptide via a peptide bond.

During translation, ribosomes convert a sequence of mRNA (messenger RNA) to an amino acid sequence. Each 3-base-pair-long segment of mRNA is a codon which corresponds to one amino acid or stop signal. Amino acids can have multiple codons that correspond to them. Ribosomes do not directly attach amino acids to mRNA codons. They must utilize tRNAs (transfer RNAs) as well. Transfer RNAs can bind to amino acids and contain an anticodon which can hydrogen bind to an mRNA codon. The process of bind an amino acid to a tRNA is known as tRNA charging. Here, the enzyme aminoacyl-tRNA-synthetase catalyzes two reactions. In the first one, it attaches an AMP molecule (cleaved from ATP) to the amino acid. The second reaction cleaves the aminoacyl-AMP producing the energy to join the amino acid to the tRNA molecule.

Ribosomes have two subunits, one large and one small. These subunits surround the mRNA strand. The larger subunit contains three binding sites: A (aminoacyl), P (peptidyl), and E (exit). After translational initiation (which is different in prokaryotes and eukaryotes), the ribosome enters the elongation period which follows a repetitive cycle. First a tRNA with the correct amino acid enters the A site. The ribosome transfers the peptide from the tRNA in the P site to the new amino acid on the tRNA in the A site. The tRNA from the P site will be shifted into the E site where it will be ejected. This continually occurs until the ribosome reaches a stop codon or receives a signal to stop. A peptide bond forms between the amino acid attached to the tRNA in the P site and the amino acid attached to a tRNA in the A site. The formation of a peptide bond requires an input of energy. The two reacting molecules are the alpha amino group of one amino acid and the alpha carboxyl group of the other amino acids. A by-product of this bond formation is the release of water (the amino group donates a proton while the carboxyl group donates a hydroxyl).

Translation can be downregulated by miRNAs (microRNAs). These RNA strands can cleave mRNA strands they are complementary to and will thus stop translation. Translation can also be regulated via helper proteins. For example, a protein called eukaryotic initiation factor-2 (eIF-2) can bind to the smaller subunit of the ribosome, starting translation. When elF-2 is phosphorylated, it cannot bind to the ribosome and translation is halted.

Post-translational Modifications

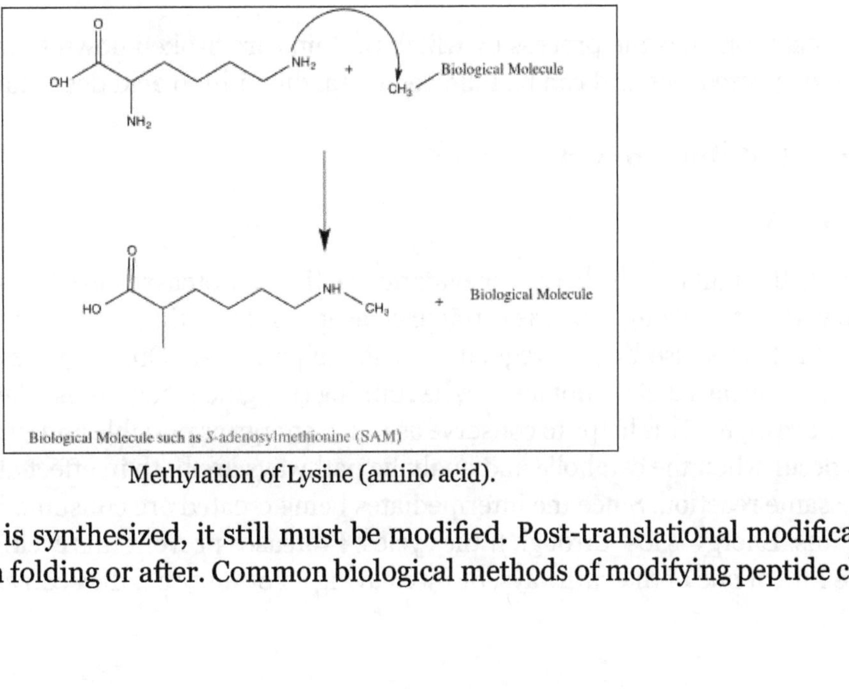

Methylation of Lysine (amino acid).

Once the peptide chain is synthesized, it still must be modified. Post-translational modifications can occur before protein folding or after. Common biological methods of modifying peptide chains after translation include methylation, phosphorylation, and disulfide bond formation. Methylation often occurs to arginine or lysine and involves adding a methyl group to a nitrogen (replacing a hydrogen). The R groups on these amino acids can be methylated multiple times as long as the bonds to nitrogen does not exceed. Methylation reduces the ability of these amino acids to form hydrogen bonds so arginine and lysine that are methylated have different properties than their standard counterparts. Phosphorylation often occurs to serine, threonine, and tyrosine and involves replacing a hydrogen on the alcohol group at the terminus of the R group with a phosphate group. This adds a negative charge on the R groups and will thus change how the amino acids behave in comparison to their standard counterparts. Disulfide bond formation is the creation of disulfide bridges (covalent bonds) between two cysteine amino acids in a chain which adds stability to the folded structure.

Protein Folding

A polypeptide chain in the cell does not have to stay linear; it can become branched or fold in on itself. Polypeptide chains fold in a particular manner depending on the solution they are in. The fact that all amino acids contain R groups with different properties is the main reason proteins fold. In a hydrophilic environment such as cytosol, the hydrophobic amino acids will concentrate at the core of the protein, while the hydrophilic amino acids will be on the exterior. This is entropically favorable since water molecules can move much more freely around hydrophilic amino acids than hydrophobic amino acids. In a hydrophobic environment, the hydrophilic amino acids will concentrate at the core of the protein, while the hydrophobic amino acids will be on the exterior. Since the new interactions between the hydrophilic amino acids are stronger than hydrophobic-hydrophilic interactions, this is enthalpically favorable. Once a polypeptide chain is fully folded, it is called a protein. Often many subunits will combine to make a fully functional protein although physiological proteins do exist that contain only one polypeptide chain. Proteins may also incorporate other molecules such as the heme group in hemoglobin, a protein responsible for carrying oxygen in the blood.

Protein Breakdown

Protein catabolism is the process by which proteins are broken down to their amino acids. This is also called proteolysis and can be followed by further amino acid degradation.

Protein Catabolism via Enzymes

Proteases

Originally thought to only disrupt enzymatic reactions, proteases (also known as peptidases) actually help with catabolizing proteins through cleavage and creating new proteins that were not present before. Proteases also help to regulate metabolic pathways. One way they do this is to cleave enzymes in pathways that do not need to be running (i.e. gluconeogenesis when blood glucose concentrations are high). This helps to conserve as much energy as possible and to avoid futile cycles. Futile cycles occur when the catabolic and anabolic pathways are both in effect at the same time and rate for the same reaction. Since the intermediates being created are consumed, the body makes no net gain gains. Energy is lost through futile cycles. Proteases prevent this cycle from occurring by altering the rate of one of the pathways, or by cleaving a key enzyme, they can stop one of the pathways.

Proteases are also nonspecific when binding to substrate, allowing for great amounts of diversity inside the cells and other proteins, as they can be cleaved much easier in an energy efficient manner.

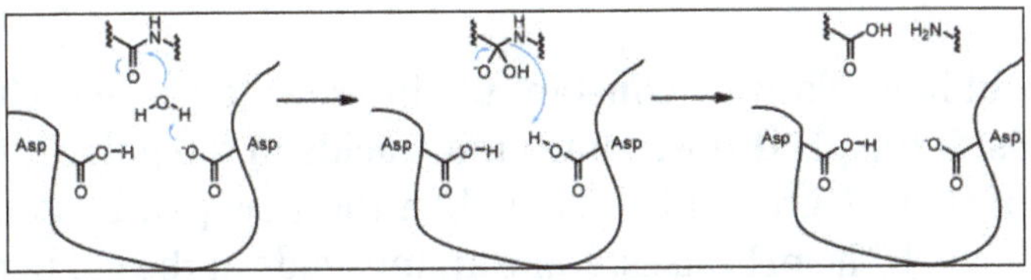

Possible mechanism for Aspartyl Protease cleaving a peptide bond. Only the peptide bond and active site are shown.

Because many proteases are nonspecific, they are highly regulated in the cell. Without regulation, proteases will destroy many essential proteins for physiological processes. One way the body regulates proteases is through protease inhibitors. Protease inhibitors can be other proteins, small peptides, or molecules. There are two types of protease inhibitors: reversible and irreversible. Reversible protease inhibitors form non-covalent interactions with the protease limiting its functionality. They can be competitive inhibitors, uncompetitive inhibitors, and noncompetitive inhibitors. Competitive inhibitors compete with the peptide to bind to the protease active site. Uncompetitive inhibitors bind to the protease while the peptide is bound but do not let the protease cleave the peptide bond. Noncompetitive inhibitors can do both. Irreversible protease inhibitors covalently modify the active site of the protease so it cannot cleave peptides.

Exopeptidases

Exopeptidases are enzymes that can cleave the end of an amino acid side chain mostly through the addition of water. These enzymes have two classes: aminopeptidases and carboxypeptidases. Aminopeptidases are enzymes that remove amino acids from the amino terminus of protein. They are present in all lifeforms and are crucial for survival since they do many cellular tasks in order to maintain stability. This form of peptidase is a zinc metalloenzyme and it is inhibited by the transition state analog. This analog is similar to the actual transition state, so it can make the enzyme bind to it instead of the actual transition state, thus preventing substrate binding and decreasing reaction rates. Carboxypeptidases cleave at the carboxyl end of the protein. While they can catabolize proteins, they are more often used in post-transcriptional modifications.

Endopeptidases

Endopeptidases are enzymes that add water to an internal peptide bond in a peptide chain and break that bond. Three common endopeptidases are pepsin, trypsin, and chymotrypsin. Chymotrypsin performs a hydrolysis reaction that cleaves after aromatic residues. The main amino acids involved are serine, histidine, and aspartic acid. They all play a role in cleaving the peptide bond. These three amino acids are known as the catalytic triad which means that these three must all be present in order to properly function. Trypsin cleaves after long positively charged residues and has a negatively charged binding pocket at the active site. Both are produced as zymogens, meaning they are initially found in their inactive state and after cleavage though a hydrolysis reaction, they becomes activated. Non-covalent interactions such as hydrogen bonding between the peptide backbone and the catalytic triad help increase reaction rates, allowing these peptidases to cleave many peptides efficiently.

Protein Catabolism via Environmental Changes

pH

Cellular proteins are held in a relatively constant pH in order to prevent changes in the protonation state of amino acids. If the pH drops, some amino acids in the polypeptide chain can become protonated if the pka of their R groups is higher than the new pH. Protonation can change the charge these R groups have. If the pH raises, some amino acids in the chain can become deprotonated (if the pka of the R group is lower than the new pH). This also changes the R group charge. Since many amino acids interact with other amino acids based on electrostatic attraction, changing the charge can break these interactions. The loss of these interactions alters the proteins structure, but most importantly it alters the proteins function, which can be beneficial or detrimental. A significant change in pH may even disrupt many interactions the amino acids make and denature (unfold) the protein.

Temperature

As the temperature in the environment increases, molecules move faster. Hydrogen bonds and hydrophobic interactions are important stabilizing forces in proteins. If the temperature rises and molecules containing these interactions are moving too fast, the interactions become compromised or even break. At high temperatures, these interactions cannot form, and a functional protein is denatured. However, it relies on two factors; the type of protein used and the amount of heat applied. The amount of heat applied determines whether this change in protein is permanent or if it can be transformed back to its original form.

Nucleic Acid Metabolism

Nucleic acid metabolism is the process by which nucleic acids (DNA and RNA) are synthesized and degraded. Nucleic acids are polymers of nucleotides. Nucleotide synthesis is an anabolic mechanism generally involving the chemical reaction of phosphate, pentose sugar, and a nitrogenous base. Destruction of nucleic acid is a catabolic reaction. Additionally, parts of the nucleotides or nucleobases can be salvaged to recreate new nucleotides. Both synthesis and degradation reactions require enzymes to facilitate the event. Defects or deficiencies in these enzymes can lead to a variety of diseases.

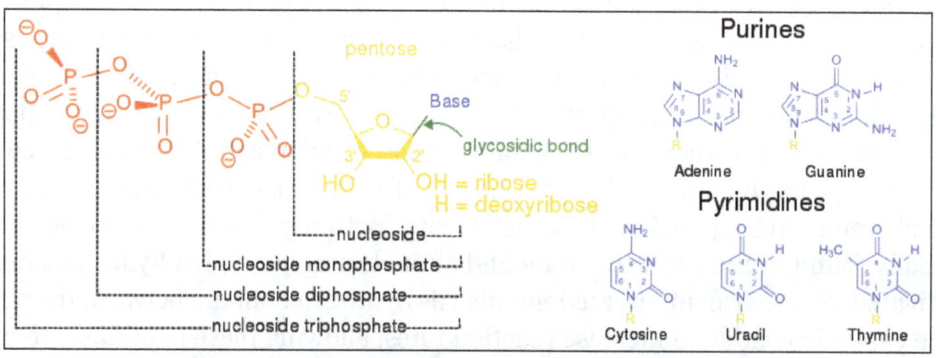

Composition of nucleotides, which make up nucleic acids.

Synthesis of Nucleic Acids

Nucleotides can be separated into purines and pyrimidines. In the more complex multicellular animals they are both primarily produced in the liver. They both contain a sugar and a phosphate, but have nitrogenous bases that are different sizes. Because of this, the two different groups are synthesized in different ways. However, all nucleotide synthesis requires the use of phosphoribosyl pyrophosphate (PRPP) which donates the ribose and phosphate necessary to create a nucleotide.

Purine Synthesis

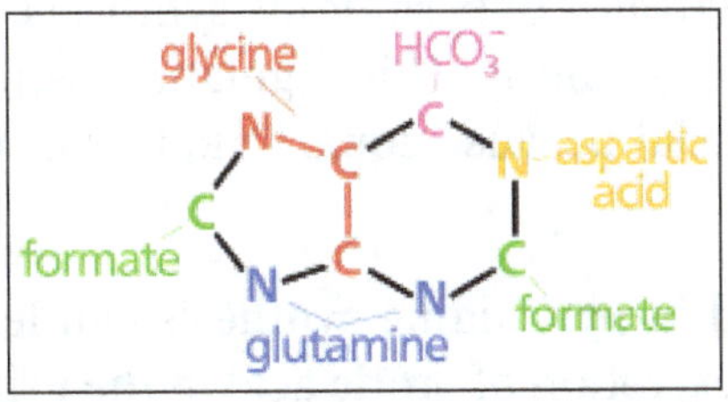

The origin of atoms that make up purine bases.

Adenine and guanine are the two nucleotides classified as purines. In purine synthesis, PRPP is turned into inosine monophosphate, or IMP. Production of IMP from PRPP requires glutamine, glycine, aspartate, and 6 ATP, among other things. IMP is then converted to AMP (adenosine monophosphate) using GTP and aspartate, which is converted into fumarate. While IMP can be directly converted to AMP, synthesis of GMP (guanosine monophosphate) requires an intermediate step, in which NAD+ is used to form the intermediate xanthosine monophosphate, or XMP. XMP is then converted into GMP by using the hydrolysis of 1 ATP and the conversion of glutamine to glutamate. AMP and GMP can then be converted into ATP and GTP, respectively, by kinases that add additional phosphates.

ATP stimulates production of GTP, while GTP stimulates production of ATP. This cross regulation keeps the relative amounts of ATP and GTP the same. Excess of either nucleotide could increase the likelihood of DNA mutations, where the wrong purine nucleotide is inserted.

Lesch-Nyhan syndrome is caused by a deficiency in hypoxanthine-guanine phosphoribosyltransferase or HGPRT, the enzyme that catalyzes the reversible reaction of producing guanine from GMP. This is a sex-linked congenital defect that causes overproduction of uric acid along with mental retardation, spasticity, and an urge to self-mutilate.

Pyrimidine Synthesis

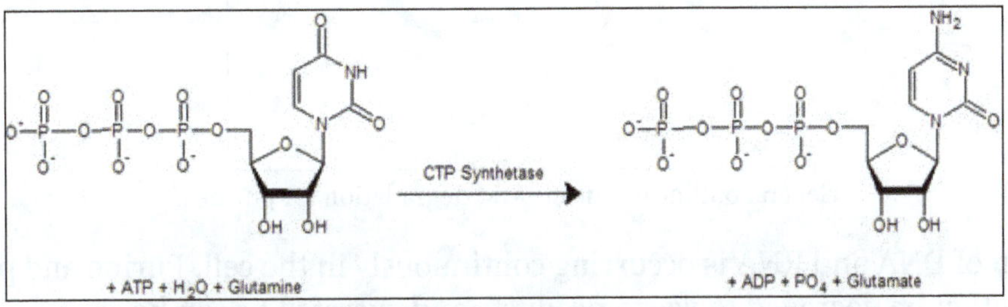

Uridine-triphosphate (UTP), at left, reacts with glutamine and other chemicals to form cytidine-triphosphate (CTP), on the right.

Pyrimidine nucleotides include cytidine, uridine, and thymidine. The synthesis of any pyrimidine nucleotide begins with the formation of uridine. This reaction requires aspartate, glutamine, bicarbonate, and 2 ATP molecules (to provide energy), as well as PRPP which provides the ribose-monophosphate. Unlike in purine synthesis, the sugar/phosphate group from PRPP is not added to the nitrogenous base until towards the end of the process. After uridine-monophosphate is synthesized, it can react with 2 ATP to form uridine-triphosphate or UTP. UTP can be converted to CTP (cytidine-triphosphate) in a reaction catalyzed by CTP synthetase. Thymidine synthesis first requires reduction of the uridine to deoxyuridine, before the base can be methylated to produce thymidine.

ATP, a purine nucleotide, is an activator of pyrimidine synthesis, while CTP, a pyrimidine nucleotide, is an inhibitor of pyrimidine synthesis. This regulation helps to keep the purine/pyrimidine amounts similar, which is beneficial because equal amounts of purines and pyrimidines are required for DNA synthesis.

Deficiencies of enzymes involved in pyrimidine synthesis can lead to the genetic disease Orotic aciduria which causes excessive excretion of orotic acid in the urine.

Converting Nucleotides to Deoxynucleotides

Nucleotides are initially made with ribose as the sugar component, which is a feature of RNA. DNA, however, requires *deoxy*ribose, which is missing the 2'-hydroxyl (-OH group) on the ribose. The reaction to remove this -OH is catalyzed by ribonucleotide reductase. This enzyme converts NDPs (nucleoside-diphosphate) to dNDPs (deoxynucleoside-diphosphate). The nucleotides must be in the diphosphate form for the reaction to occur.

In order to synthesize thymidine, a component of DNA which only exists in the deoxy form, uridine is converted to deoxyuridine (by ribonucleotide reductase), and then is methylated by thymidylate synthase to create thymidine.

Degradation of Nucleic Acids

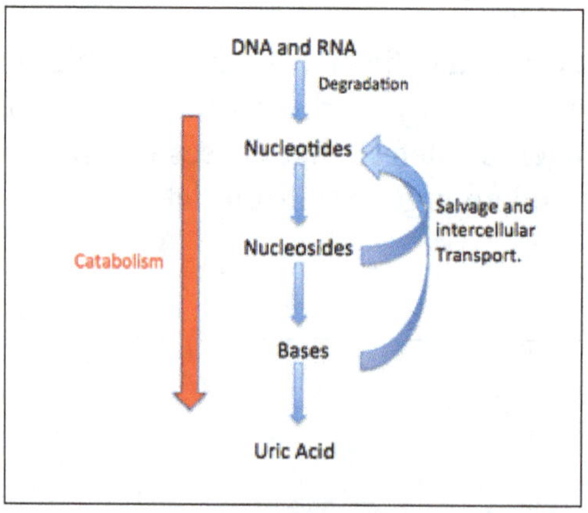

General outline of nucleic acid degradation for purines.

The breakdown of DNA and RNA is occurring continuously in the cell. Purine and pyrimidine nucleosides can either be degraded to waste products and excreted or can be salvaged as nucleotide components.

Pyrimidine Catabolism

Cytosine and uracil are converted into beta-alanine and later to malonyl-CoA which is needed for fatty acid synthesis, among other things. Thymine, on the other hand, is converted into β-aminoisobutyric acid which is then used to form methylmalonyl-CoA. The leftover carbon skeletons such as acetyl-CoA and Succinyl-CoA can then be oxidized by the citric acid cycle. Pyrimidine degradation ultimately ends in the formation of ammonium, water, and carbon dioxide. The ammonium can then enter the urea cycle which occurs in the cytosol and the mitochondria of cells.

Pyrimidine bases can also be salvaged. For example, the uracil base can be combined with ribose-1-phosphate to create uridine monophosphate or UMP. A similar reaction can also be done with thymine and deoxyribose-1-phosphate.

Deficiencies in enzymes involved in pyrimidine catabolism can lead to diseases such as Dihydropyrimidine dehydrogenase deficiency which has negative neurological effects.

Purine Catabolism

Purine degradation takes place mainly in the liver of humans and requires an assortment of enzymes to degrade purines to uric acid. First, the nucleotide will lose its phosphate through 5'-nucleotidase. The nucleoside, adenosine, is then deaminated and hydrolyzed to form hypoxanthine via adenosine deaminase and nucleosidase respectively. Hypoxanthine is then oxidized to form xanthine and then uric acid through the action of xanthine oxidase. The other purine nucleoside, guanosine, is cleaved to form guanine. Guanine is then deaminated via guanine deaminase to form xanthine which is then converted to uric acid. Oxygen is the final electron acceptor in the degradation of both purines. Uric acid is then excreted from the body in different forms depending on the animal.

Free purine and pyrimidine bases that are released into the cell are typically transported intercellularly across membranes and salvaged to create more nucleotides via nucleotide salvage. For example, adenine + PRPP --> AMP + PPi. This reaction requires the enzyme adenine phosphoribosyltransferase. Free guanine is salvaged in the same way except it requires hypoxanthine-guanine phosphoribosyltransferase.

Defects in purine catabolism can result in a variety of diseases including gout, which stems from an accumulation of uric acid crystals in various joints, and adenosine deaminase deficiency, which causes immunodeficiency.

Interconversion of Nucleotides

Once the nucleotides are synthesized they can exchange phosphates among one another in order to create mono-, di-, and tri-phosphate molecules. The conversion of a nucleoside-diphosphate (NDP) to a nucleoside-triphosphate (NTP) is catalyzed by nucleoside diphosphate kinase, which uses ATP as the phosphate donor. Similarly, nucleoside-monophosphate kinase carries out the phosphorylation of nucleside-monophosphates. Adenylate kinase is a specific nucleoside-monophosphate kinase that functions only on adenosine-monophosphate.

References

- Metabolism: britannica.com, Retrieved 16 May, 2019
- Anabolism: biologydictionary.net, Retrieved 19 August, 2019
- Catabolism: biologydictionary.net, Retrieved 23 April, 2019
- Oxidative-phosphorylation: biologydictionary.net, Retrieved 15 May, 2019
- Lipid-metabolism, anatomyandphysiology: opentextbc.ca, Retrieved 26 June, 2019
- Gluconeogenesis: Biological-Chemistry-Supplemental-Modules-(Biological-Chemistry)-Metabolism-Anabolism: chem.libretexts.org, Retrieved 28 July, 2019

5

Biochemical Techniques

Various biochemical techniques of immunoprecipitation, western blot, electrophoretic mobility shift assay, northern blot and southern blot are used for the analysis of biomolecules. The topics elaborated in this chapter will help in gaining a better perspective about these biochemical techniques.

Biochemical studies or Biochemical Techniques rely on the availability of appropriate analytical techniques and their applications. This undergraduate course deals with the different biochemical techniques such as spectrophotometry and chromatography to gain knowledge on the biomolecules such as proteins, lipids, carbohydrates, nucleic acids, and their functions.

In recent years, huge advances are made in understanding the gene structure and their expression and hence, the discipline of molecular biology overlaps with that of biochemistry in many aspects. Molecular and immunological techniques such as ELISA, radioimmunoassay, blotting, PCR, cell culture, hybridoma, and cloning protocol dealt in this course are mainly focused on understanding the diagnosis of diseases, malfunctions, and disorders in order to generate corrective measures.

Molecular and immunological techniques such as ELISA, radioimmunoassay, blotting, PCR, cell culture, hybridoma, and cloning protocol dealt in this course are mainly focused on understanding the diagnosis of diseases, malfunctions, and disorders in order to generate corrective measures.

In recent years, huge advances are made in understanding the gene structure and their expression and hence, the discipline of molecular biology overlaps with that of biochemistry in many aspects. Molecular and immunological techniques such as ELISA, radioimmunoassay, blotting, PCR, cell culture, hybridoma, and cloning protocol dealt in this course are mainly focused on understanding the diagnosis of diseases, malfunctions, and disorders in order to generate corrective measures.

Molecular and immunological techniques such as ELISA, radioimmunoassay, blotting, PCR, cell culture, hybridoma, and cloning protocol dealt in this course are mainly focused on understanding the diagnosis of diseases, malfunctions, and disorders in order to generate corrective measures.

Basics of Biochemical Techniques

Some of the Biochemical techniques are given here:

Biochemical Tests

These five tests identify the main biologically important chemical compounds. For each test take a small amount of the substance to test, and shake it in water in a test tube. If the sample is a piece of food, then grind it with some water in a pestle and mortar to break up the cells and release the cell contents. Many of these compounds are insoluble, but the tests work just as well on a fine suspension.

Starch (Iodine Test)

- To approximately 2 cm^3 of the test solution, adds two drops of iodine/potassium iodide solution.
- A blue-black color indicates the presence of starch as a starch-polyiodide complex is formed.
- Starch is only slightly soluble in water, but the test works well in a suspension or as a solid.

Reducing Sugars (Benedict's Test)

- All monosaccharides and most disaccharides (except sucrose) will reduce copper (II) sulfate, producing a precipitate of copper (I) oxide on heating, so they are called reducing sugars.
- Benedict's reagent is an aqueous solution of copper (II) sulfate, sodium carbonate, and sodium citrate.
- To approximately 2 cm^3 of test, solution adds an equal quantity of Benedict's reagent. Shake, and heat for a few minutes at 95 °C in a water bath.
- A precipitate indicates reducing sugar. The color and density of the precipitate give an indication of the amount of reducing sugar present, so this test is semi-quantitative.
- The original pale blue color means no reducing sugar, a green precipitate means relatively little sugar; a brown or red precipitate means progressively more sugar is present.

Non-reducing Sugars (Benedict's Test)

- Sucrose is called a non-reducing sugar because it does not reduce copper sulfate, so there is no direct test for sucrose.
- However, if it is first hydrolyzed (broken down) to its constituent monosaccharides (glucose and fructose), it will then give a positive Benedict's test.
- So sucrose is the only sugar that will give a negative Benedict's test before hydrolysis and a positive test afterward. First test a sample for reducing sugars, to see if there are any present before hydrolysis. Then, using a separate sample, boil the test solution with dilute hydrochloric acid for a few minutes to hydrolyze the glycosidic bond. Neutralize the solution by gently adding small amounts of solid sodium hydrogen carbonate until it stops fizzing, then test as before for reducing sugars.

- First test a sample for reducing sugars, to see if there are any present before hydrolysis. Then, using a separate sample, boil the test solution with dilute hydrochloric acid for a few minutes to hydrolyze the glycosidic bond. Neutralize the solution by gently adding small amounts of solid sodium hydrogen carbonate until it stops fizzing, then test as before for reducing sugars.

Lipids (Emulsion Test)

- Lipids do not dissolve in water but do dissolve in ethanol. This characteristic is used in the emulsion test.
- Do not start by dissolving the sample in water, but instead, shake some of the test samples with about 4 cm^3 of ethanol.
- Decant the liquid into a test tube of water, leaving any undissolved substances behind.
- If there are lipids dissolved in the ethanol, they will precipitate in the water, forming a cloudy white emulsion.

Protein (Biuret Test)

- To about 2 cm^3 of test solution adds an equal volume of biuret solution, down the side of the test tube.
- A blue ring forms at the surface of the solution, which disappears on shaking, and the solution turns lilac-purple, indicating protein.
- The color is due to a complex between nitrogen atoms in the peptide chain and Cu^{2+} ions, so this is really a test for peptide bonds.
- Qualitative Analysis: Color Reactions of Lipids
- Color reactions of Amino acids
- Color Reactions of Carbohydrates (Practical Protocol)
- Colorimetric Analysis: BRADFORD PROTEIN ASSAY

Chromatography

Chromatography is used to separate pure substances from a mixture of substances, such as a cell extract. It is based on different substances having different solubility in different solvents. A simple and common form of chromatography use filter paper.

- Pour some solvent into a chromatography tank and seal it, so the atmosphere is saturated with solvent vapor. Different solvents are suitable for different tasks, but they are usually mixtures of water with organic liquids such as ethanol or propanone.
- Place a drop of the mixture to be separated onto a sheet of chromatography paper near one end. This is the origin of the chromatogram. The spot should be small but concentrated. Repeat for any other mixtures. Label the spots with pencil, as ink may dissolve.

- Place the chromatography sheet into the tank so that the origin is just above the level of solvent, and leave for several hours. The solvent will rise up the paper by capillary action carrying the contents of the mixture with it. Any solutes dissolved in the solvent will be partitioned between the organic solvent (the moving phase) and the water, which is held by the paper (the stationary phase). The more soluble a solute is in the solvent the further up the paper it will move.
- When the solvent has nearly reached the top of the paper, the paper is removed and the position of the solvent front marked. The chromatogram may need to be developed to make the spots visible. For example, amino acids stain purple with ninhydrin.
- The chromatogram can be analyzed by measuring the distance traveled by the solvent front, and the distance from the origin to the center of each spot. This is used to calculate the Rf (relative front) value for each spot.

R_f value is characteristic of a particular solute in a particular solvent. It can be used to identify components of a mixture by comparing to tables of known R_f values.

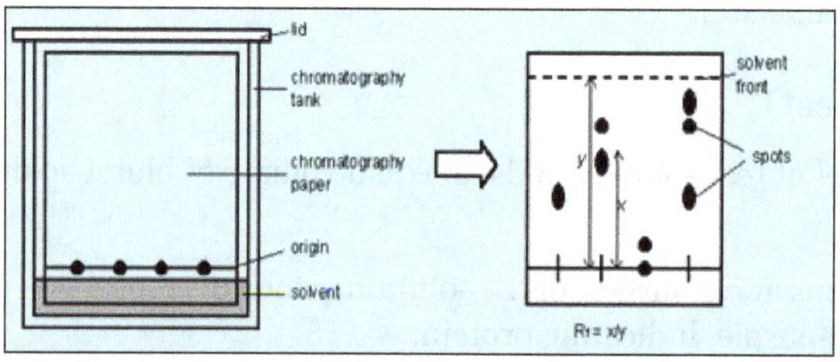

Sometimes chromatography with a single solvent is not enough to separate all the constituents of a mixture. In this case, the separation can be improved by two-dimensional chromatography, where the chromatography paper is turned through 90° and run a second time in a second solvent. Solutes that didn't separate in one solvent will separate in another because they have different solubility.

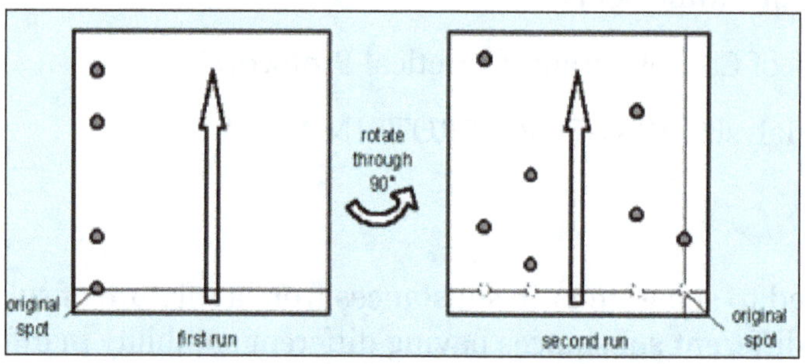

There are many different types of chromatography:

- Paper chromatography is the simplest but does not always give very clean separation.
- Thin layer chromatography (TLC) uses a thin layer of cellulose or silica coated onto a plastic or glass sheet. This is more expensive but gives much betta er and more reliable separation.
- Column chromatography uses a glass column filled with a cellulose slurry. Large samples can be pumped through the column and the separated fractions can be collected for further experiments, so this is preparative chromatography as opposed to analytical chromatography.

- High-performance liquid chromatography (HPLC) is an improved form of column chromatography that delivers excellent separation very quickly.
- Electrophoresis uses an electric current to separate molecules on the basis of charge. It can also be used to separate on the basis of molecular size, and as such is used in DNA sequencing.

Cell Fractionation

This means separating different parts and organelles of a cell so that they can be studied in detail. All the processes of cell metabolism (such as respiration or photosynthesis) have been studied in this way. This is one of the biochemical techniques. The most common method of fractionating cells is to use differential centrifugation.

A more sophisticated separation can be performed by density gradient centrifugation. In this, the cell-free extract is centrifuged in a dense solution (such as sucrose or cesium chloride). The fractions don't pellet, but instead, separate out into layers with the densest fractions near the bottom of the tube. The desired layer can then be pipetted off. This is the technique used in the Meselson-Stahl experiment and it is also used to separate the two types of ribosomes. The terms the 70S and 80S refer to their positions in a density gradient.

Enzyme Kinetics

This means measuring the rate of enzyme reactions:

- Firstly you need a signal to measure that shows the progress of the reaction. The signal should change with either substrate or product concentration, and it should preferably be something that can be measured continuously. Typical signals include color changes, pH changes, mass changes, gas production, volume changes or turbidity changes. If the reaction has none of these properties, it can sometimes be linked to a second reaction which does generate one of these changes.
- If you mix your substrate with enzyme and measure your signal, you will obtain a time-course. If the signal is proportional to substrate concentration it will start high and decrease, while if the signal is proportional to the product it will start low and increase. In both cases, the time-course will be curved (actually an exponential curve).
- How do you obtain a rate from this time-course? One thing that is not a good idea is to measure the time taken for the reaction, for as the time-course shows it is very difficult to say when the reaction ends: it just gradually approaches the end-point. A better method is to measure the initial rate – that is the initial slope of the time-course. This also means you don't need to record the whole time-course, but simply take one measurement a short time after mixing.
- Repeat this initial rate measurement under different conditions (such as different substrate concentrations) and then plot a graph of rate vs. the factor. Each point on this second graph is taken from a separate initial rate measurement (or better still is an average of several initial rate measurements under the same conditions). Draw a smooth curve through the points.
- Be careful not to confuse the two kinds of the graph (the time-course and rate graphs) when interpreting your data.

- One useful trick is to dissolve the substrate in agar in an agar plate. If a source of the enzyme is placed on the agar plate, the enzyme will diffuse out through the agar, turning the substrate into the product as it goes. There must be a way to distinguish the substrate from the product, and the reaction will then show up as a ring around the enzyme source.
- The higher the concentration of enzyme, the higher the diffusion gradient, so the faster the enzyme diffuses through the agar, so the larger the ring in a given time.
- The diameter of the ring is, therefore, proportional to the enzyme concentration. This can be done for many enzymes, e.g. a protein agar plate can be used for a protease enzyme, or a starch agar plate can be used for the enzyme amylase.

Microscopy

Of all the techniques used in biology, microscopy is probably the most important. The vast majority of living organisms are too small to be seen in any detail with the human eye, and cells and their organelles can only be seen with the aid of a microscope. Cells were first seen in 1665 by Robert Hooke (who named them after monks' cells in a monastery) and were studied in more detail by Leeuwenhoek using a primitive microscope.

Units of measurement. The standard SI units of measurement used in microscopy are:

- Metre (m) = 1 m
- Millimetre (mm) = 10^{-3} m
- Micrometre = 10^{-6} m
- Nanometre (nm) = 10^{-9} m
- Picometer (pm) = 10^{-12} m
- Angstrom (Å°) = 10^{-10} m (obsolete)

Magnification and Resolving Power. By using more lenses microscopes can magnify by a larger amount, but this doesn't always mean that more detail can be seen. The amount of detail depends on the resolving power of a microscope, which is the smallest separation at which two separate objects can be distinguished (or resolved). It is calculated by the formula.

Where lambda is the wavelength of light, and *n.a.* is the numerical aperture of the lens (which ranges from about 0.5 to 1.4). So the resolving power of a microscope is ultimately limited by the wavelength of light (400-600nm for visible light). To improve the resolving power a shorter wavelength of light is needed, and sometimes microscopes have blue filters for this purpose (because blue has the shortest wavelength of visible light).

Different kinds of Microscope

Light Microscope

- This is the oldest, simplest and most widely-used form of microscopy.
- Specimens are illuminated with light, which is focussed using glass lenses and viewed using the eye or photographic film.

- Specimens can be living or dead, but often need to be stained with a colored dye to make them visible.
- Many different strains are available that stain specific parts of the cell such as DNA, lipids, cytoskeleton, etc.
- All light microscopes today are compound microscopes, which means they use several lenses to obtain high magnification.
- Light microscopy has a resolution of about 200 nm, which is good enough to see cells, but not the details of cell organelles.
- There has been a recent resurgence in the use of light microscopy, partly due to technical improvements, which have dramatically improved the resolution far beyond the theoretical limit. For example fluorescence microscopy has a resolution of about 10 nm, while interference microscopy has a resolution of about 1 nm.

Electron Microscope

- This uses a beam of electrons, rather than electromagnetic radiation, to "illuminate" the specimen.
- This may seem strange, but electrons behave like waves and can easily be produced (using a hot wire), focussed (using electromagnets) and detected (using a phosphor screen or photographic film).
- A beam of electrons has an effective wavelength of less than 1 nm, so can be used to resolve small sub-cellular ultrastructure.
- The development of the electron microscope in the 1930s revolutionized biology, allowing organelles such as mitochondria, ER, and membranes to be seen in detail for the first time.

The main problem with the electron microscope is that specimens must be fixed in plastic and viewed in a vacuum, and must, therefore, be dead. Other problems are that the specimens can be damaged by the electron beam and they must be stained with an electron-dense chemical (usually heavy metals like osmium, lead or gold). Initially, there was a problem of artifacts (i.e. observed structures that were due to the preparation process and were not real), but improvements in technique have eliminated most of these.

There are two kinds of the electron microscope. The transmission electron microscope (TEM) works much like a light microscope, transmitting a beam of electrons through a thin specimen and then focussing the electrons to form an image on a screen or on film. This is the most common form of the electron microscope and has the best resolution. The scanning electron microscope (SEM) scans a fine beam of an electron onto a specimen and collects the electrons scattered by the surface. This has a poorer resolution but gives excellent 3-dimentional images of surfaces.

- X-ray Microscope: This is an obvious improvement to the light microscope, since x-rays have wavelengths a thousand time shorter than visible light, and so could even be used to resolve atoms. Unfortunately, there are no good x-ray lenses, so an image cannot be focussed, and usable x-ray microscopes do not yet exist. However, x-rays can be used without focussing on giving a diffraction pattern, which can be used to work out the structures of molecules, such as those of proteins and DNA.

- Scanning Tunnelling Microscope (or Atomic Force Microscope): This uses a very fine needle to scan the surface of a specimen. It has a resolution of about 10 pm and has been used to observe individual atoms for the first time.

Table: Comparison of light and electron microscopes.

	Light Microscope	Electron Microscope
illumination and source	light from lamp	electrons from hot wire
focusing	glass lenses	electromagnets
detection	eye or film	phosphor screen or film
magnification	1 500 x	500 000 x
resolution	200 nm	1 nm
specimen	living or dead	dead
staining	coloured dyes	heavy metals
cost	cheap to expensive	very expensive

Immunoprecipitation

Immunoprecipitation (IP) is the technique of precipitating a protein antigen out of solution using an antibody that specifically binds to that particular protein. This process can be used to isolate and concentrate a particular protein from a sample containing many thousands of different proteins. Immunoprecipitation requires that the antibody be coupled to a solid substrate at some point in the procedure.

Types

Individual Protein Immunoprecipitation (IP)

Involves using an antibody that is specific for a known protein to isolate that particular protein out of a solution containing many different proteins. These solutions will often be in the form of a crude lysate of a plant or animal tissue. Other sample types could be body fluids or other samples of biological origin.

Protein Complex Immunoprecipitation (Co-IP)

Immunoprecipitation of intact protein complexes (i.e. antigen along with any proteins or ligands that are bound to it) is known as co-immunoprecipitation (Co-IP). Co-IP works by selecting an antibody that targets a known protein that is believed to be a member of a larger complex of proteins. By targeting this *known* member with an antibody it may become possible to pull the entire protein complex out of solution and thereby identify *unknown* members of the complex.

This works when the proteins involved in the complex bind to each other tightly, making it possible to pull multiple members of the complex out of solution by latching onto one member with an antibody. This concept of pulling protein complexes out of solution is sometimes referred to as a "pull-down". Co-IP is a powerful technique that is used regularly by molecular biologists to analyze protein–protein interactions.

- A particular antibody often selects for a subpopulation of its target protein that has the epitope exposed, thus failing to identify any proteins in complexes that hide the epitope. This can be seen in that it is rarely possible to precipitate even half of a given protein from a sample with a single antibody, even when a large excess of antibody is used.
- As successive rounds of targeting and immunoprecipitations take place, the number of identified proteins may continue to grow. The identified proteins may not ever exist in a single complex at a given time, but may instead represent a network of proteins interacting with one another at different times for different purposes.
- Repeating the experiment by targeting different members of the protein complex allows the researcher to double-check the result. Each round of pull-downs should result in the recovery of both the original known protein as well as other previously identified members of the complex (and even new additional members). By repeating the immunoprecipitation in this way, the researcher verifies that each identified member of the protein complex was a valid identification. If a particular protein can only be recovered by targeting one of the known members but not by targeting other of the known members then that protein's status as a member of the complex may be subject to question.

Chromatin Immunoprecipitation (ChIP)

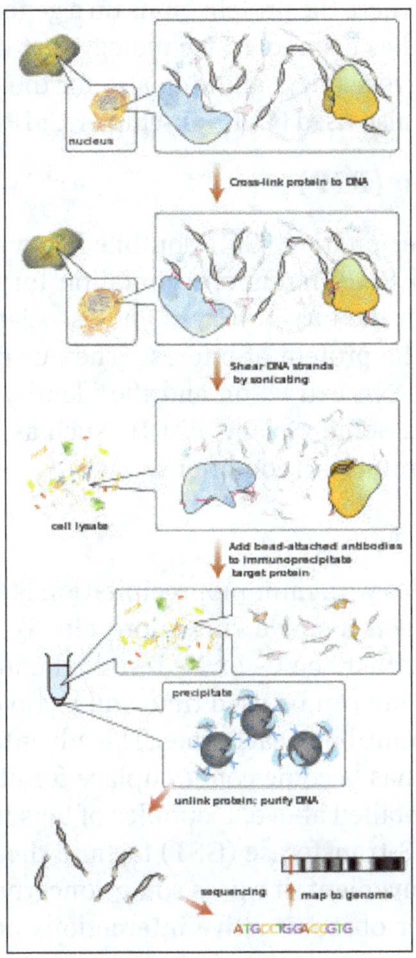

ChIP-sequencing workflow.

Chromatin immunoprecipitation (ChIP) is a method used to determine the location of DNA binding sites on the genome for a particular protein of interest. This technique gives a picture of the protein–DNA interactions that occur inside the nucleus of living cells or tissues. The *in vivo* nature of this method is in contrast to other approaches traditionally employed to answer the same questions.

The principle underpinning this assay is that DNA-binding proteins (including transcription factors and histones) in living cells can be cross-linked to the DNA that they are binding. By using an antibody that is specific to a putative DNA binding protein, one can immunoprecipitate the protein–DNA complex out of cellular lysates. The crosslinking is often accomplished by applying formaldehyde to the cells (or tissue), although it is sometimes advantageous to use a more defined and consistent crosslinker such as DTBP. Following crosslinking, the cells are lysed and the DNA is broken into pieces 0.2–1.0 kb in length by sonication. At this point the immunoprecipitation is performed resulting in the purification of protein–DNA complexes. The purified protein–DNA complexes are then heated to reverse the formaldehyde cross-linking of the protein and DNA complexes, allowing the DNA to be separated from the proteins. The identity and quantity of the DNA fragments isolated can then be determined by PCR. The limitation of performing PCR on the isolated fragments is that one must have an idea which genomic region is being targeted in order to generate the correct PCR primers. Sometimes this limitation circumvented simply by cloning the isolated genomic DNA into a plasmid vector and then using primers that are specific to the cloning region of that vector. Alternatively, when one wants to find where the protein binds on a genome-wide scale, ChIP-Sequencing is used and has recently emerged as a standard technology that can localize protein binding sites in a high-throughput, cost-effective fashion, allowing also for the characterization of the cistrome. Previously, DNA microarray was also used (ChIP-on-chip or ChIP-chip).

RNP Immunoprecipitation (RIP)

Similar to chromatin immunoprecipitation (ChIP) outlined above, but rather than targeting DNA binding proteins as in ChIP, an RNP immunoprecipitation targets ribonucleoproteins (RNPs). Live cells are first lysed and then the target protein and associated RNA are immunoprecipitated using an antibody targeting the protein of interest. The purified RNA-protein complexes can be separated by performing an RNA extraction and the identity of the RNA can be determined by cDNA sequencing or RT-PCR. Some variants of RIP, such as PAR-CLIP include cross-linking steps, which then require less careful lysis conditions.

Tagged Proteins

One of the major technical hurdles with immunoprecipitation is the great difficulty in generating an antibody that specifically targets a single known protein. To get around this obstacle, many groups will engineer tags onto either the C- or N- terminal end of the protein of interest. The advantage here is that the same tag can be used time and again on many different proteins and the researcher can use the same antibody each time. The advantages with using tagged proteins are so great that this technique has become commonplace for all types of immunoprecipitation including all of the types of IP detailed above. Examples of tags in use are the Green Fluorescent Protein (GFP) tag, Glutathione-S-transferase (GST) tag and the FLAG-tag tag. While the use of a tag to enable pull-downs is convenient, it raises some concerns regarding biological relevance because the tag itself may either obscure native interactions or introduce new and unnatural interactions.

Pull down assay using tagged proteins.

Methods

The two general methods for immunoprecipitation are the direct capture method and the indirect capture method.

Direct

Antibodies that are specific for a particular protein (or group of proteins) are immobilized on a solid-phase substrate such as superparamagnetic microbeads or on microscopic agarose (non-magnetic) beads. The beads with bound antibodies are then added to the protein mixture, and the proteins that are targeted by the antibodies are captured onto the beads via the antibodies; in other words, they become immunoprecipitated.

Indirect

Antibodies that are specific for a particular protein, or a group of proteins, are added directly to the mixture of protein. The antibodies have not been attached to a solid-phase support yet. The antibodies are free to float around the protein mixture and bind their targets. As time passes, the beads coated in protein A/G are added to the mixture of antibody and protein. At this point, the antibodies, which are now bound to their targets, will stick to the beads.

From this point on, the direct and indirect protocols converge because the samples now have the same ingredients. Both methods gives the same end-result with the protein or protein complexes bound to the antibodies which themselves are immobilized onto the beads.

Selection

An indirect approach is sometimes preferred when the concentration of the protein target is low or when the specific affinity of the antibody for the protein is weak. The indirect method is also used when the binding kinetics of the antibody to the protein is slow for a variety of reasons. In most situations, the direct method is the default, and the preferred, choice.

Technological Advances

Agarose

Historically the solid-phase support for immunoprecipitation used by the majority of scientists has been highly-porous agarose beads (also known as agarose resins or slurries). The advantage of this technology is a very high potential binding capacity, as virtually the entire sponge-like structure of the agarose particle (50 to 150μm in size) is available for binding antibodies (which will in turn bind the target proteins) and the use of standard laboratory equipment for all aspects of the IP protocol without the need for any specialized equipment. The advantage of an extremely high binding capacity must be carefully balanced with the quantity of antibody that the researcher is prepared to use to coat the agarose beads. Because antibodies can be a cost-limiting factor, it is best to calculate backward *from* the amount of protein that needs to be captured (depending upon the analysis to be performed downstream), *to* the amount of antibody that is required to bind that quantity of protein (with a small excess added in order to account for inefficiencies of the system), and back still further *to* the quantity of agarose that is needed to bind that particular quantity of antibody. In cases where antibody saturation is not required, this technology is unmatched in its ability to capture extremely large quantities of captured target proteins. The caveat here is that the *"high capacity advantage"* can become a *"high capacity disadvantage"* that is manifested when the enormous binding capacity of the sepharose/agarose beads is not completely saturated with antibodies. It often happens that the amount of antibody available to the researcher for their immunoprecipitation experiment is less than sufficient to saturate the agarose beads to be used in the immunoprecipitation. In these cases the researcher can end up with agarose particles that are only partially coated with antibodies, and the portion of the binding capacity of the agarose beads that is not coated with antibody is then free to bind anything that will stick, resulting in an elevated background signal due to non-specific binding of lysate components to the beads, which can make data interpretation difficult. While some may argue that for these reasons it is prudent to match the quantity of agarose (in terms of binding capacity) to the quantity of antibody that one wishes to be bound for the immunoprecipitation, a simple way to reduce the issue of non-specific binding to agarose beads and increase specificity is to preclear the lysate, which for any immunoprecipitation is highly recommended.

Preclearing

Lysates are complex mixtures of proteins, lipids, carbohydrates and nucleic acids, and one must assume that some amount of non-specific binding to the IP antibody, Protein A/G or the beaded support will occur and negatively affect the detection of the immunoprecipitated target(s). In most cases, *preclearing* the lysate at the start of each immunoprecipitation experiment is a way to remove potentially reactive components from the cell lysate prior to the immunoprecipitation to prevent the non-specific binding of these components to the IP beads or antibody. The basic preclearing procedure is described below, wherein the lysate is incubated with beads alone, which are then removed and discarded prior to the immunoprecipitation. This approach, though, does not account for non-specific binding to the IP antibody, which can be considerable. Therefore, an alternative method of preclearing is to incubate the protein mixture with exactly the same components that will be used in the immunoprecipitation, except that a non-target, irrelevant antibody of the same antibody subclass as the IP antibody is used instead of the IP antibody itself. This approach attempts to use as close to the exact IP conditions and components as the actual immunoprecipitation to remove any non-specific cell constituent without capturing the target protein (unless, of course, the target protein non-specifically binds to some other IP component, which should be properly controlled for by analyzing the discarded beads used to preclear the lysate). The target protein can then be immunoprecipitated with the reduced risk of non-specific binding interfering with data interpretation.

Superparamagnetic Beads

While the vast majority of immunoprecipitations are performed with agarose beads, the use of superparamagnetic beads for immunoprecipitation is a much newer approach that is only recently gaining in popularity as an alternative to agarose beads for IP applications. Unlike agarose, magnetic beads are solid and can be spherical, depending on the type of bead, and antibody binding is limited to the surface of each bead. While these beads do not have the advantage of a porous center to increase the binding capacity, magnetic beads are significantly smaller than agarose beads (1 to 4μm), and the greater number of magnetic beads per volume than agarose beads collectively gives magnetic beads an effective surface area-to-volume ratio for optimum antibody binding.

Commercially available magnetic beads can be separated based by size uniformity into monodisperse and polydisperse beads. Monodisperse beads, also called microbeads, exhibit exact uniformity, and therefore all beads exhibit identical physical characteristics, including the binding capacity and the level of attraction to magnets. Polydisperse beads, while similar in size to monodisperse beads, show a wide range in size variability (1 to 4μm) that can influence their binding capacity and magnetic capture. Although both types of beads are commercially available for immunoprecipitation applications, the higher quality monodisperse superparamagnetic beads are more ideal for automatic protocols because of their consistent size, shape and performance. Monodisperse and polydisperse superparamagnetic beads are offered by many companies, including Invitrogen, Thermo Scientific, and Millipore.

Agarose vs. Magnetic Beads

Proponents of magnetic beads claim that the beads exhibit a faster rate of protein binding over agarose beads for immunoprecipitation applications, although standard agarose bead-based immunoprecipitations have been performed in 1 hour. Claims have also been made that magnetic beads are better for immunoprecipitating extremely large protein complexes because of the complete lack of an upper size limit for such complexes, although there is no unbiased evidence stating this claim. The nature of magnetic bead technology does result in less sample handling due to the reduced physical stress on samples of magnetic separation versus repeated centrifugation when using agarose, which may contribute greatly to increasing the yield of labile (fragile) protein complexes. Additional factors, though, such as the binding capacity, cost of the reagent, the requirement of extra equipment and the capability to automate IP processes should be considered in the selection of an immunoprecipitation support.

Binding Capacity

Proponents of both agarose and magnetic beads can argue whether the vast difference in the binding capacities of the two beads favors one particular type of bead. In a bead-to-bead comparison, agarose beads have significantly greater surface area and therefore a greater binding capacity than magnetic beads due to the large bead size and sponge-like structure. But the variable pore size of the agarose causes a potential upper size limit that may affect the binding of extremely large proteins or protein complexes to internal binding sites, and therefore magnetic beads may be better suited for immunoprecipitating large proteins or protein complexes than agarose beads, although there is a lack of independent comparative evidence that proves either case.

Some argue that the significantly greater binding capacity of agarose beads may be a disadvantage because of the larger capacity of non-specific binding. Others may argue for the use of magnetic beads because of the greater quantity of antibody required to saturate the total binding capacity of agarose beads, which would obviously be an economical disadvantage of using agarose. While these arguments are correct outside the context of their practical use, these lines of reasoning ignore two key aspects of the principle of immunoprecipitation that demonstrates that the decision to use agarose or magnetic beads is not simply determined by binding capacity.

First, non-specific binding is not limited to the antibody-binding sites on the immobilized support; any surface of the antibody or component of the immunoprecipitation reaction can bind to nonspecific lysate constituents, and therefore nonspecific binding will still occur even when completely saturated beads are used. This is why it is important to preclear the sample before the immunoprecipitation is performed.

Second, the ability to capture the target protein is directly dependent upon the amount of immobilized antibody used, and therefore, in a side-by-side comparison of agarose and magnetic bead immunoprecipitation, the most protein that either support can capture is limited by the amount of antibody added. So the decision to saturate any type of support depends on the amount of protein required.

Cost

The price of using either type of support is a key determining factor in using agarose or magnetic beads for immunoprecipitation applications. A typical first-glance calculation on the cost of magnetic beads compared to sepharose beads may make the sepharose beads appear less expensive. But magnetic beads may be competitively priced compared to agarose for analytical-scale immunoprecipitations depending on the IP method used and the volume of beads required per IP reaction.

Using the traditional batch method of immunoprecipitation, where all components are added to a tube during the IP reaction, the physical handling characteristics of agarose beads necessitate a minimum quantity of beads for each IP experiment (typically in the range of 25 to 50μl beads per IP). This is because sepharose beads must be concentrated at the bottom of the tube by centrifugation and the supernatant removed after each incubation, wash, etc. This imposes absolute physical limitations on the process, as pellets of agarose beads less than 25 to 50μl are difficult if not impossible to visually identify at the bottom of the tube. With magnetic beads, there is no minimum quantity of beads required due to magnetic handling, and therefore, depending on the target antigen and IP antibody, it is possible to use considerably less magnetic beads.

Conversely, spin columns may be employed instead of normal microfuge tubes to significantly reduce the amount of agarose beads required per reaction. Spin columns contain a filter that allows all IP components except the beads to flow through using a brief centrifugation and therefore provide a method to use significantly less agarose beads with minimal loss.

Equipment

Only standard laboratory equipment is required for the use of agarose beads in immunoprecipitation applications, while high-power magnets are required for magnetic bead-based IP reactions. While the magnetic capture equipment may be cost-prohibitive, the rapid completion of immunoprecipitations using magnetic beads may be a financially beneficial approach when grants are due, because a 30-minute protocol with magnetic beads compared to overnight incubation at 4 °C with agarose beads may result in more data generated in a shorter length of time.

Automation

An added benefit of using magnetic beads is that automated immunoprecipitation devices are becoming more readily available. These devices not only reduce the amount of work and time to perform an IP, but they can also be used for high-throughput applications.

While clear benefits of using magnetic beads include the increased reaction speed, more gentle sample handling and the potential for automation, the choice of using agarose or magnetic beads based on the binding capacity of the support medium and the cost of the product may depend on the protein of interest and the IP method used. As with all assays, empirical testing is required to determine which method is optimal for a given application.

Protocol

Once the solid substrate bead technology has been chosen, antibodies are coupled to the beads and the antibody-coated-beads can be added to the heterogeneous protein sample (e.g. homogenized tissue). At this point, antibodies that are immobilized to the beads will bind to the proteins that they specifically recognize. Once this has occurred the immunoprecipitation portion of the protocol is actually complete, as the specific proteins of interest are bound to the antibodies that are themselves immobilized to the beads. Separation of the immunocomplexes from the lysate is an extremely important series of steps, because the proteins must remain bound to each other (in the case of co-IP) and bound to the antibody during the wash steps to remove non-bound proteins and reduce background.

When working with agarose beads, the beads must be pelleted out of the sample by briefly spinning in a centrifuge with forces between 600–3,000 x g (times the standard gravitational force). This step may be performed in a standard microcentrifuge tube, but for faster separation, greater consistency and higher recoveries, the process is often performed in small spin columns with a pore size that allows liquid, but not agarose beads, to pass through. After centrifugation, the agarose beads will form a very loose fluffy pellet at the bottom of the tube. The supernatant containing contaminants can be carefully removed so as not to disturb the beads. The wash buffer can then be added to the beads and after mixing, the beads are again separated by centrifugation.

With superparamagnetic beads, the sample is placed in a magnetic field so that the beads can collect on the side of the tube. This procedure is generally complete in approximately 30 seconds, and the remaining (unwanted) liquid is pipetted away. Washes are accomplished by resuspending the beads (off the magnet) with the washing solution and then concentrating the beads back on the tube wall (by placing the tube back on the magnet). The washing is generally repeated several times to ensure adequate removal of contaminants. If the superparamagnetic beads are homogeneous in size and the magnet has been designed properly, the beads will concentrate uniformly on the side of the tube and the washing solution can be easily and completely removed.

After washing, the precipitated proteins are eluted and analyzed by gel electrophoresis, mass spectrometry, western blotting, or any number of other methods for identifying constituents in the complex. Protocol times for immunoprecipitation vary greatly due to a variety of factors, with protocol times increasing with the number of washes necessary or with the slower reaction kinetics of porous agarose beads.

Steps

- Lyse cells and prepare sample for immunoprecipitation.
- Pre-clear the sample by passing the sample over beads alone or bound to an irrelevant antibody to soak up any proteins that non-specifically bind to the IP components.
- Incubate solution with antibody against the protein of interest. Antibody can be attached to solid support before this step (direct method) or after this step (indirect method). Continue the incubation to allow antibody-antigen complexes to form.
- Precipitate the complex of interest, removing it from bulk solution.
- Wash precipitated complex several times. Spin each time between washes when using agarose beads or place tube on magnet when using superparamagnetic beads and then remove the supernatant. After the final wash, remove as much supernatant as possible.
- Elute proteins from the solid support using low-pH or SDS sample loading buffer.
- Analyze complexes or antigens of interest. This can be done in a variety of ways:
    - SDS-PAGE (sodium dodecyl sulfate-polyacrylamide gel electrophoresis) followed by gel staining.
    - SDS-PAGE followed by: gel staining, cutting out individual stained protein bands, and sequencing the proteins in the bands by MALDI-Mass Spectrometry.
    - Transfer and Western Blot using another antibody for proteins that were interacting with the antigen, followed by detection using a chemiluminescent or fluorescent secondary antibody.

Electrophoretic Mobility Shift Assay

An electrophoretic mobility shift assay (EMSA) or mobility shift electrophoresis, also referred as a gel shift assay, gel mobility shift assay, band shift assay, or gel retardation assay, is a common affinity electrophoresis technique used to study protein–DNA or protein–RNA interactions. This procedure can determine if a protein or mixture of proteins is capable of binding to a given DNA or RNA sequence, and can sometimes indicate if more than one protein molecule is involved in the binding complex. Gel shift assays are often performed in vitro concurrently with DNase footprinting, primer extension, and promoter-probe experiments when studying transcription initiation, DNA replication, DNA repair or RNA processing and maturation, as well as pre-mRNA splicing.

Although precursors can be found in earlier literature, most current assays are based on methods described by Garner and Revzin and Fried and Crothers.

Principle

A mobility shift assay is electrophoretic separation of a protein–DNA or protein–RNA mixture on a polyacrylamide or agarose gel for a short period (about 1.5-2 hr for a 15- to 20-cm gel). The speed at which different molecules (and combinations thereof) move through the gel is determined by their size and charge, and to a lesser extent, their shape. The control lane (DNA probe without protein present) will contain a single band corresponding to the unbound DNA or RNA fragment. However, assuming that the protein is capable of binding to the fragment, the lane with a protein that binds present will contain another band that represents the larger, less mobile complex of nucleic acid probe bound to protein which is 'shifted' up on the gel (since it has moved more slowly).

Under the correct experimental conditions, the interaction between the DNA (or RNA) and protein is stabilized and the ratio of bound to unbound nucleic acid on the gel reflects the fraction of free and bound probe molecules as the binding reaction enters the gel. This stability is in part due to a "caging effect", in that the protein, surrounded by the gel matrix, is unable to diffuse away from the probe before they recombine. If the starting concentrations of protein and probe are known, and if the stoichiometry of the complex is known, the apparent affinity of the protein for the nucleic acid sequence may be determined. Unless the complex is very long lived under gel conditions, or dissociation during electrophoresis is taken into account, the number derived is an apparent Kd. If the protein concentration is not known but the complex stoichiometry is, the protein concentration can be determined by increasing the concentration of DNA probe until further increments do not increase the fraction of protein bound. By comparison with a set of standard dilutions of free probe run on the same gel, the number of moles of protein can be calculated.

Variants and Additions

An antibody that recognizes the protein can be added to this mixture to create an even larger complex with a greater shift. This method is referred to as a *supershift assay*, and is used to unambiguously identify a protein present in the protein – nucleic acid complex.

Often, an extra lane is run with a competitor oligonucleotide to determine the most favorable binding sequence for the binding protein. The use of different oligonucleotides of defined sequence allows the identification of the precise binding site by competition (not shown in diagram). Variants of the competition assay are useful for measuring the specificity of binding and for measurement of association and dissociation kinetics. Thus, EMSA might also be used as part of a SELEX experiment to select for oligonucleotides that do actually bind a given protein.

Once DNA-protein binding is determined *in vitro*, a number of algorithms can narrow the search for identification of the transcription factor. Consensus sequence oligonucleotides for the transcription factor of interest will be able to compete for the binding, eliminating the shifted band, and must be confirmed by supershift. If the predicted consensus sequence fails to compete for binding, identification of the transcription factor may be aided by Multiplexed Competitor EMSA (MC-EMSA), whereby large sets of consensus sequences are multiplexed in each reaction, and where one set competes for binding, the individual consensus sequences from this set are run in a further reaction.

For visualization purposes, the nucleic acid fragment is usually labelled with a radioactive, fluorescent or biotin label. Standard ethidium bromide staining is less sensitive than these methods and can lack the sensitivity to detect the nucleic acid if small amounts of nucleic acid or single-stranded nucleic acid(s) are used in these experiments. When using a biotin label, streptavidin conjugated to an enzyme such as horseradish peroxidase is used to detect the DNA fragment. While isotopic DNA labeling has little or no effect on protein binding affinity, use of non-isotopic labels including flurophores or biotin can alter the affinity and/or stoichiometry of the protein interaction of interest. Competition between fluorophore- or biotin-labeled probe and unlabeled DNA of the same sequence can be used to determine whether the label alters binding affinity or stoichiometry.

Western Blot

Western Blot (WB) is a common method to detect and analyze proteins. It is built on a technique that involves transferring, also known as blotting, proteins separated by electrophoresis from the gel to a membrane where they can be visualized specifically. The procedure was first described by H. Towbin et al in 1979 and two years later given its name by W. Neal Burnette. Towbin et al described electrophoretic transfer of proteins from polyacrylamide gels to nitrocellulose sheets where the original gel pattern was accurately obtained. The setup consists of a standard set of seven steps, figure.

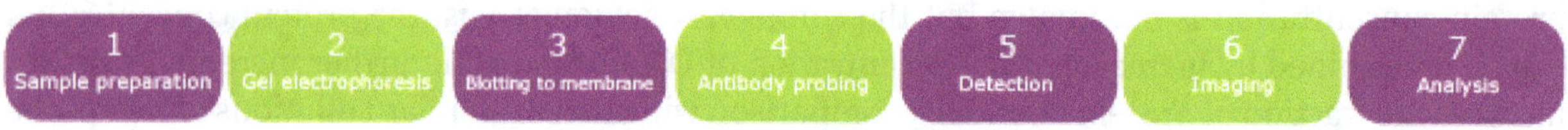

The standard steps in Western Blotting.

Samples are prepared and loaded on to a gel and during the electrophoresis the negatively charged proteins move toward the positively charged anode. In order to further analyze the proteins, they are transferred onto a membrane in a procedure called blotting. After the transfer, the membrane is blocked in order to prevent unwanted membrane-protein interaction in the following steps. To visualize the protein of interest the membrane is commonly first probed using a primary protein-specific antibody followed by a labeled secondary antibody used for detection. An image is taken of the membrane and the result is analyzed.

By adding a separate marker solution to one of the wells in the gel, it is possible to estimate the size of the protein in addition to the antibody interactions that are used to verify the specific protein. The separation on the gel is not only due to size but also to some extent depending on the molecular charge, hydrophobic regions, and degree of denaturation. The setup of the experiment can be varied in many ways to best suit the specific inquiry. When analyzing the results, variations between lanes regarding loading and transfer rates between blots, must be taken into consideration. In addition, the non-linear relation of the generated signal across the concentration range of the samples is also an aspect of consideration when interpreting the results. The outcome of a WB experiment depends on three important factors; the ability of the antibody to bind a specific protein, the strength of the interaction, and the concentration of the protein of interest itself. Moreover, the specificity of the binding to the target and a low cross reactivity are important features as well. The result form the WB is not always easy to interpret as the size of the protein may vary from the theoretical weight due to posttranslational modifications, such as glycosylation, or interactions with other proteins. However, WB is a very common method and almost all available commercial antibodies have been validated using this method.

Technology

Sample Preparation

The first step of a WB is to prepare the sample, e.g. tissue, cells, or other solution, which is going to be analyzed. Usually the tissue needs to be broken down by blending, homogenization, or sonication. Buffers are added to lyse the cells and solubilize the proteins and often an inhibitor is added to prevent denaturation or degradation. Different types of filtration and centrifugation methods are applied to further prepare the samples. It is important to determine the total protein concentration of the generated extract to be able to load a specific amount on the gel to enable comparison between samples. Usually a biochemical assay is used in order to determine the protein concentration. The extract is then diluted with loading buffer consisting of glycerol and a dye (e.g. bromophenol blue). The glycerol is used to simplify the loading by raising the density of the extract and the dye is added to visualize the sample. Heat is applied on the samples in order to break the structures of the protein, which will help keeping the negative charge from neutralization. Preferably positive and negative controls are included in the set up to confirm identity of the protein as well as the activity of the antibody.

Gel Electrophoresis

After sample preparation the extract is ready to be loaded to separate the proteins according to size by gel electrophoresis. An electric field is applied over the gel that causes the charged molecules to move. In WB polyacrylamide gels are used for protein separation and the method is therefore called polyacrylamide gel electrophoresis (PAGE) when using native condition. For denaturing conditions, sodium dodecyl sulfate (SDS) is added to the system and the method is therefore called SDS-PAGE. The SDS binds to the protein and form a negatively charged micelle around the protein regardless of inherent charge. The denaturing condition dissolves the tridimensional structure of the proteins and the charge of the protein becomes relative to its size resulting in separation of the proteins only by size. When using native conditions the mobility is depending on both charge and hydrodynamic size allowing detection of changes in charge due to chemical degradation, conformational changes due to folding or unfolding, aggregation, or other binding events.

The gel typically consists of two sections with different densities: (i) a stacking gel, and (ii) a separating gel, Figure. The differences between the two sections are in pH and gel concentration. With somewhat acidic pH and a lower concentration of acrylamide the stacking gel separates the proteins poorly but allows them to form highly defined sharp bands before they enter the separating gel. With more basic conditions and higher gel concentration, the separating gel makes the proteins differentiate by size as smaller proteins travel faster in the gel than bigger ones. Precast gels are convenient; however, it is possible to cast them by hand. The gel is immersed in buffer and the protein samples and markers are loaded to the wells in the gel. A voltage is applied on the gel and the proteins will start to travel down the gel due to their negative electrical charge. Selecting the proper voltage is important since too high voltage will overheat the gel and maybe deform the bands.

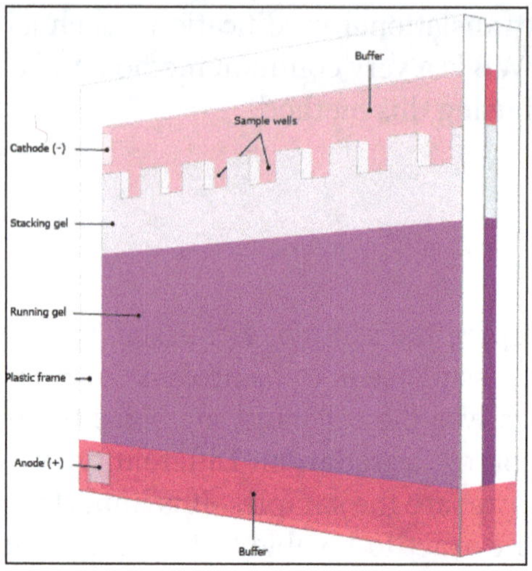

A typical gel immersed in buffer.

Blotting to Membrane

After gel electrophoresis the proteins are transferred to a solid support membrane, which is the third step of Western Blot. In the transfer process voltage is applied to transfer the proteins from the gel to the membrane. The setup includes sponges, filter papers, the gel, and the membrane, which is placed between the gel and the positive electrode, Figure. This ensures the migration of the negatively charge proteins from the gel to the membrane. There are three types of membranes: nitrocellulose, polyvinylidene difluoride (PVDF), and nylon. Even though nylon membranes are superior in several aspects, the high background binding and irreversible staining of some dyes makes this type of membrane less common than the other two alternatives. The major advantage of nitrocellulose membranes is the low background regardless of detection method. Due to a relatively large average pore size, nitrocellulose membranes should not be used for transfer of proteins with low molecular weight. Moreover, when dry, the membrane becomes brittle which makes them difficult to handle. The more stable PVDF membrane allows relabeling and is more convenient to store. The hydrophobic nature of PVDF result in high protein binding capacity, however, as a consequence the background is also higher.

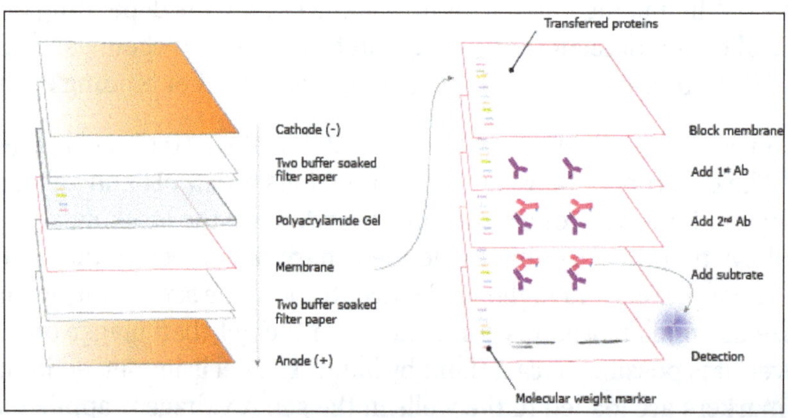

The proteins in the gel are blotted to a membrane and the sample is visualized through blocking, adding antibodies, and washing according to a certain schedule.

There are two methods for the blotting called wet and semi-dry. The wet conditions are preferred when the transfer must be efficient and give high quality regarding distinct and sharp bands. In addition, this is the better choice for transfer of larger protein complex. The gel, membrane, and filter papers are completely immersed in buffer during the transfer and there is no risk of drying out the gel. Semi-dry blotting is more rapid and less volume of buffer is needed. However, this transfer method is usually less efficient, especially for larger proteins, and there is a risk of overheating and drying the gel when using extended transfer times.

Antibody Probing

The forth step of the WB is antibody probing. In order to prevent unspecific binding of the antibodies to the membrane, rather than binding specific to the protein of interest, a substance is used to block out the residual sites on the membrane. Common substances used are dried non-fat milk, 5% Bovine Serum Albumin (BSA) diluted in Tris Buffered Saline Tween (TBST), normal goat serum, casein, or fish gelatin. Milk is easy to get hold of and inexpensive, however not suitable for all detection labels. Fish gelatin gives lower background but can mask some proteins as well as being a relatively expensive blocking buffer. BSA is inexpensive, whereas serum can contain immunoglobulins giving rise to cross-reactivity. Careful selection of the blocking agent is key since none of the blocking buffers are ideal for all different antigen-antibody interactions. The blocking procedure consists of incubating the membrane in the appropriate blocking buffer for an hour or longer. When using long incubation times, the blocking should be performed at +4 °C to rule out the risk of staining artifacts or background. Blocking is a delicate balance between reducing the background without decreasing the signal from the protein of interest.

The blocked membrane is thereafter incubated with the primary antibody. The antibody is diluted to a suitable concentration in TBST, phosphate buffered saline (PBS), or wash buffer. It is preferred to incubate the antibody with BSA if the antibody is going to be re-used. After washing the membrane, the membrane is incubated with the secondary antibody that binds to the primary antibody. The secondary antibody is labeled with a reporter. When using a polyclonal antibody as secondary antibody, it may give rise to some background. In the case of background staining, the secondary antibody may be pre-blocked with non-immune serum from the host it was generated in. Optimization of the concentration of the secondary antibody is recommended due to quite extended variations between antibodies as well as detection system used.

Detection

In the fifth step of a WB, the protein-antibody-antibody complex is detected on the membrane. There are several kinds of labeling of the secondary antibody, e.g. enzymes, fluorophores, biotinylation, gold-conjugation, and radioisotopes, as exemplified in figure. Amongst enzymes the most common is HRP used together with chemiluminescent, chemifluorescent, or chromogenic substances. HRP has a high substrate specificity giving low background, is stable, and inexpensive. In chemiluminescense the HRP enzyme catalyzes the oxidation of luminol from the luminol peroxide detection reagent. The multi-step reaction generates light emission. Certain chemicals like phenols can enhance the emitted light. A direct method is the use of fluorescence; the fluorophores emit light after being excited and no detection agent is needed. It is well suitable for quantitative Western and since different fluorophores emit light of different wavelengths it is possible to perform multiplexing and specific detection of more than one protein at the time. Using a chemical and/or an enzyme to induce the generation of an active fluorophore from a fluorogenic substrate is called chemifluorescence. To further enhance the signal intensity a two-step biotin streptavidin based system may be used. Gold conjugation is also a method where proteins stain dark red due to accumulation of gold. It is also possible to use radioisotopes but they require special handling and are quite expensive.

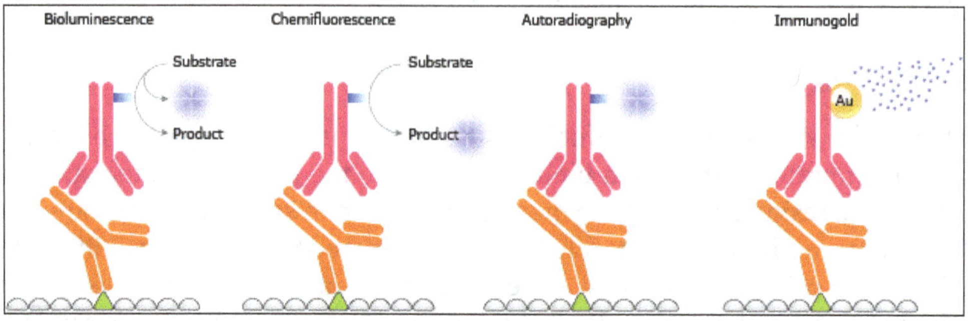

Different reporter systems.

Imaging

Imaging is the sixth step of WB and the capturing can be analogue using a film, or digitally preformed with a CCD camera or scanner capturing the different kinds of emitted signals. The CCD imaging device enables quantitation with high detection sensitivity and a broad linear range with no chemical waste or need for a dark room. It may be used to detect membranes, stained gels, or for ultraviolet light applications.

Analysis

The last step of a WB is to analyze the results. In a typical qualitative application, the presence of a protein of interest is confirmed, the amount is approximated by visual inspection, and the size is determined by comparison with a marker. Improvements and developments, especially towards highly sensitive detection reagents and advanced imaging techniques, make WB a potential tool for quantitative analysis. The quantitative applications entail a definition of the amount of protein in relative or absolute terms. Some factors are to take under consideration like sensitivity, signal stability, linear dynamic range, normalization, and the signal-to-noise ratio. The minimum of protein that can be seen in a given assay gives the limits of detection (LOD), and the limit of signal intensity that can be reliably used for precise quantification is the limit of quantification (LOQ). Factors that affect these terms are antibody quality and concentrations as well as exposure times when considering the minimum amount of protein detected. A stable signal system expands the time window for reaching high sensitivity, multiple exposures, and possibility to detect weak bands. The range that allows an even and precise quantitation where the signal intensity still is proportional to the amount of protein is called the linear dynamic range. It is important to avoid signal saturation due to excessive amounts of protein or high concentrations of antibodies. A low LOD and quantitation of both weak and strong signals gives a broad linear dynamic range. The protein of interest should be normalized to an internal reference that allows fluctuations in amount of protein loaded onto each well or different concentrations. This can be achieved with housekeeping or spiked protein. The ratio between the signal and noise is important in order to properly quantitate the protein. This is of outmost importance when detecting weak bands where a higher background is expected.

Specific Examples

Although the theoretical weight of many proteins differ from the real weight due to post translational modifications, almost all commercial antibodies are validated using WB.

Within the Human Protein Atlas project WB is used for quality control of the polyclonal antibodies generated in the project. After purification, the antibodies are used to detect bands in a setup of lysate and different tissues. The use of siRNA WB is currently implemented to further analyze the antibodies.

The entire WB protocol, including dilution of both primary- and secondary antibodies and the final detection step, is performed in a routine manner and no specific experimental optimization is made for individual antibodies. At first, total protein lysates from selected cell lines and tissues (15μg of RT-4, U-251MG, liver, and tonsil as well as 25μg of HSA- and IgG-depleted plasma) and a marker (PageRuler Plus Prestained Protein Ladder; Thermo Scientific) are loaded on precast 4–20% Criterion SDS–PAGE gradient gels (Bio-Rad Laboratories) and run under reducing conditions, followed by transfer to PVDF membranes (Bio- Rad Laboratories) using Trans-Blot turbo (Bio-Rad Laboratories) according to the manufacturer's recommendations. The Criterion SDS–PAGE gradient gels together with the marker make it possible to analyze protein in the sizes ranging from 10 to 250 kDa. Membranes are then activated in methanol before blocking (5% dry milk, 0.5% Tween 20, 1× TBS; 1 mM tris–HCl, 0.15 M NaCl) for 1 h at room temperature during constant shaking. The membranes are then incubated with primary HPA antibody diluted to ~0,6μg/ml, for 1 h, followed by washing (1 mM tris–HCl, 0.15 M NaCl, 0.05% Tween) and incubation for 45 minutes with the secondary peroxidase-conjugated antibody (swine anti-rabbit 1:4000, Dako). For commercial antibodies a dilution of 1:500 is used and the secondary antibody is either swine anti-rabbit (1:3000, Dako) or goat anti-mouse (1:3000, Dako). A CCD-camera (Bio-Rad Laboratories) is used for detection of signal from the substrate (Immobilon Western Chemiluminescence HRP Substrate; Millipore).

A typical Western Blot from HPA is seen in figure.

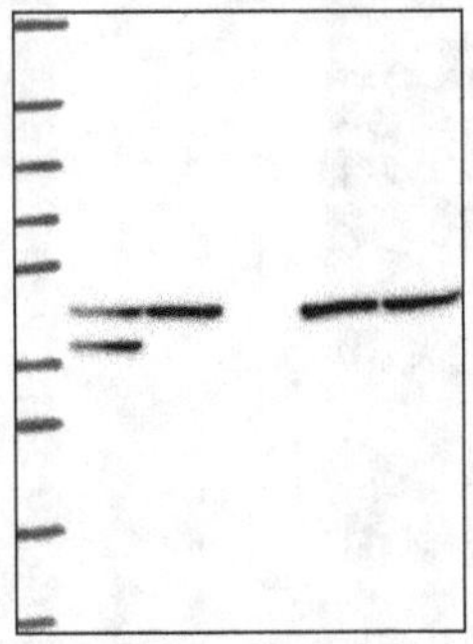

Typical Western Blot result using HRP and a CCD camera.

Southern Blot

A Southern blot is a method used in molecular biology for detection of a specific DNA sequence in DNA samples. Southern blotting combines transfer of electrophoresis-separated DNA fragments to a filter membrane and subsequent fragment detection by probe hybridization.

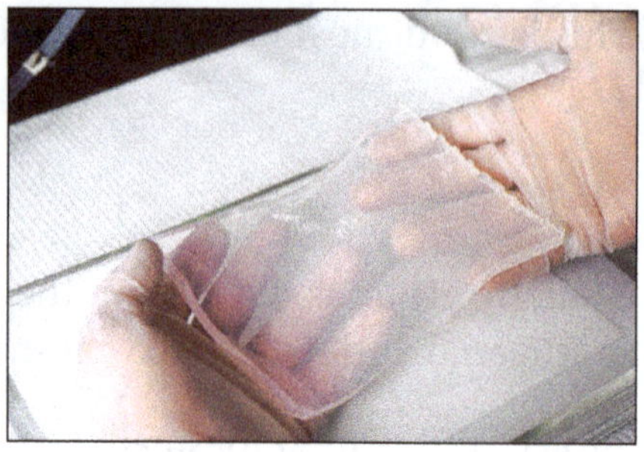

Agarose Gel

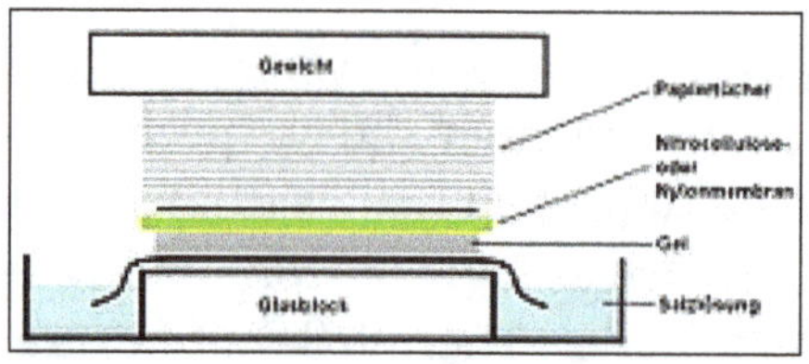

Tray with a stack consisting top down of a weight, paper towels, membrane of nitrocellulose or nylon, gel, salt solution and a slab of glass.

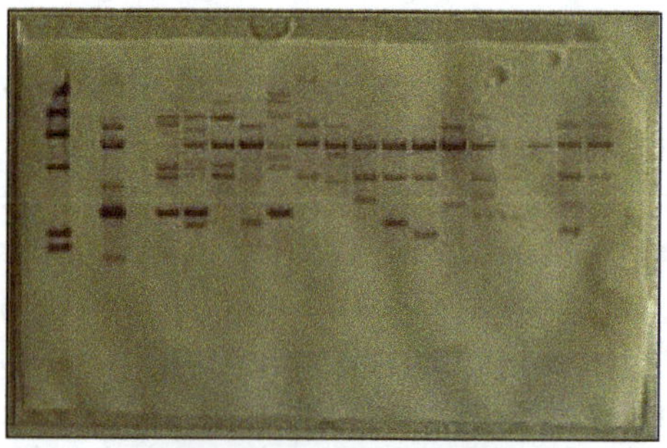

Southern blot membrane after hybridization and rinsing.

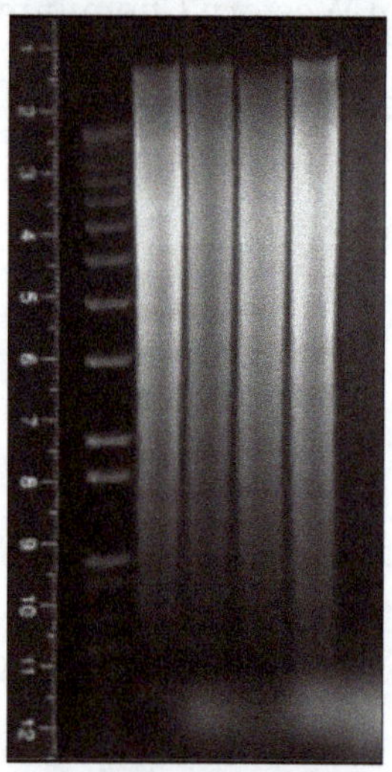

Southern blot agarose gel under ultraviolet illumination.

The method is named after the British biologist Edwin Southern, who first published it in 1975. Other blotting methods (i.e., western blot, northern blot, eastern blot, southwestern blot) that employ similar principles, but using RNA or protein, have later been named in reference to Edwin Southern's name. As the label is eponymous, Southern is capitalised, as is conventional of proper nouns. The names for other blotting methods may follow this convention, by analogy.

Method

1. Restriction endonucleases are used to cut high-molecular-weight DNA strands into smaller fragments.
2. The DNA fragments are then electrophoresed on an agarose gel to separate them by size.
3. If some of the DNA fragments are larger than 15 kb, then prior to blotting, the gel may be treated with an acid, such as dilute HCl. This depurinates the DNA fragments, breaking the DNA into smaller pieces, thereby allowing more efficient transfer from the gel to membrane.
4. If alkaline transfer methods are used, the DNA gel is placed into an alkaline solution (typically containing sodium hydroxide) to denature the double-stranded DNA. The denaturation in an alkaline environment may improve binding of the negatively charged thymine residues of DNA to a positively charged amino groups of membrane, separating it into single DNA strands for later hybridization to the probe, and destroys any residual RNA that may still be present in the DNA. The choice of alkaline over neutral transfer methods, however, is often empirical and may result in equivalent results.
5. A sheet of nitrocellulose (or, alternatively, nylon) membrane is placed on top of (or below, depending on the direction of the transfer) the gel. Pressure is applied evenly to the gel (either using suction, or by placing a stack of paper towels and a weight on top of the membrane and gel), to ensure good and even contact between gel and membrane. If transferring by suction, 20X SSC buffer is used to ensure a seal and prevent drying of the gel. Buffer transfer by capillary action from a region of high water potential to a region of low water potential (usually filter paper and paper tissues) is then used to move the DNA from the gel onto the membrane; ion exchange interactions bind the DNA to the membrane due to the negative charge of the DNA and positive charge of the membrane.
6. The membrane is then baked in a vacuum or regular oven at 80 °C for 2 hours (standard conditions; nitrocellulose or nylon membrane) or exposed to ultraviolet radiation (nylon membrane) to permanently attach the transferred DNA to the membrane.
7. The membrane is then exposed to a hybridization probe—a single DNA fragment with a specific sequence whose presence in the target DNA is to be determined. The probe DNA is labelled so that it can be detected, usually by incorporating radioactivity or tagging the molecule with a fluorescent or chromogenic dye. In some cases, the hybridization probe may be made from RNA, rather than DNA. To ensure the specificity of the binding of the probe to the sample DNA, most common hybridization methods use salmon or herring sperm DNA for blocking of the membrane surface and target DNA, deionized formamide, and detergents such as SDS to reduce non-specific binding of the probe.
8. After hybridization, excess probe is washed from the membrane (typically using SSC buffer), and the pattern of hybridization is visualized on X-ray film by autoradiography in the

case of a radioactive or fluorescent probe, or by development of colour on the membrane if a chromogenic detection method is used.

Result

Hybridization of the probe to a specific DNA fragment on the filter membrane indicates that this fragment contains DNA sequence that is complementary to the probe. The transfer step of the DNA from the electrophoresis gel to a membrane permits easy binding of the labeled hybridization probe to the size-fractionated DNA. It also allows for the fixation of the target-probe hybrids, required for analysis by autoradiography or other detection methods. Southern blots performed with restriction enzyme-digested genomic DNA may be used to determine the number of sequences (e.g., gene copies) in a genome. A probe that hybridizes only to a single DNA segment that has not been cut by the restriction enzyme will produce a single band on a Southern blot, whereas multiple bands will likely be observed when the probe hybridizes to several highly similar sequences (e.g., those that may be the result of sequence duplication). Modification of the hybridization conditions (for example, increasing the hybridization temperature or decreasing salt concentration) may be used to increase specificity and decrease hybridization of the probe to sequences that are less than 100% similar.

Applications

Southern blotting transfer may be used for homology-based cloning on the basis of amino acid sequence of the protein product of the target gene. Oligonucleotides are designed so that they are similar to the target sequence. The oligonucleotides are chemically synthesized, radiolabeled, and used to screen a DNA library, or other collections of cloned DNA fragments. Sequences that hybridize with the hybridization probe are further analysed, for example, to obtain the full length sequence of the targeted gene.

Southern blotting can also be used to identify methylated sites in particular genes. Particularly useful are the restriction nucleases *MspI* and *HpaII*, both of which recognize and cleave within the same sequence. However, *HpaII* requires that a C within that site be methylated, whereas *MspI* cleaves only DNA unmethylated at that site. Therefore, any methylated sites within a sequence analyzed with a particular probe will be cleaved by the former, but not the latter, enzyme.

Northern Blot

The Northern blot is a technique used in molecular biology research to study gene expression in a sample, through detection of RNA (or isolated messenger RNA). With Northern blotting it is possible to observe cellular control over structure and function by determining the particular gene expression levels during differentiation, morphogenesis, as well as abnormal or diseased conditions. Northern blotting involves the use of electrophoresis to separate RNA samples by size and detection with a hybridization probe complementary to part of or the entire target sequence.

The term 'Northern blot' actually refers specifically to the capillary transfer of RNA from the elec-trophoresis gel to the blotting membrane. However, the entire process is commonly referred to as Northern blotting. The northern blot technique was developed in 1977 by James Alwine, David Kemp, and George Stark at Stanford University. Northern blotting takes its name from its similarity to the first blotting technique, the Southern blot, named for biologist Edwin Southern. The major difference is that RNA, rather than DNA, is analyzed in the Northern blot.

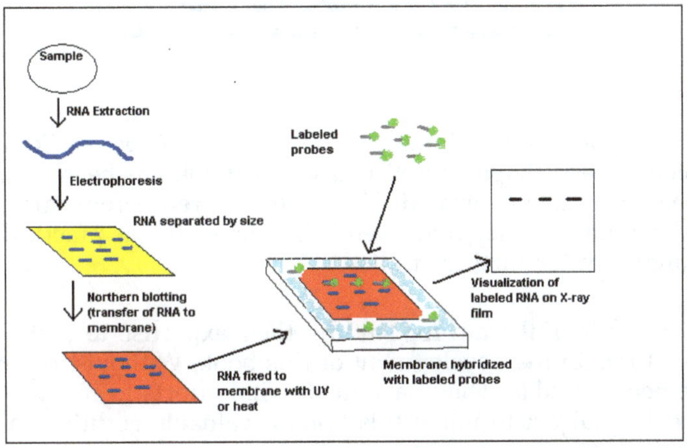

Northern blot technique: Flow diagram outlining the general procedure for RNA detection by northern blotting.

A general blotting procedure starts with extraction of total RNA from a homogenized tissue sample or from cells. Eukaryotic mRNA can then be isolated through the use of oligo (dT) cellulose chromatography to isolate only those RNAs with a poly(A) tail. RNA samples are then separated by gel electrophoresis. Since the gels are fragile and the probes are unable to enter the matrix, the RNA samples, now separated by size, are transferred to a nylon membrane through a capillary or vacuum blotting system.

References

- Biochemical-techniques-basics: biochemden.com, Retrieved 17 January, 2019
- Rosenberg, Ian (2005). Protein analysis and purification: benchtop techniques. Springer. p. 520. ISBN 978-0-8176-4340-9
- Hellman, Lance M; Fried, Michael G (2007). "Electrophoretic mobility shift assay (EMSA) for detecting protein–nucleic acid interactions". Nature Protocols. 2 (8): 1849–1861. doi:10.1038/nprot.2007.249. ISSN 1754-2189. PMC 2757439. PMID 17703195
- Mahmood T et al, 2012. Western blot: technique, theory, and trouble shooting. N Am J Med Sci. PubMed: 23050259 DOI: 10.4103/1947-2714.100998
- Smith AJ, Humphries SE (January 2009). "Characterization of DNA-binding proteins using multiplexed competitor EMSA". J. Mol. Biol. 385 (3): 714–7. doi:10.1016/j.jmb.2008.11.035. PMID 19059416

Permissions

All chapters in this book are published with permission under the Creative Commons Attribution Share Alike License or equivalent. Every chapter published in this book has been scrutinized by our experts. Their significance has been extensively debated. The topics covered herein carry significant information for a comprehensive understanding. They may even be implemented as practical applications or may be referred to as a beginning point for further studies.

We would like to thank the editorial team for lending their expertise to make the book truly unique. They have played a crucial role in the development of this book. Without their invaluable contributions this book wouldn't have been possible. They have made vital efforts to compile up to date information on the varied aspects of this subject to make this book a valuable addition to the collection of many professionals and students.

This book was conceptualized with the vision of imparting up-to-date and integrated information in this field. To ensure the same, a matchless editorial board was set up. Every individual on the board went through rigorous rounds of assessment to prove their worth. After which they invested a large part of their time researching and compiling the most relevant data for our readers.

The editorial board has been involved in producing this book since its inception. They have spent rigorous hours researching and exploring the diverse topics which have resulted in the successful publishing of this book. They have passed on their knowledge of decades through this book. To expedite this challenging task, the publisher supported the team at every step. A small team of assistant editors was also appointed to further simplify the editing procedure and attain best results for the readers.

Apart from the editorial board, the designing team has also invested a significant amount of their time in understanding the subject and creating the most relevant covers. They scrutinized every image to scout for the most suitable representation of the subject and create an appropriate cover for the book.

The publishing team has been an ardent support to the editorial, designing and production team. Their endless efforts to recruit the best for this project, has resulted in the accomplishment of this book. They are a veteran in the field of academics and their pool of knowledge is as vast as their experience in printing. Their expertise and guidance has proved useful at every step. Their uncompromising quality standards have made this book an exceptional effort. Their encouragement from time to time has been an inspiration for everyone.

The publisher and the editorial board hope that this book will prove to be a valuable piece of knowledge for students, practitioners and scholars across the globe.

Index

P

R

S

T

U

X

Z

www.ingramcontent.com/pod-product-compliance
Lightning Source LLC
LaVergne TN
LVHW080919210926
848485LV00004B/683